Lecture Notes in Earth Sciences   84

Editors:
S. Bhattacharji, Brooklyn
G. M. Friedman, Brooklyn and Troy
H. J. Neugebauer, Bonn
A. Seilacher, Tuebingen and Yale

W0018238

Springer
Berlin
Heidelberg
New York
Barcelona
Hong Kong
London
Milan
Paris
Singapore
Tokyo

Henry V. Lyatsky   Gerald M. Friedman
Vadim B. Lyatsky†

Principles of Practical Tectonic Analysis of Cratonic Regions

With Particular Reference
to Western North America

With 89 Figures, 3 Tables and 3 Foldouts

 Springer

Authors

Dr. Henry V. Lyatsky
Lyatsky Geoscience Research & Consulting Ltd.
4827 Nipawin Cr. N.W.
Calgary, AB, Canada T2K 2H8
E-mail: lyatskyh@cadvision.com

Prof. Dr. Gerald M. Friedman
Brooklyn College and Graduate School
of the City University of New York
Northeastern Science Foundation, Inc.
Rensselaer Center of Applied Geology
15 Third Street, P.O. Box 746
Troy, NY 12181-0746, USA
E-mail: gmfriedman@juno.com

Dr. Vadim B. Lyatsky†

"For all Lecture Notes in Earth Sciences published till now please see final pages of the book"

Library of Congress Cataloging-in-Publication Data

Lyatsky, Henry V., 1962–
   Principles of practical tectonic analysis of cratonic regions :
with particular reference to Western North America / Henry
V. Lyatsky, Gerald M. Friedman, Vadim B. Lyatsky.
      p.   cm. -- (Lecture notes in earth sciences ; 84)
   Includes bibliographical references.
   ISBN 3-540-65346-5 (softcover)
   1. Cratons--North America. 2. Geology, Structural--North America.
I. Friedman, Gerald M. II. Lyatsky, Vadim B., 1927-   .
III. Title. IV. Series.
QE626.L9 1999
551.8'097--dc21   98-51416
                                                      CIP

ISSN 0930-0317
ISBN 3-540-65346-5 Springer-Verlag Berlin Heidelberg New York

Typesetting: Camera ready by authors
SPIN: 10698237   32/3142-543210 - Printed on acid-free paper

PREFACE

Most of the North American continent is underlain by long-cratonized crust. However, the geologists', geochemists' and geophysicists' attention is mainly focused on the surrounding mobile megabelts. Only few publications consider the North American craton as a whole and attempt to review its entire history from the Late Archean to the present. Even in the last, as yet unfinished synopsis, created under the label *Decade of North American Geology* (*DNAG*), the geology of the craton is divided between volumes separately considering the Canadian Shield and cratonic platforms, platformal cover and basement, in the U.S. and Canada.

This separateness reflects a real disconnection between professionals in different countries, provinces or states (consider the well-known border "faults"), industries and branches of geology - hard-rock and soft-rock. Hard-rock geologists are trained in igneous and metamorphic petrology and related disciplines. Soft-rock geologists specialize in sedimentology and disciplines related to it. Geologists in these two groups concentrate on different practical tasks, use different data, read different journals. The gap between them is great, and the links weak.

Tectonics in these two fields of knowledge usually focuses on different rocks: crystalline shields and platformal basement, or unmetamorphosed sedimentary cover. Their interrelationship is often reduced to just so-called basement influences on the distribution of provenance areas and structure of sedimentary basins. Worse, tectonics is commonly equated with just the structure of the crust or its parts, ignoring other manifestations of tectonism. The dynamic causes of tectonic events are often presumed to be located away from cratons and even away from continents.

In such a partition, tectonics often loses its identity as a specific geological discipline, and regional tectonics misses the principal focus of its exploratory effort. Tectonics relies on systematic knowledge of all endogenic geodynamical processes, which are expressed in their four principal aspects: tectono-sedimentological, tectono-magmatic, tectono-metamorphic (including all kinds of alteration), and tectono-deformational. These aspects of tectonism are observable through geologic mapping of the crust. Geologic mapping is more than just compilation of geologic maps (which are the usual final product). It includes direct observation and description of rock types and their areal distribution, as well as observation, measurement and description of rock bodies and their spatial and temporal relationships. This information is supplemented with the results of drilling in many areas of soft and hard rocks.

Tectonics synthesizes the findings from field mapping and from many related specialized disciplines - petrology, paleontology, etc. A tectonist's job is to combine into an internally consistent concept a broad range of facts and data from different fields of industrial, academic and scientific activity. From a huge volume of diverse information, he/she must select that which permits to evaluate the observed phenomena induced by endogenic processes in the lithosphere. Tectonic analysis is carried out in the four aspects: tectono-sedimentary, tectono-magmatic, tectono-metamorphic, and tectono-deformational.

In this book, we use this integrated approach in some detail, primarily with examples from the Northern Hemisphere, the North American craton, and particularly the Alberta Platform in it. This approach is rock-based and fact-driven, and that gives it realistic predictive power. This book is thus not about plate tectonics, nor terrane tectonics, nor tectonic modeling, but rather it is about non-speculative tectonics of cratons, which are the largest parts of continents. Cratonic regions are the most suitable areas for human habitation, agriculture, and industry. They contain enormous natural resources - oil and gas, coal and salt, bauxite, titanite sands, and various construction materials. These regions provide clues to the understanding of the entire continental and global tectonics.

This book is intended for those whose job is to manage the existing resources and find new ones, maintain and develop the urban and rural infrastructure, transportation, and environment. It is also intended as a tutorial for young geologists interested in studying regional tectonics and practical applications of tectonic knowledge. Of course, proper understanding of the tectonic fundamentals would enhance their use for prognostic purposes. To help the practitioners and students inspired by practical societal needs is the main purpose of this book.

A crucial point is that *geologic field mapping was, is and will remain the principal source of regional tectonic information.* Deep drilling has added a lot of information to what is known from surface studies. Deep continental drilling through crystalline rocks provides invaluable information about the endogenic regimes, putting unshakable constraints on speculations about the crust. We are now also blessed with many auxiliary techniques - geochemical, isotopic, radiometric, geophysical, remote-sensing, experimental - as well as with a strong mathematical apparatus and computer capability for data processing and modeling useful for some tectonic tasks. These techniques and tools put a new burden on the modern tectonists - to be familiar not only with a wide variety of relevant factual geologic data and analytic results, but also with the potential and limitations of a great variety of methods, techniques and procedures used in obtaining geochemical and geophysical information.

Tectonics is in fashion these days. Paleontologists and volcanologists, hydrogeologists and stratigraphers, geophysicists and geochemists try to leave their mark by applying their findings to tectonics. The pursuit of tectonic applications may, however, distract the specialists from their work, causing them instead to enter the realm of uncertainty and speculation. Only deep knowledge of all four aspects of tectonics, and skills to apply them properly to particular areas, provide the antidote against groundless speculation conferred by healthy skepticism.

Tectonics, like geology itself, embraces practice and theory, local descriptions and regional generalizations. Tectonics, like all of geology, is the practice of mineral and petroleum exploration, as well as a science dealing with the evolution of geologic systems. Like geology, tectonics, as a science, is descriptive and restricted to rock-made bodies (the largest of which is the lithosphere) Tectonics, like all of geology, is a science of principles, i.e. empirical generalizations no less meaningful than the principles which are mathematically described and physically modeled. In this book, the reader will find little speculation, and lots of skepticism about one-sided physical and chemical solutions to the problems of regional tectonics. The reader will find examples of careful use of these principles in some of the best-studied areas of the North American craton: parts of southern Canadian Shield and western Interior Platform.

Not least, this book is aimed at readers who would stay with our logical development from the beginning to the end. It would require time and patience. But we will try to make the readers' efforts worthwhile.

ACKNOWLEDGMENTS

This book has arisen from our decades-long experience of practical research and exploration in western Canada and elsewhere. Our work has benefited from many helpful discussions with our exploration-oriented colleagues in the oil industry, at the Universities of Calgary and Saskatchewan, and at the Calgary, Ottawa and Vancouver offices of the Geological Survey of Canada. These colleagues have brought to our attention many relevant facts and scientific ideas.

Particular thanks go to Dr. Andrew Holder (formerly, Vice-President, Home Oil Co. Ltd., Calgary) and Profs. James Brown and Don Lawton (University of Calgary), whose support and encouragement are deeply appreciated. Technical support with the production of horizontal-gradient vector and other potential-field maps has been provided, at various times, by Home Oil Co. Ltd., University of Calgary (where our horizontal-gradient vector software currently resides) and Geological Survey of Canada. Garth Keyte and Four West Consultants (Calgary) kindly provided printing, drafting and reproduction facilities.

TABLE OF CONTENTS

3 - REGIONAL TECTONIC ANALYSIS OF THE WESTERN CANADA SEDIMENTARY PROVINCE IN ALBERTA, SASKATCHEWAN AND ADJACENT PARTS OF THE U.S.

4 - METHODOLOGICAL REASONS FOR USING STRUCTURAL-FORMATIONAL ÉTAGES IN REGIONAL TECTONIC ANALYSIS

5 - RECONSTRUCTION OF STRUCTURAL HISTORY OF A CRATON BY USING STRUCTURAL-FORMATIONAL ÉTAGES IN ITS COVER AND PRE-COVER VOLCANO-SEDIMENTARY BASINS

6 - FINDING AN ADEQUATE REGIONAL TECTONIC INTERPRETATION CONSTRAINED BY MAPPABLE PROPERTIES OF ROCKS

7 - PLACE OF NEOTECTONIC AND CURRENT TECTONIC CRUSTAL MOVEMENTS IN REGIONAL TECTONIC STUDIES

8 - CONCLUSIONS: ADVANTAGES OF PRACTICAL TECTONICS

LIST OF TABLES AND FIGURES

1 BASIC NOTIONS AND DEFINITIONS

Intrinsic difficulties in describing geologic objects

Geology was born from mankind's practice of using the stones underfoot: first to make stone ornaments and tools, then to construct dwellings and extract metals. Gradual recognition that some useful rock properties are associated with certain kind of rocks, and later that the distribution of rock types is regular, has led to the rise of geology as a field of knowledge and, eventually, a science. Geology at first concerned itself with studying rocks at the surface; later it went deeper, and the 20th century enabled geology to expand its scope to the entire rock-made outer shell of the Earth. But at all times, geology has dealt with rocks (i.e. mineral aggregates) and with bodies made of rocks.

The modern ability to picture some characteristics of the deeper geospheres should not distract geology from its main focus as a rock-related practice and science. Geology must limit its attention to the Earth's rock-made perisphere, where its own means are sufficient to meet its tasks. Geology, as a human social practice (including geologic mapping and resource exploration in the mining and petroleum industries) and a science, deals primarily with minerals and their aggregates (rocks) in a solid state and with geologic (rock-made) bodies, restricted to the lithosphere. Geology has its own objects and methods for the study of those Earth layers that are subject to direct observation and permissible close extrapolation. Geology is an observational and descriptive science, whose knowledge is obtained chiefly through geologic field mapping.

Studies of the entire Earth are a job of other disciplines: planetology, geodesy, "big" geophysics, global geochemistry. Their results, when reasonably constrained, may be useful for understanding the geologic evolution of different geologic regions. But material in the sublithospheric mantle is in different physico-chemical states than in the lithosphere. Although the above definition puts it outside the scope of geology, considerations of interaction of the lithosphere with the sublithospheric mantle as well as the atmosphere constrain geologic generalizations.

Tectonics is an intrinsic part of geology, and as such it deals with the lithosphere and its constituents. In this definition, tectonics mostly has the same material objects as geology per se, but with regards only to internally generated, endogenic kinematics and dynamics (geology, by contrast, is much broader, including also exogenic processes and their products). Geology (like astronomy) is essentially an observational, descriptive and empirical discipline (e.g., Frodeman, 1995), with relatively little room for experimentation. This is particularly true of tectonics, whose processes are rooted at great depths, at long-lived enormous temperatures and pressures; they proceed over long periods of time that cannot be reproduced experimentally. Visible mappable manifestations of endogenic (i.e. tectonic) processes may be specific for each region, but all of them fall into four classes: tectono-sedimentological, tectono-magmatic, tectono-metamorphic, and tectono-deformational.

Observations obtained from geologic field mapping, determination of rock types and their spatial distribution, as well as description of relationships between rock-made bodies - all that forms the basis of tectonics. Any analysis of rocks may contain tectonic information. A core part of tectonics, as of geology itself, is regional stratigraphy of stratal lithic bodies (those whose lateral extent is much bigger than their thickness) and petrology of discordant bodies (chiefly intrusive igneous ones). Lithostratigraphy permits to assess the temporal succession and spatial distribution of stratified bodies, sedimentary, volcanic and metamorphic, and the place of intrusive bodies in a time-stratigraphic continuum.

In the geologic record, rock bodies are arranged in space and time in a way that permits to discern regional tectonic units. This, in turn, makes it possible to restore the penecontemporaneous regional tectonic regimes, interpreted from characteristics of rocks and rock bodies (rock composition, structural interrelationships, etc.). Because the division of a standard stratigraphic column into tectonic units requires some ideas about their tectonic genesis, uncertainty inevitably occurs in the descriptions of regional geology in terms of tectonics. But geologic, i.e. rock-made, bodies are the only source of objective geological (including tectonic) information. Their main characteristics are their material composition (chemical, mineralogic, petrologic) and structure, be it internal (i.e. organization of their constituents) or external (pertaining to their form and shape). Recognition of geologic bodies corresponding to certain tectonic regimes in the past and reconstruction of their evolution as structural-compositional systems is the main job of tectonics.

Tectonic definition of continents, and of cratons as their fundamental constituents

Confusingly, words like *ocean, oceanic, continent* and *continental* are now often used in various senses. It may be geographical, referring to huge bodies of seawater (oceans) and continuous landmasses (continents); geophysical, referring to the types of Earth's crust and lithosphere possessing or lacking specific physical characteristics; or geological, referring to global-scale domains made up of certain types of rocks. Continents as tectonic features are large continental-crust bodies of felsic (andesitic) composition, in the geologic past intermittently emergent or covered by epicontinental seas, at present standing mainly above sea level and demarcated by the lower (outer) continental slopes and, where present, deep-sea troughs or trenches. Geologic boundaries of the North American continent are not everywhere distinct (Fig. 1): there are no clear boundaries with South America and Asia (the reader is encouraged to examine this and following figures before reading further, to familiarize himself/herself with geologic regions, localities and place names which will be mentioned in the text). The most apparent are the Atlantic and Pacific margins in the U.S., but in Canada the geologic outline of the continent is in many areas still in dispute (Grant, 1987; Lyatsky, 1996).

Continents as tectonic units are quite different from the surrounding oceanic-lithosphere domains. They are distinguished by their average seismic velocity structure (crustal thickness in the tens of kilometers); geochemistry (enrichment in incompatible trace elements), and petrology (andesitic composition); presence of all rock types known on Earth, in various associations and combinations and with rocks of all ages starting with >4,000 Ma (Anderson, 1995; Rudnick, 1995). Regions of oceanic lithosphere are, by contrast, dominated by thin crust (a few kilometers in thickness) of uniform composition (mostly mafic).

Continental crust is strikingly variable vertically and laterally, and a great variety of rock types and rock-body structures is found on continents. Modern geochemistry and petrology offer many ways to associate the elemental and isotopic systems and petrologic characteristics of igneous and metamorphic rocks with their melt sources and protoliths. Genetic hypotheses about the origin of continental rock types and their links with specific tectonic settings are abundant and usually controversial; also disputable are many physical and theoretical models of rock behavior at great depths. Nonetheless, some fundamental ground truth about rocks can be established. There are igneous rocks, which originate at depth by hot melting, and sedimentary rocks of surficial exogenic origin. The latter may be clastic or biologically or chemically precipitated. Rocks of all types are altered after they are formed, in places reaching very high grades of metamorphism.

It is possible that the base of the lithosphere is related to phase changes in the material. Transformations of rocks in the uppermost mantle and crust are probably less drastic, although the state of rocks at the lower-crustal levels is still known poorly. However, in places rocks that were previously at lower-crustal depths are now exposed at the surface, where they are available for observation. Subcrustal and deep crustal samples are available in xenoliths.

From the distribution of rock types and rock-made bodies, the continents as geologic entities consist of two principal types of lateral whole-crust (and whole-lithosphere) tectonic units: the vast, relatively stable cratons, and the big mobile belts (sometimes called loosely "orogens"). The biggest parts of continents are the cratons. Most of the stratigraphic rock record was formed when the cratons were covered by shallow epeiric seas. Water depths in these seas have at different times in the geologic past ranged from several tens to several hundred meters

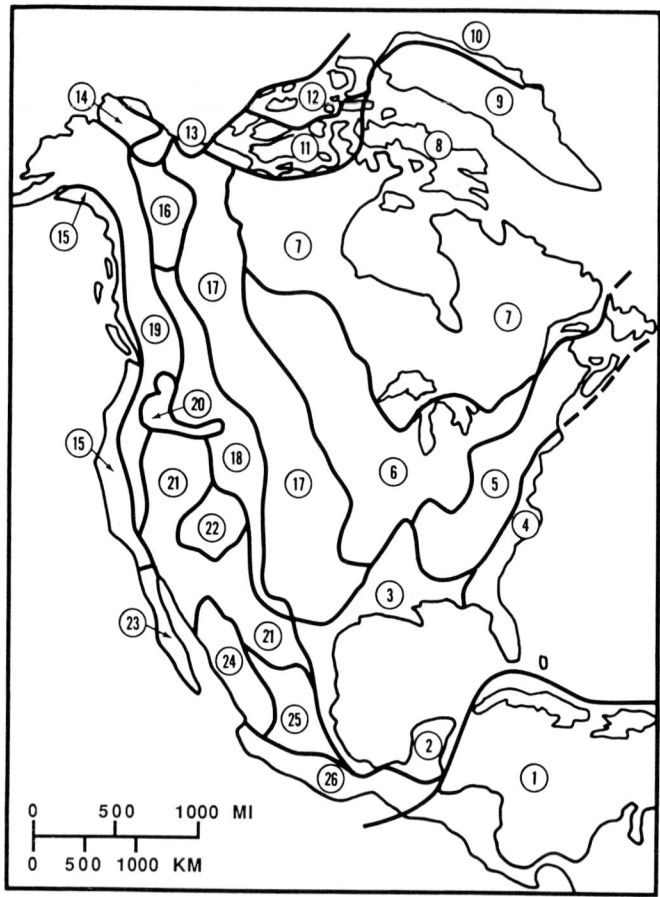

Figure 1. Main geomorphologic provinces of the North American continent and vicinity (modified from Bally et al., 1989). Geomorphologic provinces: 1 - Central America and Caribbean; 2 - Yucatan platform; 3 - Gulf of Mexico coastal plain; 4 - Atlantic coastal plain; 5 - Appalachian mountains and plateaus; 6 - Central lowland; 7 - Canadian Shield; 8 - Baffin Island; 9 - Greenland ice cap; 10 - West and North Greenland mountains and fiords; 11 - Arctic lowland; 12 - Innuitian; 13 - Arctic coastal plain; 14 - Brooks Range; 15 - Pacific Rim and Pacific Coast Ranges; 16 - Mackenzie; 17 - Great Plains; 18 - Rocky Mountains; 19 - Interior plateaus and mountains; 20 - Columbia Plateau; 21 - Basin and Range; 22 - Colorado Plateau; 23 - Baja California; 24 - Sierra Madre Occidental; 25 - Sierra Madre Oriental; 26 - Sierra Madre del Sur.

(~300 m in the Cretaceous); the present time is characterized by predominant emergence.

On the North American continent, the craton continues from the Appalachian mobile megabelt to the Cordilleran, and from the Arctic Ocean almost to the Gulf of Mexico (Figs. 1, 2). The boundaries and extent of the modern North American craton and its main parts are often defined in the literature only loosely, and with some confusion between geographical and geological definitions (King, 1977). In a physiographical sense, Flint and Skinner (1974) restricted the craton to only the Canadian Shield, and regarded the Canadian Shield as an opposite of the Central Lowlands and Great Plains. In more geological descriptions, cratons are frequently defined as rather flat areas, whether blanketed by a cratonic sedimentary cover or not, contrasting with deformed mountainous areas (cp. Sloss, 1988a). However, some mountainous areas, such as the very broad Rocky Mountains in the U.S., can in tectonic terms be cratonic. King (1969) described the Canadian Shield as an assemblage of Precambrian foldbelts, and distinguished the cratonic platforms as structurally bipartite entities, with a cover overlying a crystalline basement of different ages.

In Europe, the cratons were originally distinguished as a stable and rigid mass, as opposed to the intensely folded and faulted Alpides. The term *kratogen* (which means "strength-generating") has been used since the early 20th century to describe tectonic entities strong enough to resist the forces that caused orogenic deformation. Stille (1924, 1941) simplified this term to *kraton* (with the accent on the second vowel). Kay (1951), who applied these ideas to North America and Anglicized the German spelling of this word to *craton*, though he still pronounced it the German way. As this word entered the mainstream North American geologic literature, it soon acquired its current English pronunciation. For Stille (1941) and his contemporaries, cratons were whole-crust entities consisting of shields and platforms. Shields, for them, were merely regions where the crystalline basement of cratons was exposed, and tectonic platforms (see also Suess, 1906) were simply regions complementary to shields (some authors referred to platforms as *tables*). The commonly accepted defining characteristic of platforms was the presence of a thick, flat-lying sedimentary cover.

Indeed, the etymology of "platform" is from the Greek *platys* (flat) or the French *plate-forme* (flat form) (Suess, 1906). In terms that are largely stratigraphic, platforms are considered to be those parts of the cratons which are covered by flat-lying or gently tilted strata (see Friedman et al., 1992). To King (1969, 1977), platforms were largely structural entities, morphologically smooth and flat, as opposed to geosynclines. As the classical geosyncline theory declined, the very essence of this notion lost its clarity. The limits of cratons, particularly of the modern North American craton (e.g., Sloss, 1988a), became uncertain.

Figure 2a. Tectonic index map of North America, showing main tectonic provinces (simplified from Bally et al., 1989). Abbreviations of tectonic provinces: BR - Basin and Range; BRO - Brooks Range; F.B. - fold belt; MK - Mackenzie Mountains; MA - Marathon uplift; OU - Ouachita Mountains.

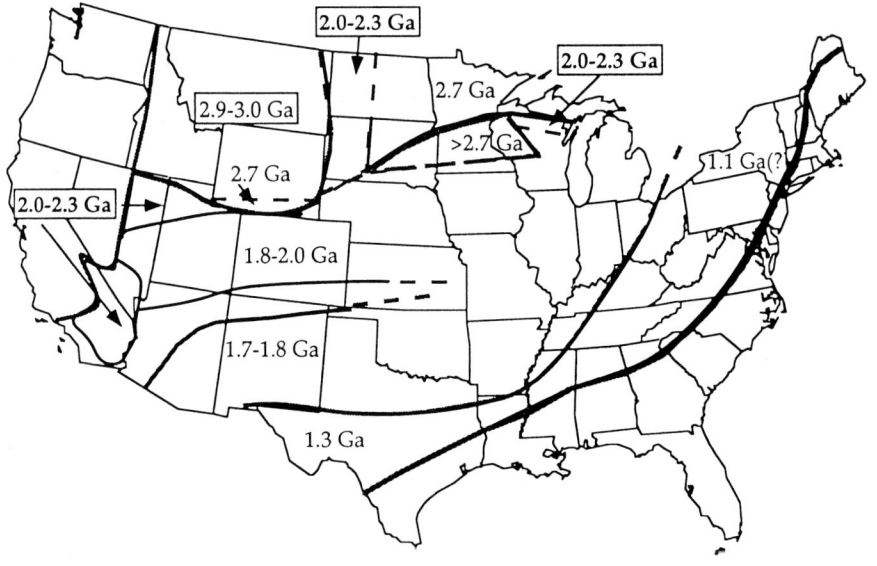

Figure 2b. Crystalline-crust age provinces in North America (modified from Reed et al., 1993).

8

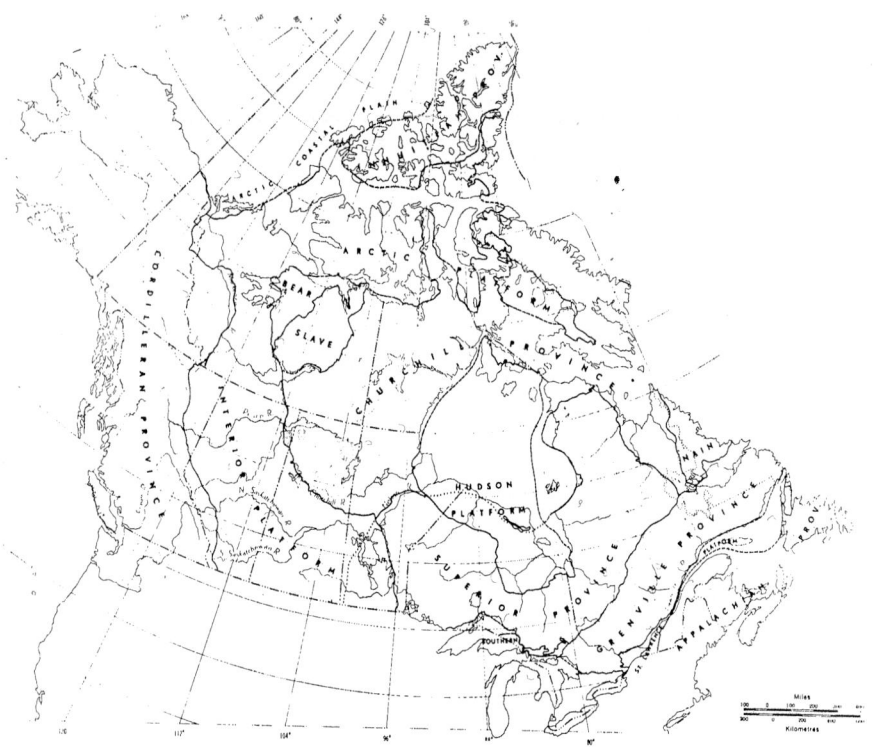

Figure 2c. Main tectonic provinces of Canada (modified from Ruzicka, 1987).

Still, contrasting tectonic regions do exist on continents: mobile megabelts are mostly coincident with mountainous regions; relatively immobile platforms mainly underlie areas of low relief. These two types of region are distinguished not only by their landforms, but also by their mappable geologic characteristics and by the properties of their entire crust and lithosphere.

Geographical continental lands account for only about one-third of the Earth's surface. The geologically defined crustal (and lithospheric) continent masses cover about 40% of the globe. The total continental crust, including submerged and exotic-looking continental-crust blocks in oceanic-crust regions (such as Rockall Plateau or Seychelles Islands block), occupies almost half the globe (cp. Cogley, 1984; White, 1988; Anderson, 1989; Kuvaas and Kodaira, 1997). On several lines of geochemical and geophysical evidence, the continents, including North America, have roots deeper than 400 km (e.g., Gossler and Kind, 1996; also re-view by Lyatsky, 1994a), much deeper than the elastic layer required to support even isostatically balanced mountains (estimated to be >100 km thick; Anderson, 1995). From seismic velocity studies, Silver and Chan (1988) found evidence for deep-seated material of high viscosity under the Canadian Shield. Pinet et al. (1991) also suggested that lithospheric structure under the Canadian Shield is different than under the extensive cratonic platforms to the south.

Seismic anisotropy in the deep roots under the Canadian Shield matches the Archean geologic grain of the Superior province (Silver and Chan, 1988). Armstrong (1991) argued that most continental-crust masses appeared in the Late Archean. Rudnick (1995) agreed that ~70% of present-day continental masses probably existed since the end of the Archean. Continental cratons appear to have been rooted deep in the mantle for most of their history. The most stable and compositionally distinct cratonic roots are common under shields (e.g., Anderson, 1995) which tend to stand higher than platforms for protracted periods of time. This implies that within cratons, long-lived, stable, deep roots are a fundamental peculiarity that distinguishes shields from platforms, and initial cratons from younger mobile megabelts. Cratons and their principal constituents, shields and cratons, are not just structural but wholly tectonic units intrinsically typical for continents.

Interaction of indigenous and external tectonic forces acting on evolving cratons

The recognizable thermo-tectonic, tectono-magmatic, tectono-metamorphic and tectono-deformational processes that are conventionally combined under the name *orogeny* have affected the North American continent since Archean time (the Kenoran orogeny around the Archean-Proterozoic boundary). The next such episode, dated as Early Proterozoic, is called Hudsonian. Huge bodies of granitoid and anorthosite composition, accompanied by a wide variety of metallic mineral de-

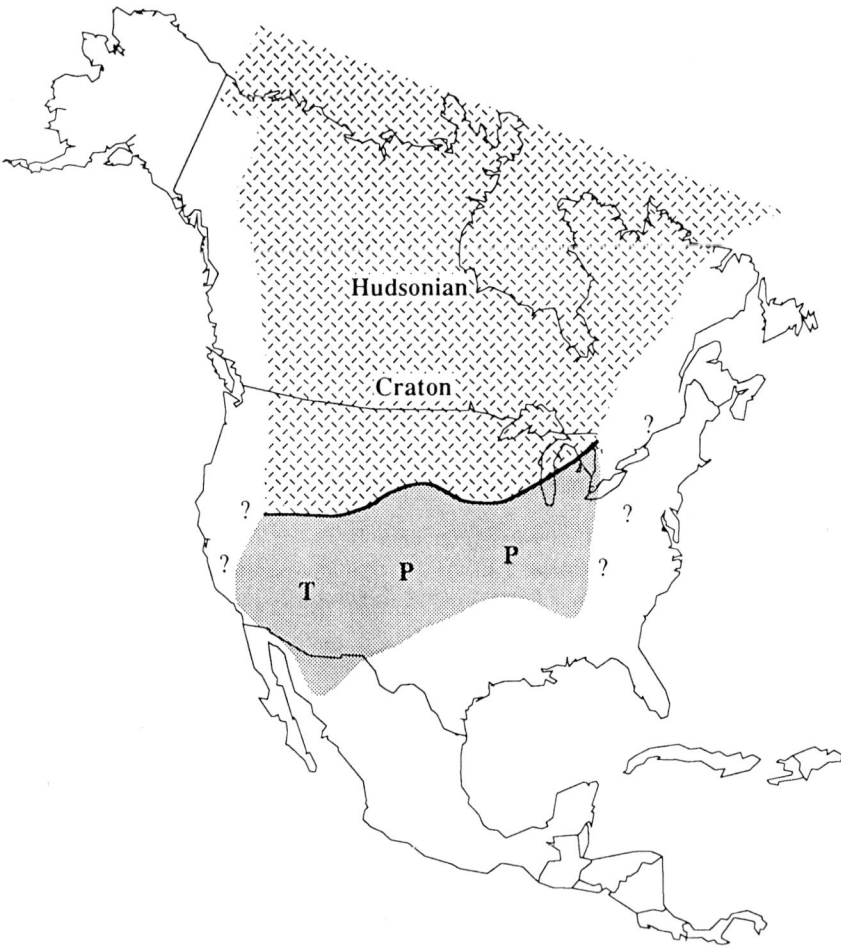

Figure 3a. Approximate known extent of the Early Proterozoic Laurentian craton (after 1,800 Ma) in North America, showing also the location of the Transcontinental Proterozoic provinces (TPP). Modified from Van Schmus et al. (1993).

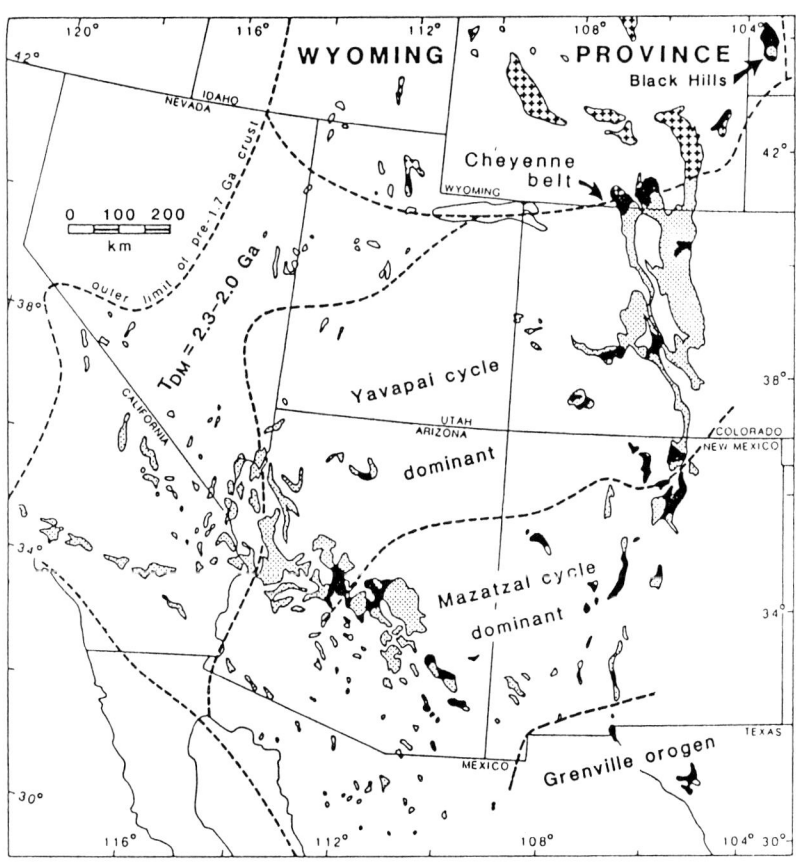

Figure 3b. Distribution of Early Proterozoic volcanic and sedimentary rocks (solid) in the cratonic crystalline basement of the Phanerozoic Midcontinent and Interior cratonic platforms. Plutonic rocks are shown in stippled pattern. Modified from Hoffman (1988).

posits now mined in the Canadian Shield, intruded into the Archean geologic framework during that time.

These main orogenies caused extensive areas of cratonized crust to appear across much of the continent (Fig. 3). Cratonization of lithosphere in North America continued in the Mesoproterozoic and early Neoproterozoic. Later orogenies partly reworked the Proterozoic cratons, mostly in two mobile megabelts, Appalachian and Cordilleran, on the eastern and western rims of the continent. Evolution of these orogenized mobile megabelts continued into the Paleozoic to early Mesozoic (in the Appalachians) and even Cenozoic (in the Cordillera).

As components of a single tectonic entity - the continental-crust mass - cratons and their rimming mobile megabelts do not develop in isolation from one another: mutual influences left traces in their geologic record. These influences, understandably, were the strongest in zones of contact between these contrasting lateral tectonic units. Existence of transitional miogeosynclinal belts between craton and megabelt hinterlands is vivid evidence for interaction of cratonic and mobile-megabelt tectonic processes. Miogeosynclinal belts are found in the western Appalachians and eastern Cordillera. The most impressive of the transitional tectonic features are grandiose thrust belts verging from the mature mobile megabelts cratonward. On the other hand, intermittent expansion of cratonic sedimentation settings occurred all across one of the most impressive such zones in the world - the Rocky Mountain Belt of folding and thrusting, which runs from Mexico in the south to Alaska in the north (Fig. 4).

Definition of tectonic boundaries of ancient and modern cratons with the mobile megabelts during each particular time interval has important practical implications in locating the areas of interest for petroleum and mineral exploration (e.g., Blockley et al., 1989; Finlayson et al., 1989). Proper understanding of the timing of craton-sourced and orogen-sourced influences in the transition zones between cratons and megabelts would improve the accuracy of resource predictions and reduce the risk inherent in exploration efforts.

Recognition of vertical and horizontal tectonic units in cratons

Any tectonic unit in the crust and lithosphere is volumetrically and laterally restricted. Vertically, they are limited by the brittle-ductile rheological transition in the mid-crust, by the base of the crust or lithosphere, or other subhorizontal discontinuities. Laterally, these tectonic units are delineated by the breaks in their main defining characteristics.

The physical concept of lithospheric layering comes from the basic laws of physics, and is supported by considerations from seismic, gravity and magnetic data. Recognition, from gravity data, that somewhere in the Earth there must be a level

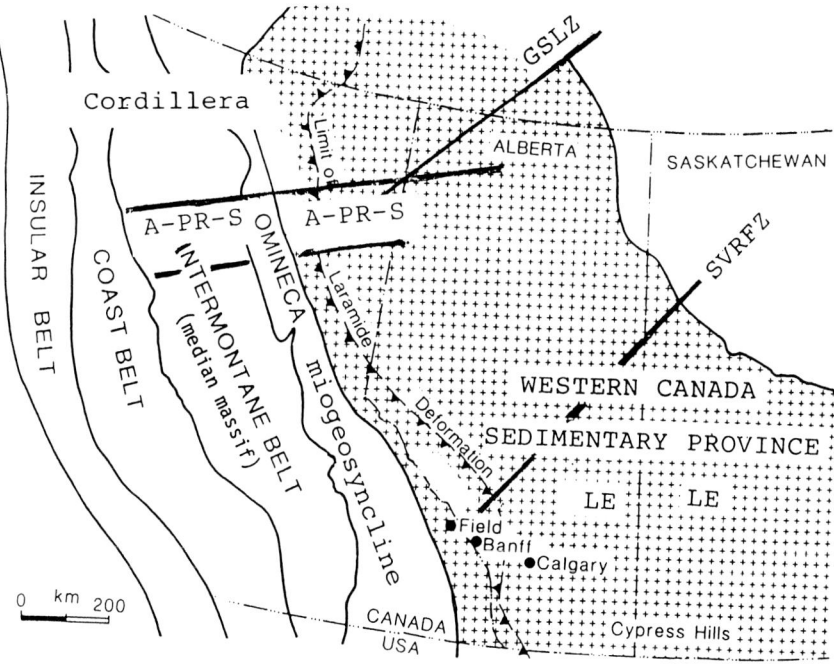

Figure 4a. Location of the Cordilleran miogeosyncline in Canada, in relation to other major tectonic features in the craton and the Cordilleran mobile megabelt. A-PR-S - Athabasca-Peace River-Skeena zone of crustal weakness, which continues from the craton across the grain of the Cordilleran mobile megabelt. LE - early Paleozoic Lloydminster embayment.

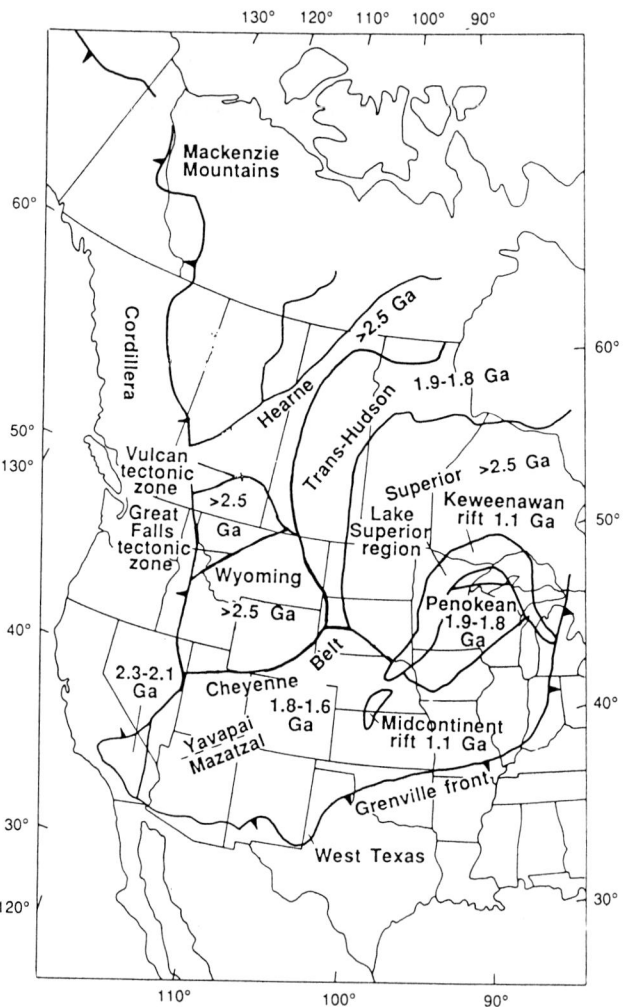

Figure 4b. Precambrian tectonic zoning of the central and western part of the North American craton, according to Link et al. (1993, with modifications). An updated zonation is proposed in this book. Additional details are shown in Figure 48a.

of isostatic compensation has led to the notion of a mechanically strong lithosphere floating on a weaker asthenosphere (Barrell, 1914). Thermal demagnetization of rocks deeper than the Curie isotherm has led to the notion of the magnetic layer. Other simple considerations and experiments point to rheological breaks, finite effective thickness, and so on, depending on the properties of particular rock types at different pressure and temperature conditions in the lithosphere.

Vertical variations in the lithosphere, recognized as layering, arises also due to changes in rock type: Archean and post-Archean, crystalline basement and sedimentary cover, and so on. The most apparent and geologically mappable kind of layering is that visible in stratal rocks. Bodies of stratified rocks predominate in many regions. Gravitation and heat are crucial factors that govern magma differentiation, fluid flow, spread of lavas, chemical precipitation from solution, clast transport, deposition of sediments. Many of these processes result in the formation of *strata*. This general collective term is usually (and loosely) applied to all layers that were spread out one at a time. A stratum is an elementary tablet-like body. Conventionally, the smallest such bodies (centimeter-thick strata, for example, are found in medium- and fine-textured sands, silts and clay-size sediments) are not regarded as separate strata and are instead treated as structural or textural characteristics of beds; likewise, cross-bedding is treated as being an internal characteristic of a clastic rock body.

A rock layer, like any rock-made body, is bounded by breaks in rock characteristics. For the purposes of geologic mapping in English-speaking countries, principal unit of stratal rocks of the same age and composition is called a *formation*. Formations are sometimes combined together into *groups* and *supergroups*, and their subordinates are *members* (North American Stratigraphic Code, 1983). In the Russian/Soviet Stratigraphic Code (Zhamoida, ed., 1979), equivalents of formations are stratal *suites*. The most important characteristic of formations (suites) is their mappability based on their distinct lithologic identity, easily distinguishable from that of underlying and overlying formations. Such a lithologic criterion enables the definition of formations right in the field, even before their rock ages are determined. Formation (suite) is the most common formal unit used to subdivide the whole stratigraphic record into named units on the basis of their lithostratigraphic characteristics.

A regional vertical rock record can be divided based on any criteria: depositional, metamorphic, deformational, and so on. The first reflects a commonality of sedimentation settings, the second a metamorphic overprint of older rock bodies, the third internal structural characteristics of stratal units. Correlation of rock units from unmetamorphosed to metamorphosed domains may be very complicated, because while in low-grade metamorphic successions the original stratification is well recognizable, in high-grade rocks the imposed metamorphic zoning ignores the older rock-body boundaries and only new homogenized entities can be mapped. In most Precambrian high-grade metamorphic-rock terrains on cratons, the original character of formations is often obliterated almost completely. Though in many cases protoliths can be established, the genetic nature of the protoliths is usually questionable.

Early approaches to subdividing the crystalline metamorphic basement of the North American craton

Post-metamorphic boundaries of tectonic blocks can be delineated from sharp lateral discontinuities in the structural-metamorphic patterns of the mapped terrains. In contrast, pre-metamorphic boundaries of crustal blocks may be difficult to determine. Especially difficult is correlation between different blocks, particularly in the crystalline basement of tectonic platforms. The common sparseness of information necessitates great caution in the interpretations of the basement. Collerson et al. (1990) noted that regional metamorphism in western Canada had obliterated many original rock characteristics in the cratonic basement even in its relatively young parts. Importantly, through geologic mapping (also Lewry and Collerson, 1990) it is now possible to recognize several levels in the Canadian Shield. There is a level of Archean basement, and a level of Early Proterozoic Hudsonian (~1,900-1,800 Ma; Table IA) orogenic grain. These levels are in places covered by Proterozoic platformal and basinal successions of different ages, some of which bear an overprint of Hudsonian tectonism. Currently, there is a tendency to treat the distinct structural-compositional (structural-formational) packages as detached and displaced thrust sheets. But in many areas, Early Proterozoic platformal assemblages of sedimentary and volcanic rocks overlie the Archean and earliest Proterozoic basement in a normal, upward-younging stratigraphic succession.

Platformal covers of late Paleoproterozoic and Mesoproterozoic age (at least 1,300 Ma) are now recognized over much of the North American craton in Canada and the U.S. (e.g., Campbell, ed., 1981; Reed et al., eds., 1993; Burwash and Muehlenbachs, 1997). The Proterozoic volcano-sedimentary cover successions are commonly intruded by granitic plutonic bodies described as non-orogenic. These granitic bodies are comagmatic with abundant rhyolitic volcanic strata Proterozoic platformal cover of different ages.

The nature of boundaries between the contrasting packages of metamorphic rocks may be variable: original stratigraphic, structural (in the case of thrust sheets), rheological. Where vertical stacks of rock complexes are exposed, their original boundaries are often recognizable despite the metamorphic overprint; sometimes it is possible even to map depositional unconformities between them. But mainly, compositional and structural contrasts, considered together, allow to separate dissimilar successions into a regional geologic record. Different primary tectono-sedimentary and tectono-magmatic characteristics, as well as variable secondary tectono-magmatic and tectono-metamorphic characteristics with tectono-deformational overprints of several generations permit to distinguish tectonic units in a regional record.

A method of tectonic analysis based on integration of compositional and structural characteristics of stratal complexes intruded by discordant paragenetic bodies was developed by a Swiss geologist Wegmann (1935, 1956), who discerned and classified big complexes of metamorphic rocks in Greenland. He designated such complexes as *stockworks*, emphasizing their complicated nature. But the term *stockwork* was and still is used to denote mineral deposits of very complex shape, and to utilize it for vertical tectonic units is awkward.

Since the principal time of cratonization in the Archean and Paleoproterozoic, tectonism in the continental-crust regions has found an expression in the sedimentological and volcanic characteristics and subsequent burial-related alterations of the cover. Where the burial was not deep, regional metamorphism of the cover rocks was minor, cryptic or even absent. Folding in the cover is also restricted, usually to narrow marginal zones along the flanks of mobile megabelts. In this sense, cratons may be defined broad stable parts of continental masses; *hedreocraton* is a long-persistent craton. But at deeper levels in the cratonic basement, metamorphic processes and fold/fault deformation never ceased.

Traditionally, since the 19th century, cratonic platforms have been considered as tectonically rather inert. For example, Sloss (1988a, p. 49) stated that in the North American craton "...the cratonic margins were largely passive and the interior was riven by extensional forces; flexural cratonic downwarpings were rare except as sequels to rifting..." This dull scenario was thought to account for the visible rather uniform properties of the platform cover and landscape, but it is not a realistic description of the tectonism of the North American and other cratons. Throughout their >2,800-Ma history, many generations of cratons on the North American continent were nothing like inert. In the modern North American craton outside the Canadian Shield, during the last ~600 m.y. long-lived subsidence made room for the vast Phanerozoic sedimentary cover. Epeirogenic movements are broad vertical movements of the lithosphere, unaccompanied by crumpling of stratal rocks in the cover and producing crustal upwarps and downwarps of different sizes (see Friedman, 1987a-c; Friedman et al., 1992). Downward epeirogenic movements enable thick sedimentary successions to accumulate. Upward movements produce raised continental features, such as plateaus.

Typical for the modern cratons are low-relief topography, broad structural upwarps, and slightly disturbed stratal-rock bedding. But these gently and laterally persistent near-surface manifestations of whole-lithosphere platformal tectonism recorded in the sedimentary cover obscure the ancient zoning in the metamorphic grade and patterns of partial melting and structural deformation of crystalline rocks in the basement.

Presence of a platform does not necessarily mean platformal structures exist below the basement-cover contact. Since cratonization, common for the basement and cover have been vertical tectonic displacements: upwarping, downwarping and block movements. But tectono-metamorphic, -magmatic and -deformational processes differed as a function of depth and corresponding pressure and temperature conditions. Nonetheless, the manifestations of tectonic platforms are not just shallow. Platforms are whole-crust or whole-lithosphere tectonic units, distinguished from others as specific spatio-temporal geodynamic systems.

Nomenclature of lateral tectonic units relevant to tectonic platforms

As geology evolved, it was diversified into many interrelated disciplines: mineralogy and petrology, sedimentology and structural geology, geological (geo)chemistry and (geo)physics, etc. Each discipline acquired its own thesaurus, partly distinctive but all too often overlapping. From this, semantic inconsistencies arise between distinct branches of geology. Terminological confusion ensues, with the same words meaning different things to specialists in different branches. Example are terms like *facies, sequence, rift* and others. In geology today, this problem is compounded by the common lack of knowledge of the history of ideas, notions, and theories, and the corresponding terminological systems and their epistemological roots. Besides, there is a temptation for modern researchers to appear original by inventing new names even for known natural phenomena, or for rediscovered old ideas. As a result, the terminological jargon mess gets bigger, but not necessarily healthier. Besides, many tectonic terms are often treated loosely, such as *craton, orogen, platform, sedimentary basin*, etc. This compels us to formalize the terminology used in this book.

Cratons are first-order lateral tectonic units of continents; so are mobile megabelts. Cratons are resistant to rapid and extreme vertical movements of the type common in mobile megabelts, and to folding of rocks at the level of brittle upper crust. Fundamental components of cratons are *shields* and *tectonic platforms*, distinguished by the absence or presence of a flat-lying sedimentary blanket or cover over the metamorphic crystalline rocks in the basement. In the modern cratonic platforms, the "basement" is taken to be everything beneath the Late Proterozoic and Phanerozoic unmetamorphosed, subhorizontal, continuous (except for some vertical offsets) sedimentary cover. Though the volume of the cover was different at different stages of a craton's evolution, the basic distinction between a metamorphosed and folded basement and an unmetamorphosed and continuous cover remained.

Pulses of intense tectonic reworking sometimes turned the older platformal cover into a basement for younger platformal successions, and the basement-cover contact shifted upward in the crust. Relative to a specific reworking episode, tectonic rock units may be "pre-orogenic", "syn-orogenic", and "post-orogenic".

Tectonic platforms can be of two types: (a) fully cratonic, part of an ancient craton overlain by subhorizontal (volcano-)sedimentary cover (e.g., North American Midcontinent and Interior, Russian, Siberian, Korean, Yangtze platforms), and (b) part of a leveled and subsided younger (usually no older than Mesoproterozoic or Neoproterozoic) mobile megabelts, where orogenic processes ceased and the terrains have been covered by undeformed and unmetamorphosed sedimentary successions (North German Variscan, Central Asian Hercynian, South Midcontinent Acadian to Alleghenian in North America). Cratonic tectonic platforms are different from tectonic platforms formed over the younger extinct orogenic belts: they are stiffer, and have a sharper basement-cover discontinuity representing an unconformity of long duration.

In this book, clear distinctions are maintained by frequent resort to descriptive adjectives: tectonic platform vs. carbonate platform, cratonic platform vs. younger one.

Cratonic platforms pass through several distinct evolutionary stages. Most of them occurred before the deposition of the cover. Punctuation of the sedimentary cover by unconformities reflects alternating episodes of tectonic subsidence and uplift of various areal extent. Due to eustatic changes in global sea level, tectonic uplift and subsidence are sometimes difficult to recognize. Marine transgressions easily spread epeiric sea conditions over the flat platforms. But relative to mobile megabelts, lithostratigraphic record of tectonic platforms, especially those of cratonic origin, reveals considerable tectonic quiescence during the lifespan of the sedimentary cover.

The platformal cover itself consists of many semi-separate bodies with greatly variable lithofacies and thicknesses. Such bodies may be terminated by lateral and vertical facies changes or thickness reductions and complete wedge-outs. The largest such rock bodies are conventionally designated *sedimentary basins*. (By contrast, depressions in which sedimentary basins form are referred to as *basins of sedimentation*.) The rimming areas, marked by reduced thicknesses and wedgeouts of stratified rocks, are usually distinguished as elongated *arches* or round *domes*. There, stratigraphic horizons may be thin or absent, forming condensed sections or stratigraphic hiati, due to tectonic elevation of the bedrock (be it crystalline basement or older parts of the cover).

Platformal sedimentary or volcano-sedimentary blankets, lying on top of an orogenically-formed Precambrian crystalline basement, were laid down after cratonization and peneplanation of pre-existing paleotopography. The first blankets were volcano-sedimentary basins, discontinuous and usually separate. The later cover

over the cratons was more laterally continuous and extensive, and predominantly sedimentary.

All the sedimentary and volcano-sedimentary basins can be regarded as similar phenomena from the sedimentological perspective. There must be areas of sediment provenance, and topographic depressions and/or catchment areas (e.g., at the foot of the mountains) where deposits can accumulate and be preserved. Huge basins are created in deltaic and continental-slope sedimentation settings. Unconsolidated sediments there can be re-transported by tides and sea-bottom slumps, with or without direct tectonic influences.

Transformations of the notion of *miogeosyncline*

Kay (1951) lumped different basins - platformal and non-platformal, epicontinental, shelfal, deep-water, orogenic, tectonic and non-tectonic, unmetamorphosed or metamorphosed - together and called them geosynclines (of various types). His classification was popular at first, but was soon found impractical and sensibly abandoned, especially after the general acceptance of plate tectonics. Also abandoned were many of the terms he had borrowed from his predecessors (notably Stille, 1924, 1936a-b, 1941). The term *eugeosyncline* (in its original sense, inner zone of a mobile megabelt, where volcanic rocks are present and granitic batholiths were intruded during the orogeny) has been avoided by most modern geologists because of its confusing associations with assumed exotic-terrane regions. The term *miogeosyncline* (or its more-recent variants *miogeocline* and *miocline*; e.g., Dietz, 1966, 1972; Dietz and Holden, 1974 and references therein) is still used commonly but very loosely. The idiosyncrasy against the original geosynclinal theory has been extended even to the word *miogeosyncline*, and many modern authors prefer to use *miocline* instead. But this variant bears only structural (rather than tectonic) meaning, and even that meaning is incorrect because these features are neither underdeveloped (*mio*) nor inclined in any gentle or simple fashion (the inclinations are actually considerable, and in different directions). Unsurprisingly, Bally (1989) still preferred the original complete spelling, miogeosyncline.

Sloss (1988a, p. 2) noted that "with the enhanced clarity of vision afforded by plate tectonics, we now identify miogeoclines (also known as *shelf prisms* or *shelf wedges*) as subsident passive (or divergent) continental margins" (italics his). To us, however, a miogeosyncline is a transitional zone between a craton and a mobile megabelt, which has been affected by orogenic processes. This is closer to the initial designation of *miogeosynclines* as outer zones of *orthogeosynclines* (Stille, 1924, 1941). Stille placed miogeosynclines along the margins of mobile megabelts, cratonward from the megabelt hinterlands, and characterized them by a lesser degree of magmatism and metamorphism than in eugeosynclines (which lie in core zones of megabelts) but typically by intense deformation including nappes.

Kay (1951) missed the essence of Stille's definition of all geosynclines as whole-crust tectonic units rather than just near-surface basins. Misuse of the term *miogeosyncline* still goes on. Dietz and Holden (1974) linked miogeosynclines with the Atlantic continental margin of the U.S., and since then many workers have used this term simply to denote the marginal zones of continents.

Semantically, the term *geosyncline* has never been very revealing. Originally, the Appalachian mobile megabelt (orogen) was considered the type example of geosyncline, but its distinction from the Midcontinent Platform (initially regarded as just plains) was based on the increased thickness of stratal rocks to the east; folding was at first thought to be a consequence of syn-depositional subsidence (Hall, 1859) but was later ascribed to a separate set of processes and events (Dana, 1873). The Plains were later assigned to the North American craton, the mainly sedimentary geosynclines of the Appalachians and the Cordillera to miogeosynclines, and the heavily magmatized axial zones of these mobile megabelts to eugeosynclines (e.g., Kay, 1951; Rich, 1951; Gilluly, 1963; King, 1977).

By their geologic properties and position, miogeosynclines have long been regarded as transitional tectonic entities between cratons and mobile megabelts, but this was never strongly formalized. Mobile megabelts across Eurasia (Fig. 4c) generally have a more complex and less regular temporal and spatial distribution of orogenic zones with miogeosynclinal and eugeosynclinal characteristics than do such zones in North America. Often, these entities occur in vertical as well as lateral stacks, with non-systematic repetitions (e.g., in the Eastern Sayan). In the hinterlands of the Urals, for example, ophiolitic formations developed twice, in the late Precambrian and in the Ordovician. On the other hand, in foldbelts like Verkhoyansk and Salair in Siberia, mafic magmatism was not extensive at all. Substantially felsic are many orogenic belts in Kazakhstan and Central Asia (e.g., Northern Tien-Shan), which contain predominantly carbonate and clastic rocks, plutonized with only a moderate amount of granitoids. In the eastern Canadian Cordillera, many extensive carbonate platforms conventionally assigned to a miogeosyncline (see below) are reliably correlated with similar and coeval rock successions in the tectonic Interior Platform on the craton.

Incomplete orogenic development of miogeosynclines has to do with the only partial reworking of the cratonic basement on which these miogeosynclines developed. It is evident that ex-cratonic basement underlies not just the miogeosynclines in the Appalachians and Cordillera, but also some interior zones of these mobile megabelts as well. It is a commonplace that an ex-cratonic basement, only partly reworked, underlies the Canadian Cordillera as far west as the Omineca and even Intermontane belts (Brown et al., 1992; Marquis et al., 1995). Basement windows in uplifted blocks are known in the Alps, Caucasus, Altai, Sayans, Appalachians and other orogenic areas in various parts of the world (Spizharsky, ed., 1973; Huang, 1980; Krasnyi, ed., 1980; Rodgers, 1995). Little-reworked

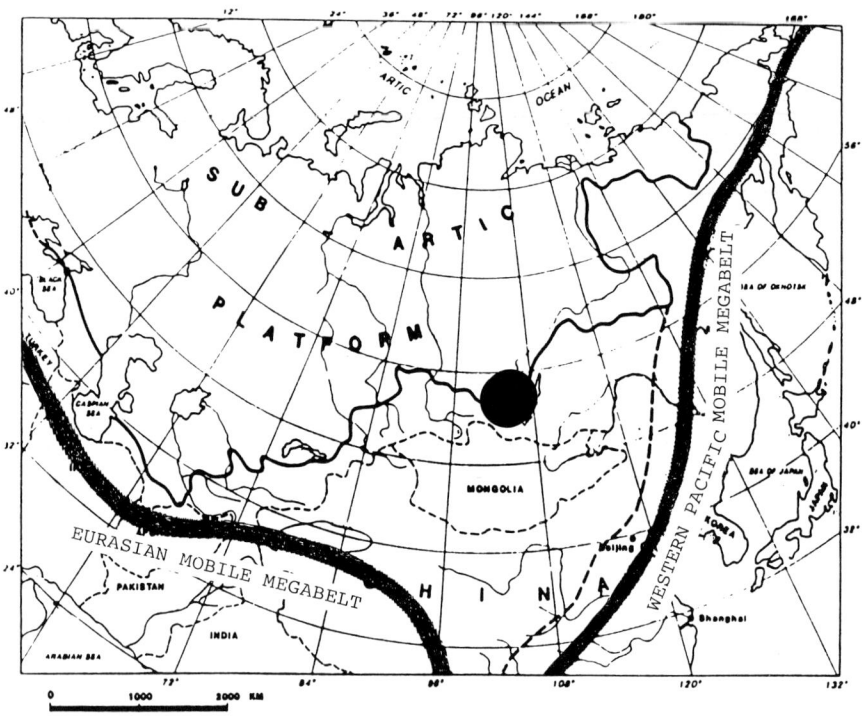

Figure 4c. General tectonic index map of northern Asia, showing the main platforms and mobile megabelts (modified from Lyatsky et al., 1988). Area marked by black circle is shown in more detail in Fig. 78b.

basement in megabelts occurs in median massifs (e.g., Intermontane and Yukon-Tanana in the Canadian and Alaskan Cordillera). In miogeosynclinal orogenic zones as well, old cratonic basement remains only partly reworked.

Review of previous approaches to discrimination of tectonic units in platformal sedimentary cover in the North American craton

The crystalline basement of cratonic platforms may be same age as or even older than the basement in the adjacent mobile megabelts. Archean basement underlies the Early Proterozoic platforms delineated in the Canadian Shield, and Early Proterozoic crystalline basement lies beneath the volcano-sedimentary basins of later Proterozoic and younger ages. The continuous sedimentary platformal cover of the North American craton rests on a basement whose age and nature are heterogeneous. But all its pre-Phanerozoic vertical and lateral tectonic units are distinct from the cratonic cover by their metamorphism and folding. This provides a reliable criterion to differentiatethe basement and the cover as two principal vertical tectonic units of modern platforms. But each is divided into vertical tectonic units of a second or third order. For practical purposes, such division is particularly important in the sedimentary cover, which in many regions is the object of resource exploration and environmental studies.

Depending on the perception and criteria, a regional stratigraphic record can be regarded as continuous or discontinuous. The criteria for its division are chosen for the specific purpose of a particular classification. They may be purely lithic, reflecting rock types or rock suites of the same setting, paleontological, rock-mechanical, hydrogeological, tectonic. Lithostratigraphy provides an apparent natural segmentation of rock successions based on visible lithological contrasts in stratified rocks. Strata are grouped into sequences (and sequence sets), series, suites, formations, groups or members which on the basis of their lithological characteristics can be similar or dissimilar to various degrees. As the science of geology advanced, paleontological studies provided another reliable basis for definition of units in the rock record. For a long time, it was a sole clue for correlation of the numerous regional stratigraphic records. Faunal correlations were established over broad areas with very differentlithologies but within the same faunal provinces. Biostratigraphic boundaries are also natural, and major unconformities with large time gaps between the rocks below and above were used as the most apparent breaks dividing units in the geologic records. "Naturalness" became a principle used to formalize the most fundamental time-stratigraphic sequences, denoted as *systems*. All the original systems in their stratotypes were bounded by unconformities.

The need to standardize worldwide the correlation of fundamental biostratigraphic taxonomic units was discussed as early as in the 1880s, at the Second International Geological Congress in Bologna. With growing urgency, these discussions went on after that, because detailed mapping demonstrated complications in correlation between natural faunal provinces. The development of radiometric dating techniques confirmed that biostratigraphic geologic breaks in different provinces commonly occurred at different times. A dichotomy developed between biostratigraphy and chronostratigraphy. Ager (1984, p. 98) described the continuing confusion as follows: "...The best level at which to place a boundary is, paradoxically, the level at which it is least obvious". The principle of "naturalness" had to be rethought.

Making a distinction between chronostratigraphy (which is truly universal, in the sense that it is applicable worldwide), biostratigraphy (which is in essence provincial) and lithostratigraphy (which depends even more on local paleoenvironmental conditions) brought new thinking on what constitutes a natural division of a stratigraphic record. After many discussions, at least one fundamental principle to emerge was that "...all chronostratigraphical units should be defined by means of their bases only, the tops being defined automatically by the base of the succeeding unit..." (Cope, 1996, p. 107). The same, of course, applies to the designation of tectonic units, particularly those in a platformal sedimentary cover.

Even though such radical revision of "naturalness" has been under way since the mid-20th century, a group of U.S. geologists continued to look for "natural" boundaries to divide the platformal stratigraphic record in North America (Sloss et al., 1949). They accepted as natural only those unit boundaries which were related to unconformities, and efforts continued to establish a stratigraphic-system scale that would supposedly be "natural" for North America (Sloss, 1963). The general geologic time scale whose stratigraphic systems were historically based on European standards was deemed unacceptable across the Atlantic, as Sloss missed the role of a standard geologic time scale as an independent reference frame for interprovincial correlations. Rather, he promulgated an alternative "Sequence-stratigraphic" scale, which was presumably natural for the North American continent because of its derivation from the unconformity-bounded stratigraphic successions in the Midcontinent Platform. Sloss (1988a-b) filled the time span from the latest Precambrian to the Tertiary with numerous Sequences and Subsequences (all named and capitalized) and canonized for the entire continent. These Sequences became almost ubiquitous in the descriptions of the North American geology (see, e.g., Salvador, 1987).

Sloss was also wrong to treat all unconformity-bounded sequences or successions as tectonic units. Indeed, in the 1930s von Bubnoff had described widespread unconformities in the cratonic geologic record in Europe, which are related to great marine transgressions and regressions. Many such events were products of tectonic epeirogenic movements of big parts of cratons, but some of them were non-tectonic results of changes of the global sea level (Lyatsky, 1974; Vail et al.,

1977). These eustatic sea-level fluctuations could have been produced by vertical movements of the ocean floor or by episodes of glaciation and deglaciation. At least some of the big unconformities in the Mesozoic western seaway in the Interior Platform adjacent to the Cordillera could have been caused by variations in sea level (Caldwell and Kauffman, eds., 1993). Some impressive unconformities in the large Tertiary Beaufort-Mackenziebasin were also induced by eustatic sea-level changes (Lyatsky, 1988; see also Dott, ed., 1992). The nature of many unconformities all across the North American continent and craton is a topic of ongoing discussions (Vail et al., 1977; Whittaker et al., 1991; Vail, 1992).

Ambiguities in the definition of Sloss' Sequences

Sloss (1988a, p. 27) stated that "...interregional unconformities place unambiguous bounds on time slices for [tectonic] analysis of cratonic evolution". Bally (1989, p. 415), however, held a differentopinion and cautioned that "the inception and cessation of tectonic regimes [across North America]... often are not correlative with any of the sequence boundaries that have been proposed". Worse was his reduction of tectonic analysis of the North American craton to examination of unconformities in the platformal cover (mainly in the Midcontinent region). Geologic units and their boundaries are actually not always obvious. Division of a rock record into tectonic, hydrogeologic, chemical and other units must be based on specific criteria relevant to the objectives of the study. If the purposes are not clearly formulated, or the purposes from differentdisciplines are lumped together, there is a risk of undermining the meaning and utility of the resulting unit definition.

In his criticism of the attempts to formally include regional unconformity-bounded units into the North American Stratigraphic Code, Murphy (1988, p. 155) stressed that an unconformity is just "an interpretation of the relations between stratigraphic units". He emphasized the error of the idea that "unconformity-bounded units... are not lithostratigraphic" (Salvador, 1987, p. 234), because in fact they are. Murphy (1988, p. 155) concluded emphatically that "there is no value in adding words to our jargon that are already synonyms, especially those that masquerade as being different from what we already have".

Unconformities may in fact be no more important than other boundaries in a lithostratigraphic record. Attempts to formalize unconformity-bounded bodies (Chang, 1975; Salvador, 1987) as standard units would undermine the significance of the geologic time scale. Causes of unconformities can be tectonic and eustatic, and these two influences are usually hard to disentangle. Unconformity-bounded bodies are usually diachronous, and they are limited to certain regions. Provincial successions of unconformity-bounded bodies, useful as they are for local and regional correlations, should not be used to contest the already-established international geologic time scale. Because such bodies are of differentorigins, to regard them as tectonic units is misleading.

Since the time of Steno in the 17th century, stratigraphers dealing with strata of distinct rocks have recognized that bedded rocks are divided by discontinuities in their characteristics. The most pronounced changes commonly occur at unconformities. At first, unconformities were noted between metamorphosed and unmetamorphosed rock complexes, as well as between compacted and consolidated rocks and unconsolidated deposits. Historically, these complexes were named, first in northern Italy and then elsewhere in Europe, as primary, secondary, tertiary and quaternary (for details, see Friedman and Sanders, 1978). Finer division, dating, and designation of lithostratigraphic rock sequences from differentperspectives came later. For a long time, the primary sequence was associated with the undifferentiatedcrystalline basement. The secondary sequence was found already in the 19th century to contain many distinguishable stratigraphic units: they were named Cambrian, Devonian, Mesozoic and so on. Only Tertiary and Quaternary, capitalized, are still used as names for stratigraphic systems.

In Britain and later elsewhere, it was found that such stratified rock sequences are separated by big regional unconformities, which came to be used as tools for regional stratigraphic correlation. The first formalization of stratigraphic units was thus lithological. Later, definition of dissimilar fossil assemblages permitted to formalize the stratigraphic systems we know today. But even then, a practice persisted to treat some of these new units as essentially lithological sequences, echoed in formal names such as Carboniferous or Cretaceous. The boundaries of these units were for a while still put on regional unconformities, which continued to be seen as natural boundaries.

Soon, however, it became obvious that the boundaries natural for Western Europe were not so apparent in Eastern Europe, Asia or North America. A pronounced natural, unconformity-related stratigraphic boundary was found in the U.S. within the Carboniferous, which in Europe is a single system. Mismatches such as this gave the North American geologists an impulse to search for a geologic time scale more native to their continent's geology. Attempts were made to revise the European biostratigraphic scheme to fit North America. Notably, for example, the Carboniferous was split up into two systems - Mississippian and Pennsylvanian (though the Carboniferous is now regaining its standing).

From extensive geologic mapping in the Midcontinent cratonic platform, Sloss et al. (1949) concluded that many other lithologic sequences have unconformity-related boundaries, and they are differentthan those in Europe. This served as a starting point for the reasoning that for the North American cratonic platformal cover, another time scale is required. Sloss (1963) explicitly formulated the challenge to erect a new geologic scale natural for North America. Several regional unconformities segmenting the cover of the Midcontinent Platform were raised to a rank of phenomena universal to the whole continent, and these broad unconformities were interpreted as fundamental, and tectonic, breaks in the evolution of the entire North American craton (Sloss, 1963, 1988a-b). Sloss thus identified some important provincial evolutionary stages, which are mainly but not entirely tectonic.

Sloss (1963) recognized six Sequences in the Phanerozoic cover of the Midcontinent Platform. He named them with words borrowed from the local - American Indian and Spanish - languages: Sauk (late Vendian-Early Cambrian to Early Ordovician); Tippecanoe (late Early or Middle Ordovician through Silurian to Middle Devonian); Kaskaskia (late Early or Middle Devonian to Late Mississippian); Absaroka (latest Mississippian through Pennsylvanian, Permian and Triassic to late Early Jurassic); Zuni (late Early Jurassic through Cretaceous to early Paleogene); and Tejas (the rest of the Tertiary). The bottom of each sequence was fixed exactly in time: 586-540 Ma for Sauk, 488-470 Ma for Tippecanoe, 401 Ma for Kaskaskia, 330 Ma for Absaroka, 186 Ma for Zuni, 60 Ma for Tejas (also Menning, 1988), and this scheme was canonized as universal for the entire continent (Sloss, 1988a-b). Sloss acknowledged that difficulties arise in certain areas from a lack of sufficiently clear breaks between some of his Sequences (e.g., Sauk or Tippecanoe), but all the same, he pushed to standardize not only his six large Sequences but also 16 subdivisions within them.

Early on, Sloss (1963, p. 95) claimed that his Sequences, where preserved, can be identified "virtually everywhere on the [North American] craton", and he (Sloss, 1988a, his Figs. 1-7) extended them as far west as Idaho, Nevada and southern California. Yet, he noted that even the most profound unconformities that are clearly mappable in most of the Midcontinent Platform (e.g., Early Devonian Kaskaskia) pass into continuous stratigraphic successions at the craton's margins, in the miogeosynclinal belts of the Appalachians and the Cordillera. With this approach, he found no clear boundaries between the North American craton and the Appalachian and Cordilleran mobile megabelts (Sloss, 1988a). He mused that a tectonic challenge lies in the "delineation of the cutoffs between geosynclinal and cratonic units" (Sloss, 1963, p. 110).

Still, the habit to force-fit the local geologic records into the Sloss scheme has remained strong in North America. However, as more detailed geological studies progressed, the boundaries and extent of Sloss' Sequences came under criticism (e.g., Bally, 1989). In many places, the proposed Sequence boundaries could not be distinguished (e.g., in the Michigan Basin and the Western Canada Sedimentary Province; Podruski et al., 1988; Sloss, ed., 1988). The assumed "broadly correlatable" Sequences were found to be inconsistent in terms of their lithostratigraphic and chronostratigraphic content in different platforms on the North American craton (e.g., in Stott and Aitken, eds., 1993).

Other sequence stratigraphies

Even etymologically, the word *sequence* is hardly appropriate for a formalized geologic unit. In the English-speaking countries, this word has been used freely for centuries: any kind of stratal rock succession has been described as a sequence (e.g., in the famous cross-sections produced in England by William Smith in 1819). The use of *sequence* to mean cluster, bunch, stack or succession of rocks remains commonplace to this day. Sloss et al. (1949) were inconsistent from the beginning, applying the word *sequence* (which he capitalized) variously to distinct rock assemblages that are biostratigraphic, or lithostratigraphic, or tectono-stratigraphic. Generally, Sloss perceived his Sequences as mostly lithostrati-graphic, defining them as unconformity-bounded units "higher than supergroup", thus putting them in the same hierarchy as formations, but at the same time he treated them as "tectonic units". In the 1970s, during the rise of a new practice of geophysically discriminating rock assemblages, seismic sequence stratigraphy arose, with particular emphasis on depositional paleogeographic environments (Payton, ed., 1977). At present, "sequence stratigraphy" means little more than lithofacies analysis (in much the same sense as was recognized long ago by the Swiss geologist Gressly, 1838; see also Cross and Homewood, 1997) of stratal rock complexes (Van Wagoner et al., 1988).

Dramatic improvements in the resolution of seismic reflection data in the 1970s permitted to display a wide variety of geometries in seismically defined rock pack-ages. Mitchum et al. (1977) applied this method (called "seismic stratigraphy") to sedimentary wedges primarily on continental shelves, where distinct bodies were treated as "depositional sequences", bounded by "unconformities or their correlative conformities" (also Van Wagoner et al., 1988 and references therein). These sequences were distinguished on "a single objective criterion - the physical relations of the strata" (Mitchum et al., 1977, p. 53). Yet, all disconformities and unconformities were attributed to one process only - eustatic (by definition, worldwide) changes in sea level.

Specific reflection geometries and signatures can indeed be used to correlate spe-cific lithologic packages with certain depositional paleoenvironments. But genetic connections are usually ambiguous. Even lithology itself is often hard to infer reliably from seismic sections, to say nothing of lithofacies sets. For these rea-sons, a straightforward use of seismic images to determine the sedimentological framework is nowhere near full-proof (see, e.g., the critique of Hubbard et al., 1985a,b by Lyatsky, 1988). As a decade ago, it is worth restating (Lyatsky, 1988, p. 148) that seismic methods are still the most powerful available "tool for looking into sedimentary basins ahead of drilling. This tool allows the tracing of correlatable horizons from one well to another or to a new location. This is the principal source of our knowledge of the spatial extent of some [unexposed] physi-cal bodies which can sometimes be interpreted geologically..." However, "seismic data commonly contain multiple possibilities for very different geologic solutions. From the geologic point of view, seismically identified domains... are not fully reliable as units of the regional... tectonic framework." They can be, and have been, used to restore the tectonic history of many areas, but "this must be done with caution, and only after the drilling of a number of wells. The main problem is, first of all, the real meaning of seismic unit boundaries... An unconformity is a function of three factors: vertical tectonic movements, sedimentation rates, and sea level changes. An unconformity may or may not depend on any particular combination of these three [factors], which together define the depositional environments and the formation of different geologic bodies. One of the best examples is available from the Beaufort Shelf, where a significant tectonic event of a Late Eocene age is represented only by a disconformity, while the most pronounced Mid-Late Miocene unconformity does not have any actual tectonic importance, being merely an expression of a major drop in seal level. Tectonically, both the Late Paleogene and Neogene sequences there have to be considered as a single unit distinct from the tectonic unit of the Late Cretaceous-Early Tertiary age".

The wide discussions that accompanied the rise of seismic stratigraphy increased the interest in interpreting stratigraphic successions in terms of lithologies and facies changes in lithologic complexes (e.g., Van Wagoner et al., 1988). To predict the variations of lithofacies on a fairly small scale is of great service to the petroleum industry. Today, the term *sequence stratigraphy* is applied to local sophisticated sedimentological studies combining outcrop examination, drilling results, and modern high-resolution seismic techniques. Yet, it has no connection with Sloss' Sequences, and little in common with the original seismic sequence stratigraphy.

Tectonic significance of unconformities in the geologic record

Unconformities may have different origins, caused by subaerial or underwater erosion of land or sea-floor surfaces (e.g., Aubry, 1995) or just periods of non-deposition. The exact causes of any specific mapped unconformity may be hard to determine, but the result is the same: a stratigraphic gap between rock packages. Unconformities can be recognized easily if they separate rock packages with sharply contrasting lithologic and structural characteristics. But in a succession of subhorizontally layered rocks deposited in similar paleoenvironments (which happens often on cratons), disconformities rather than angular unconformities are more common, and they can be hard to identify. Importantly, in stacked stratified rocks, each contact between even small layers (strata) is a potential unconformity (Murphy, 1988).

The most pronounced are unconformity surfaces separating rock bodies with contrasting grade of metamorphism or structural characteristics. For example, James

Hutton's (1726-1797) famous unconformities at Siccar Point or at Jedburgh in Scotland truncate folded underlying rocks at a high angle. Also structural are the so-called Great Unconformity at the base of the Cambrian in the Grand Canyon in Arizona, the unconformity at the base of platformal sedimentary cover over the highly metamorphosed and severely folded Archean and Early Proterozoic rocks, etc.

These structural unconformities represent big chronostratigraphic gaps in the geologic record, between the top of the underlying succession and the bottom of the overlying sequence. Erosion bevels the exposed ground surface, which is often weathered to produce paleosols. During the time of these gaps, many geologic events could occur, including tectonic ones.

In platformal areas, cratonic or otherwise, subdued tectonism produced only slightly angular unconformities in the platformal cover. Many of them are recognized only by thorough geologic mapping of large regions (azimuthal unconformities). Disconformities, which run parallel to bedding and separate strata with almost the same attitudes, usually represent periods of non-deposition. They are often recognizable only by the contrast in rock ages above and below.

The terminology used to describe different types of unconformities is complicated and sometimes confusing: paraconformity, stratal unconformity, etc. Diagnostic criteria to identify unconformities can be very elusive. This underlines the great variety of unconformities and disconformities, and the big difficulties in detecting and tracing them in many areas. Importantly, however, their expression in the geologic record is not necessarily a measure of their geologic significance.

Approaches to rock-unit classification in the *North American Stratigraphic Code*

To formalize mappable geologic formations (or suites, as they are called in Russia) and their boundaries, and to standardize the criteria of their correlation, a number of national stratigraphic codes in different countries have been produced (for details, see, e.g., Harland et al., 1990; Harland, 1992). The first official U.S. guide to stratigraphic classification (Hedberg, ed., 1976) reasonably defined a mappable *stratigraphic unit* as a stratum or set of strata distinguished from adjacent units by its physical properties. Codified successions of layered rocks, i.e. units, can be observed and traced directly on the ground or from dillholes. In a similar line of reasoning, the *North American Stratigraphic Code* (1983, p. 847) states in its preamble: "Stratigraphic classification promotes understanding of the *geometry* and *sequence* of rock bodies" (italics in original). The *Code* thus used the word *sequence* in its original meaning, to denote merely a rock succession. But the 1983 *Code* included also intrusive tectonic rock bodies. It recommended to designate such non-stratified rock bodies of "plutonic and tectonic origin" as *lithodemic*, as opposed to *lithostratigraphic*. In the *Code*, the definition of lithodemic bodies includes rock complexes that are highly metamorphosed (to the point of losing their primary structure) or intensely deformed, although contacts of these bodies (complexes) are non-stratigraphic: intrusive, metamorphic or tectonic. Such unit boundaries can cut across the primary stratigraphy (though there can be exceptions: contacts of stratiform igneous sills subparallel to country-rock bedding, or low-angle fault zones sometimes follow bedding planes).

Many authors have complained that inclusion into the scope of stratigraphy of non-sedimentary and depositionally non-stratified rocks dilutes stratigraphy as a discipline that specially concerns itself with stratal bodies. As a counter-argument to this complaint, Hedberg (ed., 1976) retorted that the entire Earth is, in a sense, stratified, and all rocks and rock classes (sedimentary, igneous and metamorphic) are within the scope of stratigraphy and stratigraphic classification. This rebuttal, however, fails to address the fact that such a broad approach still leaves the discipline of stratigraphy diluted and unfocused. If we include igneous, metamorphosed or fault-related tectonic bodies, why not "geophysical bodies" (which are actually just images of rocks and physical discontinuities within them)? The Soviet Code (Zhamoida, ed., 1979) did so. Salvador (1987, p. 440) acknowledged the relativity of many definitions in the 1983 *Code*: "There is general agreement that stratified volcanic rocks and bodies of metamorphic rocks that can be recognized as of sedimentary and/or extrusive volcanic origin can be treated in every respect as lithostratigraphic units because their characteristic lithologic features, original layering, and stratigraphic relationships are readily distinguishable. There is, on the other hand, considerable disagreement concerning nonlayered intrusive rocks and the bodies of metamorphic rocks that have suffered such deformation and/or such drastic recrystallization that their original layering and stratigraphic (chronologic) succession may no longer be ascertained".

The imprecisions in the *North American Stratigraphic Code* may be confusing for practical geologists (e.g., Edwards and Owen, 1996). Revisions might be warranted, especially with regard to the so-called "allostratigraphic" (op. cit., p. 1158), "sequence-stratigraphic" and "unconformity-bounded" units whose inclusion into the *Code* is still under discussion (also Salvador, ed., 1994).

We reiterate that the only stratal rocks are the object of stratigraphy, which considers the stratal rock record in all its aspects. Units into which this record is divided are designated differently depending on the purpose of the division. Thus, the lithostratigraphic record may be divided based on different internal boundaries: rock-mechanical, hydrogeological, biostratigraphic, tectonic, and so forth.

2 STRUCTURAL-FORMATIONAL, TECTONIC ÉTAGES AS FUNDAMENTAL UNITS OF REGIONAL TECTONIC ANALYSIS

Difficulties in extracting tectonic information from rock characteristics

Rocks and rock-made bodies are characterized by specific physical, chemical, mineralogical, petrological, and other properties. In combination, they are used as bases for particular classifications (by rock type, paragenesis, magnetization, etc.) and serve as criteria to map rock bodies. Each body has its own internal structure (texture), as well as boundaries at which the continuum of the particular rocks ends.

At the highest hierarchical order of geologic bodies is the lithosphere. Its base, as well as the boundary between the lithospheric upper mantle and the crust, are largely inferred from changes in the physical properties of material, detected remotely by geophysical methods. But because the relationship between specific recorded rock properties ands rock type is not unique, the problem of non-uniqueness in the interpretation of geophysical data leaves room for a lot of ambiguity. A interpretation of geophysical data is obtainable only with the benefit of direct geologic observations of rock bodies.

The top levels of the Earth's crust are indeed available for observation by surface mapping and drilling. The principal units in geologic mapping are geological formations, which are "bodies of rock strata" (Hedberg, ed., 1976). By definition, formations are the fundamental units of regional lithostratigraphy, because they are the most convenient for geologic mapping. Several formations make up units of higher taxonomic level: group, supergroup, entire cover.

Only those rock bodies that are exposed at the surface or sampled by drilling can be arranged into a temporal succession, a geochronologic record containing the recognized rock bodies as well as the hiati between them. Observable characteristics of rocks and rock-made bodies are essentially non-interpretive and objective. But grouping of rocks into formations and other units inevitably requires abstraction and generalization. It actually begins during field mapping, at the outcrop, and continues later, as the rock bodies are classified in a taxonomic hierarchical order.

Lithologic properties: carbonate or non-carbonate; coarse-grained or fine-grained; white or colored; bedded, laminated or massive crystalline or non-crystalline; etc. provide the most obvious criteria to discern units. From lithologic properties, it is possible to find some apparent indications of the regional tectonic history: rocks are magmatic or sedimentary; marine or non-marine, fluvial or deltaic, shallow or deep marine, and so forth. Such discriminations are often approximate and not unique, especially in hybrid or transitional cases, but this is all we have available to infer the paleoenvironments in which the rocks were produced: high-energy or low-energy currents, approximate sea depths, etc. To complicate things, the original rock characteristics are often altered later, by various grades of diagenesis and even metamorphism.

Genetic interpretation of rocks, geologic bodies and their complexes is one of the most important and most difficult jobs. It is rarely unique. Lithological criteria, on which the definition of formations is based, are inevitably rough or incomplete ("a predominantly limestone unit with a few shale interbeds"). Plenty of internal lithologic and mineralogic variations are averaged out in such descriptions, even though small-scale changes may be useful as a basis for reliable genetic classifications (e.g., in terms of facies). Detailed sedimentological studies provide other clues how the rocks and rock bodies were created, revealing paleogeographic environments, provenance areas, approximate distances and directions of sediment transport, etc.

Still, because various geologic processes can account for similar rocks, interpretations of lithofacies (and fossil assemblages in them) in terms of paleogeologic situations and especially in terms of tectonics, are often suspect. For this reason, it is prudent to consider multiple interpretation options and working hypotheses. It is sometimes difficult even to tell if a rock sequence was formed in a continental or marginal-marine environment, on a deep shelf or a continental slope, and so on. The common reference to depositional paleoenvironments as *deep-water, shallow-water* and *nearshore* are loose and may mean very different things to different specialists. Detailed lithofacies patterns are commonly local in extent, and to extrapolate them away from a particular area of study is often problematic. To derive tectonic information from the characteristics of rocks and rock bodies is a difficult challenge.

Regional tectonic restructuring episodes as a basis for division of a geologic record into tectonic units

Division of a lithostratigraphic record into purely tectonic units involves correlating them with regional episodes of crustal reworking. This requires integrated, multidisciplinary geologic analysis of large regions (Lyatsky and Lyatsky, 1990; Lyatsky, 1994a), in order to determine the important tectonic processes that affected the region's crust. Ironically, in the tectonically complex orogenic belts, to recognize vertical tectonic units and to connect them with certain tectonic episodes is often easier than in cratons. Ophiolitic and flysch formations are usually created during early orogenic stages, reflecting processes as grand as subduction. Molassic formations are typically created during later, mountain-building stages. Intrusive and extrusive magmatism of particular types accompanies specific orogenic stages. Regional metamorphism, typical in mobile megabelts, is one of the strongest indicators of regional tectonism. Igneous and metamorphic rocks may be linked to specific tectonic stages, permitting these stages to be dated precisely. Unconformities between rock complexes of distinct tectonic stages are abundant, deep, and usually angular, which makes them easy to identify and trace.

It is harder to depict the boundaries of vertical tectonic units in the geologic record of cratons, where crystalline rocks are very old and deeply reworked, and the platformal cover is rather uniform in composition and structure. This uniformity developed because during the accumulation of the sedimentary cover, crustal movements were less dramatic and the associated structures more subtle. The unconformities and disconformities in the sedimentary cover are less pronounced. Alteration of rocks is more cryptic; metamorphism and magmatism occurred seldom, if at all. In contrast to mobile megabelts, where dramatic restructuring was often accompanied by intense metamorphism, folding and faulting, restructuring on cratons was associated chiefly with crustal block movements.

No thrusts developed in the vast cratonic regions, except in the peripheral areas adjacent to mobile megabelts. Tectonic folds are rare and localized along regional fault zones or smaller faults that bound horsts and grabens. Phenomena diagnostic of block movements are usually obvious: sharp contrasts across block boundaries stand out in thickness and lithofacies variations of the sedimentary strata and in the topography. Typical for cratonic tectonic regimes are gentler vertical displacements of the crust, in the form of large-diameter (up to thousands of kilometers) buckling and smaller (hundreds of kilometers) warping.

Miall (1987) attempted to consign the entire concept of epeirogeny to obsolescence, but Friedman (1988b) countered this attempt effectively by presenting extensive field and laboratory data on the regional alteration patterns of carbonate sedimentary rocks. He showed that flat-lying Phanerozoic carbonates and terrigenous deposits in large areas of the North American craton in the U.S. had been buried to depths of several kilometers prior to their exhumation. In Eurasia, epeirogenic tectonic origin of many extensive platformal unconformities in the northern parts of the Russian and Siberian cratons, as well as in the much younger West Siberian platform, was demonstrated by statistical analysis of scores of paleogeographic and lithofacies-distribution maps covering the entire Phanerozoic eon (Ainemer and Lyatsky, 1972; Vinogradov, ed., 1974; Lyatsky, 1978). It was shown that some of the biggest changes in the vertical distribution of lithofacies (which were formally parametrized to perform the computations) corresponded to movements of large blocks all along the Russian Arctic. Each such block, from the Fennoscandian Shield in the west to the Chukchi Peninsula in the east, is characterized by its own, unique set of lithofacies and breaks in the sedimentary cover. Based on the similarities and differences in the paleoenvironmental settings and in the timing of breaks in different areas, movements of large crustal blocks were reconstructed. Periods when the epeirogenic and eustatic processes could have operated, in tandem or not, were also well displayed.

The time intervals in a regional chronostratigraphic record, whether represented by preserved stratal rocks or falling into stratigraphic gaps, were products of stationary regional regimes. Restructuring of the tectonic distribution of lithofacies usually occurred comparatively sharply, though it was not instantaneous and in some cases was slow and took a long time. These variations between more-or-less stationary regimes and their fairly rapid changes provided a region with the tectonic settings characterizing different stages of its development.

The importance of such nuances in linking stratigraphic characteristics to their possible tectonic causes has been recognized for many decades. During the past century, in particular, it led to a considerable interest in cyclicity. Cyclical coal-bearing paralic successions were studied in detail, and paragenic sequences of different ranks were mapped. The notion of sedimentary cycles was used mainly in a lithostratigraphical sense (Weller, 1960), related to seasonal climatic fluctuations as well as tectonic pulsation, particularly in flysch settings (Vassoyevich, 1959; Friedman et al., 1992). At present, mapping of cyclothems competes with the more traditional biostratigraphy. Regional-scale megacyclicity, which pertains to a whole platform or its large parts, is supplemented with more local mesocyclicity in the studies of smaller areas. Cyclicity is common in many peripheral zones of cratons.

Sets of tectonic lithofacies, their lateral distribution, and vertical and temporal extent provide the only objective criteria for dividing a regional stratigraphic record into tectonic units. This was noted by Salvador (1987, p. 234), though his choice of terminology reflects some conceptual inconsistency. He stated: "Unconformity-bounded units... have sometimes been considered to be equivalent to 'sedimentary cycles' or tectonically controlled stratigraphic units: stratotectonic, tectostratigraphic, tectono-stratigraphic, or tectogenic units, tectonic cycles, tectosomes, structural or tectonic stages, and so on... Calling a unit a 'tectonic stage', for instance, implies that the unconformities bounding the unit are the result of tectonic events; unconformity-bounded units, on the other hand, are established and recognized without any regard to the cause of their bounding unconformities, whether they are the result of orogenic events, epeirogenic episodes, eustatic sea-level changes, or any combination of them."

Not all rearrangements of sedimentation settings are due to tectonics. To discriminate those rearrangements whose causes were indeed tectonic is a matter for regional tectonic analysis. To do this credibly, a tectonist must correlate these rearrangements with the changes in the regional structural pattern from one tectonic stage to another.

The concept of rock-made tectonic étages and tectonic stages

In a geological study of a region, we deal not with rock-making processes but with their results, expressed as rocks of different types and rock-made bodies of various forms. The fundamental unit of a regional stratigraphic record in North America is a formation (Committee on Stratigraphic Nomenclature, 1933), and it is now accepted in most countries. A formation is defined based on its lithostratigraphic, biostratigraphic and chronometric characteristics independent of the specific objectives of a particular type of geologic analysis. A tectonic lithostratigraphic unit, on the contrary, must be distinguished as a product of a particular tectonic regime during a particular interval of time in a particular region. Abrupt changes in the regional tectonic regime are recorded as breaks in the regional geologic record, demarcating tectonic units of different taxonomic levels.

Stille (1924, 1936a-b, 1941) was probably first to consider the geologic reality purely tectonically, by identifying those its aspects which were products of purely tectonic processes. In sedimentation, magmatism, metamorphism, deformation he was able to discern manifestations of tectonics, and he interpreted these manifestations in a single concept. Less realistic was Stille's idea to establish and canonize a global scheme of tectonic pulses manifested in extensive orogenic and cratonic deformation and major unconformities, presumably of worldwide significance.

To avoid terminological ambiguity, in this book we use for vertical stratigraphic tectonic units a term common in Eurasia - *étages*. These units have been used since the 1950s in the legends of many tectonic maps of Russia, Europe and entire Eurasia (produced by Shatsky, Bogdanov, Spizharsky, Yanshin). Tectonic étages are structural-formational, and are defined by their bases only (cp. Cope, 1996). The time span of such a unit, together with the time separating it from the younger étage, is directly related to the timing of a particular tectonic regime in a region. The time span of the corresponding rock-made étage, combined with the chronostratigraphic hiatus above it, is designated as a tectonic stage (Fig. 5). Ideal for geologic mapping are situations where structural-formational, i.e. tectonic, étages are bounded by pronounced unconformities across which major lithologic and structural changes are vividly manifested. Such clear breaks are extrapolated into adjacent areas where the tectonic regimes and restructuring episodes were the same but the étage boundaries are less clear.

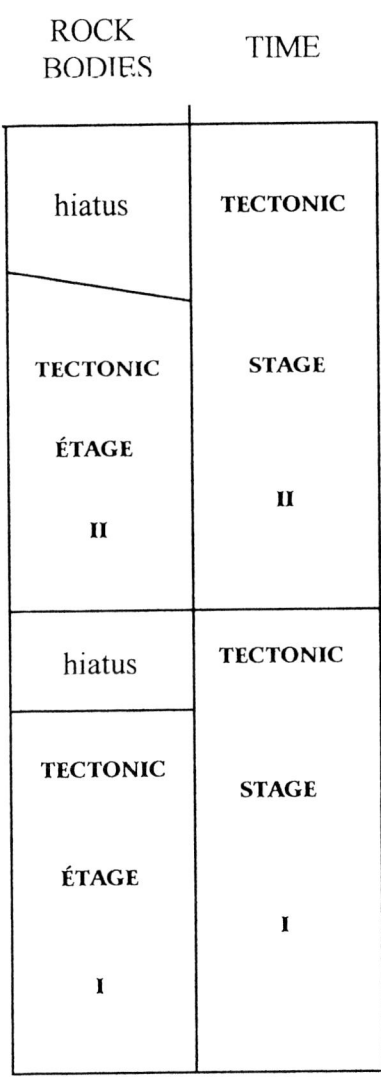

Figure 5. Tectonic etages and stages, and their time spans. The span of a tectonic time stage includes the span of its corresponding tectonic (rock-made, structural-formational) etage, plus the span of a hiatus at the etage's top. The beginning of the next stage is set at the base of the next etage.

In the 1930s, a German geologist von Bubnoff distinguished vertically stacked units in the platformal sedimentary cover of continental northern Europe, which were a response to large marine transgressions and regressions that he related to epeirogenic uplift and subsidence. His Russian contemporaries - Shatsky, Bogdanov - took this approach critically and made it more interpretive, by combining together the analysis of platformal and orogenic (geosynclinal) regions. Such tectonic units were invoked in many regional maps of the U.S.S.R. and Eurasia, including the well-known *International Tectonic Map of Europe*.

Von Bubnoff identified major units in the stratigraphic record of the sedimentary cover of ancient and young platforms, and Wegmann, Shatsky and others expanded this approach to orogenic areas that surround such platforms. In the U.S., Chang (1975) proposed a special term for big transgressive-regressive cycles in the sedimentary rock record - *synthems*. Referring to concordant rock complexes together with paragenetic rock bodies that intrude them, Wegmann (1935, 1956) used the term *stockworks*. The meaning of von Bubnoff's and Chang's units was essentially paleogeographic, but Wegmann's and Shatsky's units were purely tectonic.

In the U.S., Sloss et al. (1949) promoted an "integrated facies analysis" that was predominantly sedimentological. They distinguished "operational units" ("sequences") on the basis of "marked discontinuities in the stratal record of the [North American] craton" (Sloss et al., 1949, p. 109). Krumbein and Sloss (1951, p. 380) tried to formalize the term *sequence*, and defined it as "the rock record of a major tectonic cycle". Yet, Sloss (1963) himself was inconsistent in his interpretation of sequences as units. On the one hand, he argued that sequences should be regarded "as distinct and different type of stratigraphic [units]", but on the other hand, he treated them as just "high-rank rock-stratigraphic (lithostratigraphic) units" (Sloss, 1963, p. 93, 111). Later, he (Sloss, 1988a-b, 1990) stated firmly that the origin of stratigraphic sequences "must be attributed to *tectonic* subsidence" (Sloss, 1988a, p. 25, italics his). But this idea has for years been challenged. Vail et al. (1977) linked unconformity-bounded sequences to eustatic sea-level changes (see also Haq et al., 1987; Vail, 1992). Friedman et al. (1992) substituted the term *mesosequence* for the *sequence* of Vail et al. (1977). These authors defined sequences emphasizing their sedimentological and paleoenvironmental aspect. Galloway (1987) promulgated a sedimentological understanding of sequences, which he called genetic.

Whittaker et al. (1991), Harland (1992) and others rightly pointed out that the North American sequence-stratigraphic units as outlined above (also Van Wagoner et al., 1988) are essentially conceptual, and "...their conceptual aspect is likely to develop with each practitioner" (Harland, 1992, p. 1233). Confusion in the understanding and differentiation of tectonic and non-tectonic units in stratigraphy in general and in the sedimentary cover of various North American cratonic platforms in particular still plagues the practitioners (e.g., Murphy, 1988 vs. Salvador, 1987, 1988; Harland, 1992; Edwards and Owen, 1996).

In this book, the term *étage* (from a French word for storey, or floor, in a building) is sometimes used with the adjective *structural-formational*, to emphasize its essence as a rock-made body of certain composition and structure, and sometimes with the adjective *tectonic*, to stress its fundamental meaning. Each structural-formational étage is tectonic in its genetic essence. Each of them is related to a regional tectonic stage, i.e. a period of time between pulses of tectonic restructuring, during which a particular étage was formed.

Étages are purely tectonic structural-formational, i.e. material, units related to particular time intervals when particular tectonic regimes occurred. An étage spans the time from the beginning of its corresponding tectonic regime, typically marked by the étage's bottom, to the beginning of the next regime. Usually the étages are preserved incompletely, having been eroded prior to the creation of the next étage. Thus, temporally, tectonic stages are commonly larger than their corresponding étages. A stage includes the time span of the étage itself, as well as the time prior to the next étage. Whereas the lower boundary of an étage is usually more or less diachronous, the beginning of its corresponding tectonic stage is fixed by the age of the oldest rocks in the étage. This should be kept in mind when making a nomenclature: an Aptian to Campanian stage may be represented by an étage which is only Aptian to Cenomanian.

Levorsen's (1934, 1943, 1960, 1967) practice of reading a regional lithostratigraphic record as a layer cake is best justified for successions of units bounded by tectonically induced unconformities. A sharp change in the regional tectonic regime ends the tectonic stage, commonly producing an étage-bounding unconformity in the rock record. Each structural-formational étage is separated from the preceding and following étages by episodes of general tectonic restructuring and unconformities. Although tectonically induced regional unconformities are just one of many possible manifestations of restructuring, they have a major advantage of being objective, laterally extensive breaks in the stratigraphic layer cake.

It is easy to point to the profound unconformities between the Archean-Proterozoic crystalline basement and the sedimentary cover, which separate the main, first-order tectonic étages (basement and cover) in cratonic platforms. And yet, each of these mega-étages contains its own constituent étages, with disparities between them corresponding to episodes of tectonic restructuring. Platformal tectonism may result in general subsidence and uplift, as well as tilting of different amplitudes, rates and configurations. Each tectonic stage is characterized by its own movement patterns and block configuration. Tectonic movements affecting cratonic platforms may be subtle, and the variety of sedimentary settings may be limited to changes in provenance areas. Sedimentary rocks, especially their thickness and lithofacies patterns, commonly bear the marks of tectonic influences, but they have to be compared with results from mineralogy, petrology, organic-matter content, paleocurrent directions, facies relationships. Syn-depositional and post-depositional differential movements must be taken into account.

A *tectonic stage* is a time interval characterized by a consistent tectonic regime, whether quiescent or more active. Each rock-made tectonic étage corresponds to a tectonic time stage. In the cratons, tectonic stages are usually characterized by relative tectonic quiescence, followed by episodes of tectonic restructuring when the regional structural pattern changes more or less sharply to provide a new background for the next time stage and rock-made étage. But some stages are distinguished by higher general levels of tectonic activity than others.

In the rock record, restructuring episodes bounding regional tectonic stages are often marked by apparent unconformities over uplifted areas, but they may be less apparent in subsiding areas of continuing sedimentation. Another difficulty arises from the fact that each structural-formational étage is laterally restricted and may not be represented everywhere even within a tectonically single region. The extent of such regions also changes from one tectonic time stage to another. To delineate these regions for each tectonic stage is another challenge for regional tectonic analysis.

Regional unconformities in platformal sedimentary cover are of different types: structural, azimuthal, stratigraphic, disconformity, paraconformity, and so on. Unconformities are often diachronous at both base and top (Aubry, 1995), though the erosional bases tend to be more diachronous than the tops. Where the information about rock ages is detailed enough, the age of an unconformity can be bracketed reliably between the youngest stratified or intrusive rocks below and the oldest stratified rocks above. Tectonic stages are defined to last from the bottom of one étage to the bottom of the next, and major unconformities can be their markers.

In regional tectonic analysis, unconformities may serve as preliminary guides, but they should be viewed critically, not so much because of their diachronism as because many unconformities occur within étages. Their span is usually small, and their role in regional tectonic analysis is no bigger than an internal characteristic of a particular étage. The above makes it cleat that because the time span of a particular tectonic stage might have been longer than the geologic time interval recorded in its preserved rock-made étage, to equate a tectonic stage with an étage (cp. Salvador, 1987, 1988) would be a mistake.

Definition of tectonic stages is based on the mapping and definition of the structural-formational étages. Episodes of sharp restructuring of the regional tectonic pattern from one stage-étage to another mark the principal changes in the spatio-temporal geologic continuum. Mappable unconformities play a big role in defining étages and stages, but the main source of regional information is the sharp regional structural rearrangements of the crust.

Cope (1996, p. 107) formulated two principles (i.e. empirical rules) in identifying stratigraphic breaks: (a) boundaries segmenting the geologic record must be "... identified by sound criteria and not because of a local break in the [rock] succession which could cut out important events"; and (b) all these units "...should be defined by means of their bases only, the top being defined automatically by the base of the succeeding unit..." Structural-formational étages are defined in just such a way. Real tectonic events, expressed sedimentologically, magmatically, metamorphically (including rock alteration of any sort) or deformationally, imprinted in the observable rocks, provide the most reliable information to constrain speculations.

Structural-formational étages as tectonic units in a regional stratigraphic record

Any aspect of regional rheology can be important for tectonic analysis. Based on rock characteristics and the distribution of rock-made bodies in time and space, it is possible to realistically decipher the framework of the region's tectonic evolution. Methods that are more remote from the real rocks are auxiliary, providing valuable but usually not uniquely interpretable information. One must resist the temptation to make too-speculative hypotheses and models from this auxiliary information.

Sedimentary strata and their assemblages have been recognized to be responses to important geologic events for three centuries, since at least the time of Steno (Niels Steensen, 1638-1687). James Hutton (1726-1797) came to regard unconformities as natural boundaries recording evolutionary events in a region's history. The subsequent practice of geologic mapping showed that the more pronounced an unconformity, the likelier it is to be of tectonic origin, especially if it is structural. As tectonics became a separate discipline in geology, the search for purely tectonic units, vertical and lateral, became a centerpiece of regional tectonic analysis. Stille (1941) canonized some unconformities as tectonic markers of worldwide significance, but these cannons were rejected (Gilluly, 1949, 1950, 1963). Sloss (1963, p. 111) stressed that "it has long been recognized that the classical [European-based] systems are *not natural* groupings of strata on the North American craton" (italics ours). He tried to invent an alternative scale of stratigraphic systems (Sloss, 1963, 1988a-b), but did so without sufficient distinction of tectonic and non-tectonic geologic units.

Confusingly, in what appears to be a contradiction, Sloss stated that "although sequences have a greater time-stratigraphic significance than classic rock units, there is no implication in the sequence concept of an attempt to establish a North American, as opposed to a Western European, time scale" (op. cit.). But in fact there was such an attempt, and later Sloss (1988a, p. 26-27) revealed it more explicitly, saying that "the adherence to time-honored chronologic and chronostratigraphic subdivisions is a consequence of the necessity to conform to stratigraphic classifications approved by government agencies". To obtain a standard geologic time scale for North America, Sloss assigned "first-order importance" to unconformity-bounded units recognized in the cover of the Midcontinent Platform of the North American craton, and he tried to apply his Sequences for mapping on a continental scale. A bigger extrapolation has been an application of this scheme to other continents - for example, in the Middle East (Alsharhan and Nairn, 1997), where headings in sedimentary-basin overviews refer to the Sauk, Tippecanoe, Kaskaskia, Absaroka, Zuni, and Tejas "cycles" in Iran, Iraq, Saudi Arabia, and neighboring areas.

In reality, of course, unconformities are not required for defining stratigraphic boundaries (Ager, 1984; Cope, 1996), including tectonic ones. The internationally accepted geologic time scale did originate in Eurasia (though not just in Western Europe - for example, the Permian system or the Bashkirian stratigraphic stage are Russian contributions). But it is long since globalized, or at least major portions of it are. Not unconformities but major changes in faunal assemblages (related often to extinction events, not necessarily tectonic) served as a primary guide for the establishment of the standard time scale of worldwide applicability.

Since the middle of the 20th century, attempts have been made to produce specifically tectonic maps for large parts of continents, highlighting specifically the tectonic units. In these maps, rocks were grouped into assemblages related mainly to orogenic episodes. This required a sophisticated new approach to representing the regional geologic records. The relatively simple practice to show rock types discriminated chiefly by their age, as is done in conventional geologic maps, had to be supplemented with other criteria of unit discrimination. In contrast to general geologic maps, specialized tectonic maps used tectonic units, étages: for example, Shatsky's early (1956) *Tectonic Map of the U.S.S.R.*, and Shatsky and Bogdanov's (1964) *International Tectonic Map of Europe*. The *Tectonic Map of North America* (King, 1969) relied on the same principles, although whereas Shatsky and Bogdanov had explicitly used the term *étage*, Philip B. King (1903-1987) did not. In these tectonic maps of the 1950s and 1960s, the focus was mostly on orogenic regions, and rock bodies were grouped together based on the main epochs of their regional folding: the Alpides, Hercynides or Variscides, Caledonides, Taconides, and so on. Spizharsky's (1973) tectonic maps of Siberia and U.S.S.R. distinguished three main rock complexes in mobile megabelts: basement; post-inversional complexes bearing a strong orogenic grain; and the overlying cover of young post-orogenic platforms. Cratonic areas in his maps were mainly distinguished by thickness contours of their sedimentary cover, but in it structural-formational étages were also identified. These were the sound elements in his approach, but he erred in trying to force-fit all lateral tectonic units into an *a priori* template of lateral tectonic zoning which he assumed was stationary in time.

These outstanding contributions have not been picked up as a methodological standard for other tectonic maps. The main reason was the extreme difficulties encountered in portraying the regional geology from a multi-aspect tectonic perspective including sedimentation, magmatism, metamorphism and deformation. Phenomenological expressions of tectonism are extremely variable from time to time and from region to region. Even to outline rock bodies of the same age, as required in conventional geologic maps, became hard as the detail of field mapping grew; and enlarging the map area greatly complicated the legend. To correlate rock bodies over large distances may be problematic due to insufficient information, diachronism, age overlap of geologic bodies, and so on. To be informative and useful, tectonic maps should contain the information of conventional geologic maps, as well as the information on regional tectonics with a highlighted set of tectonic units. The initial efforts to make multi-aspect tectonic maps ended largely due to the difficulties with legibly representing on the same sheet all four aspects of tectonism: sedimentological, magmatic, metamorphic and deformational.

So-called tectonic maps produced today often take an easier route and sacrifice observable information about rocks in favor of some assumed geodynamical models and genetic speculations. High-grade metamorphic rocks in shield areas are often shown in terms of their assumed syndepositional paleogeographic settings, which are usually subjective. Pre-conceived, speculative plate-tectonic, and especially terrane-tectonic, reconstructions often take the place of real geologic configurations, and the rock bodies are defined based on these pre-fabricated templates. Because such maps do not show the real relationships of observed rock bodies, they provide little useful information about rock types, rock-body relationships, their historic changes, present-day state, or economic deposits; for this reason, they are not practical.

Tectonic maps are inevitably interpretive to some degree. But to be credible, they must be based on the real, observable regional facts. They have to be multi-aspect, showing structural-formational étages as rock-made structural complexes.

Epistemological roots of current hypotheses about the cratonic areas of North America

The leading role in distinguishing Precambrian tectonic units now belongs to radiometric methods. This adds reliability to the previous reliance on visible contrasts in metamorphic and structural styles of rocks. But despite the modern improvements and refinements of the dating techniques, many uncertainties regarding the strongly deformed and metamorphosed ancient rock bodies remain. Even dating techniques sometimes produce equivocal results. In high-grade metamorphic terranes, dating is complicated due to reworking of the mineral and chemical systems (e.g., Kröner and Jaeckel, 1995; Roberts and Finger, 1997). Cores and overgrowths in crystals of reference minerals (commonly, zircon and monazite) may be hard to separate even though their ages differ. In non-metamorphic rocks, element and isotope ratios vary. Definition from rock geochemistry of primitive vs. depleted original magma, exact protoliths of metamorphic rocks and cooling history of regions requires great caution (Issler et al., 1990; Woodsworth et al., 1991; Krogstad, 1995). Still harder is to specify the tectonic settings in which the rocks formed and were altered. Even in regions of apparent modern subduction, the influence of subduction on the properties of the youngest, and least reworked, Cenozoic Cascade basalts in Oregon and Washington is neither clear nor direct (e.g., Sherrod and Smith, 1990; see discussion in Lyatsky, 1996).

Another popular technique in the studies of Precambrian rocks is paleomagnetic. The underlying idea is that remanent magnetization of mafic minerals recorded the geomagnetic field at the time these rocks were formed. In undisturbed, untilted, unaltered, unremagnetized extrusive and intrusive igneous rocks, magnetization vectors are indeed correlatable with penecontemporaneous positions of geomagnetic poles. But many technical difficulties during field sampling, preparation of samples for laboratory analysis, and laboratory procedures themselves degrade the reliability of paleomagnetic determinations. Rock bodies are rarely undisturbed and unaltered, especially Precambrian ones. Many secondary alterations are subtle. In many areas, unrecognized tilts of rocks have been shown to invalidate the paleohorizontal determinations. There are also contaminations of rock magnetization by undesirable components, and many other complications (e.g., Beck et al., 1997), and determinations of the ancient position of blocks and entire regions based on paleomagnetic studies often contradict the regional geology (e.g., Butler et al., 1989; Dunlop, 1995; Monger and Price, 1996).

The hypothesis of so-called Wilson Cycle (a sequence of events beginning with rifting and rupture of a continent, followed by opening of an ocean between its parts, and later closing of this ocean to restore a single continent) assumes that continents were repeatedly assembled from distinct crustal blocks and then separated again, time after time. This idea underlies many other hypotheses postulating former oceanic basins, zones of continental collision and subduction on the site of modern continents and cratons. The entire North American craton has been pictured as arrangements of geologic domains assembled in a random collage of exotic pieces lacking genetic links.

The concept of a Paleozoic supercontinent Pangea that often infects the geological reasoning is now supplemented with even more speculative hypotheses. Powell et al. (1993) supposed that another supercontinent, Rodinia, existed in the Late Proterozoic. The assumed Rodinia's breakup around 700 Ma was followed by the drifting apart of the Laurentian (essentially, North America without the Grenville Belt) and Australian continents (see also Dalziel, 1995). Later, Powell et al. (1995) considered the possible existence of yet another Neoproterozoic supercontinent, Pannonia, located mostly in the Southern Hemisphere.

Breakup of each supercontinent presumably resulted in the appearance of new oceans, some as big as the Indian or Pacific oceans today. Then, for unclear reasons, the motion of these continental fragments reversed, the oceans closed at suture zones, the continental fragments collided, and new supercontinents were assembled. In such a hypothetical ultra-mobilistic framework, it has been proposed that around 1,850 Ma, a 5,000-km-wide Manikewan ocean existed in the middle of the modern Canadian Shield between the Archean Superior, Slave and Wyoming cratonic masses (cp. Stauffer, 1984; Symons, 1991). Symons et al. (1996) placed this hypothetical ocean between the Archean Slave-Rae-Hearne craton, previously situated in the tropics, and the Superior craton, located at that time near the North Pole. These cratons supposedly rotated and came together rapidly, squeezing the Manikewan ocean into a suture which is yet to be found somewhere in the "Himalaya-style" Trans-Hudson orogenic province.

Not just many geologists but also some paleomagnetists warn that grand tectonic conclusions drawn from ambiguous data may be premature (Dunlop, 1995; Halls, 1995): in the Canadian Shield, remanent magnetization in samples is often established imperfectly, paleohorizontal is misinterpreted, field associations of samples are mapped inadequately, and so on. Age calibration of the reconstructed apparent polar wander paths is sensitive to secondary remagnetization of rocks, and radiometric dating is not everywhere adequate. Paleomagnetic samples are commonly obtained from mafic igneous rocks, such as dikes, whose dating is suffers from the "notorious" difficulties with application of isotopic methods to such rocks (Reed et al., eds., 1993; Berggren et al., 1995). Many "about-coeval" or "broadly correlatable" rocks in Precambrian domains were shown by detailed studies to actually be different. Many assumed collisional sutures in fact lack suture-like characteristics (cp. Schermer et al., 1984; Schulz et al., 1993). "Confirmatory evidence for collision" is less than definitive around the Superior and Archean cratons of the North American continent (Schulz et al., 1993, p. 64).

Various hypotheses have been advanced in the last several decades about presumed paleo-positions of continents and continental fragments on the globe at different times in the geologic past (e.g., among others, Scotese et al., 1985; Scotese, 1997; Kirschvink et al., 1997). They are based not on systematic, comprehensive analysis of rock evidence but usually on paleomagnetic studies and paleontological (paleobotanic, tetrapod) correlations. It is now apparent that these hypotheses suffer from weaknesses in tectonic interpretation of paleomagnetic data, lack of clarity about mechanisms of lifeform dispersal, and so on. The purpose of this book is not to pursue that speculative approach but to examine and analyze rock-based evidence providing hard constraints on the tectonic evolution of the North American craton.

Numerous Archean blocks were proposed previously in the Churchill province of the Canadian Shield (Stockwell et al., 1970). Later, most of this province was included into the Trans-Hudson Orogen of Early Proterozoic age; that area, delineated by Hoffman (1988), contains a number of median massifs and zones of more-intense crustal reworking. In recent years, the existence of Archean blocks was confirmed in many windows and outliers in the orogen. Many Archean remnants are only partly overprinted by the Early Proterozoic orogeny. It is now widely accepted that these Archean blocks occupy much of the Trans-Hudson Orogen, and another Archean block, named Saskatchewan or Sask, has been delineated in southern Saskatchewan (e.g., Collerson et al., 1990; Lewry et al., 1996; Chiarenzelli et al., 1998).

The idea of far-traveled large continental blocks pulled apart or amalgamated in an accidental, random fashion into continents arose at the time when the relationships between lithospheric plates were understood poorly and the deep roots of cratons were not well recognized. Modern deep seismic studies (e.g., among others, Jordan, 1975; Anderson and Dziewonski, 1984; Su et al., 1994) undermine the foundations of some earlier ideas and serve to constrain speculations. For example, although ancient continental cratons have roots deeper than 400 km (also Gossler and Kind, 1996), the 410-km seismic discontinuity is on average 4-5 km shallower under continents than under oceans, but the relief of this discontinuity corresponds poorly to that of the global seismic discontinuity at ~660 km. The discontinuity at 520 km, observed beneath the oceanic-lithosphere parts of the globe, is absent under the cratonic shields of continents. All this suggests that under the continents and cratons, the mantle is distinct from that under adjacent oceanic regions to depths exceeding 520 km. Below ~660 km, mantle processes and material properties must be strongly different than above.

On the other hand, some models assume united whole-mantle dynamics between the top of the core and the base of the lithosphere. Some modelers (e.g., Grand et al., 1997) believe that patterns of whole-mantle convection are correlative with the distribution of subduction zones. Indeed, these similarities are suggested by independently derived tomographic P-wave and S-wave models of the mantle, predominantly in the well-studied Northern Hemisphere. One zone of above-average mantle velocity (probably indicating cold material) runs NW-SE from the southwestern Pacific along the modern Eurasian megabelt from the Pamir and Himalaya towards the European Alps and as far west as the Rockall Plateau area. Several Mesozoic and early Cenozoic subduction zones did exist in this belt. But another zone of high velocity is modeled from Ecuador across the Gulf of Mexico and Great Lakes region towards Hudson Bay and beyond, into the eastern Canadian Arctic. This high-velocity band is not correlatable with modern or recent subduction zones. This observation is hard to reconcile with the idea that "the cold regions appear to be continuous as a function of depth, suggesting the descent of slabs of oceanic lithosphere at the edge of tectonic plates. Indeed, the features both in the America and South Asia coincide with regions that have a long history of subduction..." (Levi, 1997, p. 18). The inference that "...heat flow from the core into the mantle helps drive plate tectonics, volcanism, earthquakes and other geological processes in the Earth's crust" (Jeanloz and Romanowicz, 1997, p. 25) also seems at odds with the above observations.

In the models of Su and Dziewonski (1995), roots of mid-ocean ridges reach depths of hundreds of kilometers. Roots of continental masses are also very deep, exceeding 400 km (Jordan, 1975; Anderson and Dziewonski, 1984; Gossler and Kind, 1996). Composition of these roots is different, but their size is very large. All this evidence points to greater stationarity of these fundamental features than is sometimes presumed.

Many models assume easy movements of numerous plates, continents and their fragments all over the globe, but often these models lack supporting physical evidence. The current uses of magnetic anomaly lineations as a reference frame to reconstruct such motions are under critical revision (e.g., Yañez and LaBrecque, 1997). On the other hand, geochemists and geologists are reporting a growing amount of new evidence about long-time connections between supposedly "exotic terranes" in the North American Cordillera (e.g., Ernst, 1988; Woodsworth et al., 1991).

With regards to the Early Proterozoic Mojave Desert province, Anderson et al. (1993, p. 187) stated: "...We disagree with the paleomagnetic interpretation (Morris and others, 1986) that transport has been on the scale of 1,000 or more kilometers". In the Canadian Cordillera, the birthplace of ideas about accreted suspect terranes, earlier paleomagnetic interpretations have been found to be flawed and in contradiction with geologic field evidence (Butler et al., 1989; Monger and Price, 1996). Elston and Link (1993, p. 575) put it strongly: "We do not understand the details and especially the timing of the fragmentation of the inferred Late Proterozoic supercontinent... The identity of the continental fragments (if any) that rifted away from western North America are contested. Furthermore, there is little agreement on the Proterozoic apparent polar wander paths for the various continents... The several reconstructions are mutually incompatible and are based on different data sets. There is disagreement about age, validity, and possibility of remagnetization of various poles..."

The lack of "confirmatory evidence" in the classical, rock-based geological practice should inspire investigators to search for such evidence among the verifiable data. In the inductive reasoning, which runs from observable facts to more abstract generalizations, evidence should be cross-checked with other valid facts. Deductive (as opposed to inductive) reasoning helps to identify the facts fitting a working hypothesis, while also noting the facts which contradict it. This makes it possible to assess the value of the observed facts for improving the existing hypotheses. A new, updated, internally non-contradictory conceptual system must be erected, incorporating all the facts. Used properly, meticulous inductive and deductive analysis is time-consuming, but such thorough reasoning prevents exuberant proclamation and publication of immature ideas.

To construct a realistic geological hypothesis or geophysical model, one has to first of all study the regional geology. Geology, as a science, is not exact but rather empirical, observational and descriptive. Current attempts to "improve" geologic reasoning by numerical models based on unrealistic assumptions are often more damaging than useful (Oreskes et al., 1994; Frodeman, 1995). Some modern beliefs that geology is becoming an exact science are plain wrong, although constraints on geological hypotheses provided by relevant numerical modeling based on geophysical or geochemical data may be invaluable. Even the most skilled mathematician cannot produce an adequate model of a phenomenon (object or process) which s/he does not understand. A good knowledge of geophysics, geochemistry, physics, chemistry, mathematics is necessary for working in geology, but none of it makes up for a lack of knowledge of geology itself. It is healthy to remind those non-geological experts who try to solve the extremely difficult geologic and tectonic problems how many such solutions have failed in the past. In all countries that carried out deep crustal deep drilling, results proved largely inconsistent with predictions derived from prior geophysical models (e.g., Kozlovsky, ed., 1984; Clauser and Huenges, 1993). Such blatant failures, born of misuse of otherwise beneficial techniques, only undermine the public confidence in science as a whole.

The common belief that "the challenge for field geologists... is to test the validity of these models" (Sinclair, 1997, p. 324) is wrong in its very essence. The job of field geologists is to gather observable facts and interpret them in an internally consistent concept, primarily by making maps. A subsequent challenge is to parametrize some of these data objectively, under a geologist's oversight. To predict the behavior of a system, computations must employ an algorithm based on a strongly formalized geologic concept. Properly done, numerical modeling can become beneficial for practical tectonics, but no model can replace an educated, experienced and creative tectonist operating with objective information.

Use of tectonic étages and stages to improve the objectivity of regional tectonic analysis

Regional tectonic analysis must first of all be founded on information obtained by geologic field mapping and drilling. Sampling rocks and measuring rock bodies provides objective, observable and verifiable information to constrain a tectonist's reasoning. Although many concept fail to explain numerous observable facts in various continental and oceanic regions, those of them which are derived from objective sources can serve as a springboard for a fruitful search for new approaches. Uniformitarian ideas in geology (Lyell, 1830-33) and specifically in tectonics (Stille, 1924, 1936a-b, 1941) are obsolete in many of their aspects, but they still provide some useful approximations. To criticize the weaknesses of older concepts became fashionable in the mid-20th century. But useful and productive parts of old methodologies should not be discarded along with the old tectonic concepts themselves.

Two main classes of tectonic regime were distinguished by Hans Stille (1876-1966) (1924, 1936a-b, 1941) as typical for geosynclinal-orogenic and craton-platformal provinces: he called them, correspondingly, *Alpinotype* and *Germanotype*. The Alpinotype tectonic regime was proposed to involve the accumulation of thick and laterally very variable stratal successions, formation of numerous igneous intrusions, high-grade metamorphism, and high degree of deformation: complex folds, synorogenic and post-orogenic nappes, abundant faults of different types. Subtypes of this regime were established: in orogenic internides (eugeosynclines), where all four manifestations of tectonism are well displayed; in orogenic externides, where thrusting is the most common type of deformation; in median massifs, where tectonic manifestations may be subdued. The Germanotype tectonic regime was distinguished as characterized by subvertical movements of blocks bounded by steep faults, which influenced the patterns of sedimentation, specific types of magmatism and moderate deformation of stratal rocks. Tectonotypes for this regime were found in the block-shaped Harz and Schwarzwald (Black Forest) Mountains and the Rhinegraben (Germany).

Stille also distinguished initial, simatic orogenic magmatism; mostly granitic synorogenic plutonism; and post-orogenic magmatism of mostly small intrusions. Such tectono-magmatic cycles of various types were identified all over Eurasia (Bilibin, 1955; Moskaleva, 1989). The initial magmatism, typified by spilite-diabase and gabbro-peridotite formations commonly associated with ophiolites, was linked to the pre-inversion stage of orogenic evolution; the voluminous granitoid plutonism was linked to the inversion stage (orogeny per se); the post-orogenic (post-inversion) stage was associated with extensional, chiefly alkaline magmatism.

It was also recognized that there is a wide spectrum of tectonic regimes with many transitional cases (paraplatformal, parageosynclinal, etc.), that the formation of mountains may be rapid or slow, that orogenic processes may or may not lead to mountain-building, that epochs of building of mountains may be orogenic or non-orogenic (e.g., Kober, 1925; Gilluly, 1949; Schultz, 1964; King, 1977). There was much discussion about how to pinpoint the beginning and the end of oro-genic regimes and their mappable expressions. It was also established that the pre-existing background on the site of the mobile megabelt, sometimes labeled vaguely as "structures of the orogen's basement", may be very complex and in-clude pieces of older orogens as well as platforms, even cratonic ones (Labazin, 1963; Lyatsky, 1967), and the pieces of cratons may be of very different ages (e.g., in Lewry and Stauffer, eds., 1990).

The initial continental-crust patches appeared sometime in the Early Archean (Tables IA, IB). They were later welded together to produce the first proto-cratons, well distinguished in contrast with mobile megabelts. On the site of the modern North American craton, the preserved original Archean cratonic masses are the Slave, Superior, Wyoming and Sask cratons. Epi-Archean platforms, such as Hearne, Rae and others, reworked by Proterozoic orogens, are also well recog-nized. The Trans-Hudson Orogen, distinguished in the central Canadian Shield as an Early Proterozoic feature, itself became cratonized, and a new, Laurentian craton appeared (Hoffman, 1988). Its extent was large, including much of the re-gions that later became the Cordilleran and Grenvillian-Appalachian mobile me-gabelts.

The Grenvillian megabelt is conspicuously colinear with the Paleozoic Appala-chian-Ouachita megabelt. Together they form a single, composite Grenville-Appalachian mobile megabelt containing elements of two distinct tectonic cycles: Grenvillian and Appalachian (including the Marathon-Ouachita orogenic prov-ince). This complex megabelt, late Mesoproterozoic to late Paleozoic in age (its cessation occurred in different regions at different times), overprinted the eastern and southern parts of the Laurentian craton. Crustal blocks detached from the pre-Grenvillian craton were incorporated into the Grenvillian orogenic belts. Such blocks are known in abundance east of the Grenville deformation front in Canada. Yet, the Grenvillian crustal blocks, including partly reworked Archean ones, are recognized also as the basement for the Appalachians. Similar tectonic relation-ships exist in the long-lived but single-cycle Cordilleran orogenic megabelt.

In the time between the Hudsonian (~1,900-1,800 Ma) and Grenvillian (~1,100-900 Ma) orogenies, the Transcontinental orogenic province was formed on the site of most of the modern Midcontinent Platform as well as in some areas in the

51

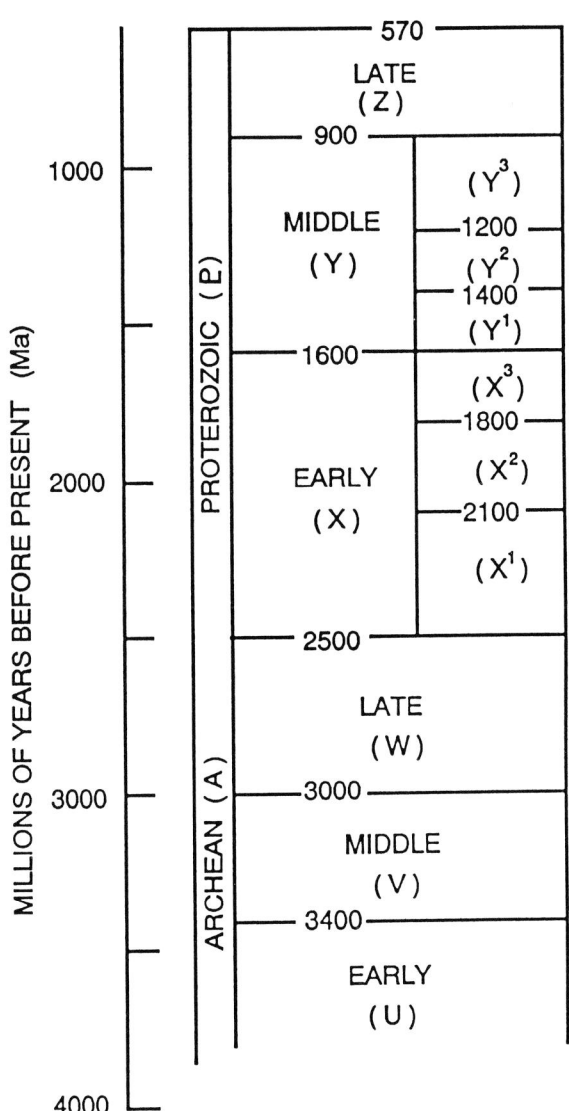

Table IA. Geologic time scale, adopted by the Geological Society of America: Precambrian (simplified from Reed and Harrison, 1993).

52

Table IB. Geologic time scale, adopted by the Geological Society of America (1983): Phanerozoic.

western Appalachians and Cordillera (in California; Van Schmus and Bickford, 1993). The Yavapai and Mazatzal orogenic belts with overlapping ages (chiefly, 1,700-1,620 Ma for the former and 1,660-1,600 Ma for the latter; Figs. 2-4; Karlstrom and Bowring, 1993) are parts of a single Transcontinental mobile megabelt. Like the older Trans-Hudson megabelt, in the Midcontinent Platform it is mainly hidden under the Phanerozoic sedimentary cover (Van Schmus et al., 1993). The E-W-trending orogens of the Transcontinental mobile megabelt cut the N-S branch of the Trans-Hudson Orogen which runs from Saskatchewan into the Dakotas. After the Transcontinental mobile megabelt died, around 1,300 Ma, a craton different than Laurentian developed on the site of the modern North American craton (sometimes it is referred to as Hudsonian).

A great number of tectono-magmatic events nonetheless occurred in Middle Proterozoic time across this new craton. The Granite-Rhyolite province in the Midcontinent region was a result of Middle Proterozoic extension which produced a grandiose episode of unorogenic magmatism, mostly between 1,500 and 1,300 Ma, partly predating the Grenville mobile megabelt. Diachronously, continental-scale crustal Mesoproterozoic extension and associated anorogenic magmatism spread over one-third of North America, from Mexico to Labrador. The Grenvillian orogeny around 1,100 Ma delineated the third-generation craton in North America, which succeeded the previous (Laurentian and Hudsonian) cratons but had its own extent and specific tectonic regimes, exhibited in particular in the formation of extensive Phanerozoic platforms and platformal sedimentary cover.

But even that was not yet the modern North American craton. Prior to the obvious commencement of the Cordilleran mobile megabelt around 800 Ma (with the appearance of the Windermere rift and its correlatives), the Proterozoic Grenvillian craton had extended much farther west, beyond the modern Rocky Mountain fold-and-thrust belt, Omineca Belt and Intermontane median-massif belt (some workers believe it might have reached even the present-day Coast Belt; e.g., Cook, 1995a-c). A surficial manifestation of the incipient modern North American craton was its differentiation into broad subsidence-prone platforms (mostly but not exclusively on the pericratonic periphery) and uplift-prone Canadian Shield; their boundaries shifted greatly in Paleozoic time (Sloss, 1988a) before they became stabilized in a position similar to the one we know today. Still, shifts in their configuration continued during the Mesozoic and later, until the Tertiary. The modern North American craton was thus delineated.

In its present-day boundaries, the modern North American craton contains a set of big and small structural-formational étages of several generations in the basement and in the volcano-sedimentary covers of different ages.

The first, Proterozoic, volcano-sedimentary basins on the cratons were discontinuous, having accumulated in separate depocenters surrounded by broad elevated areas of exposed pre-basin crystalline rocks. In what is now the Canadian Shield, such initial basins began to appear on top of the Archean crust as early as in the Early and Middle Proterozoic (Campbell, ed., 1981). In the Russian cratonic platform, similar basins seem to have developed later, in the Middle to Late Proterozoic (Igolkina, ed., 1981). Many of the early basins are located over large, subsided fault-bounded blocks or elongated zones of crustal weakness. These well-defined ancient basins were subjected to a certain amount of folding and metamorphism, but they are distinct features discordant with the underlying basement, which is much more metamorphosed and deformed.

Reflecting the shapes of underlying subsided blocks, Proterozoic volcano-sedimentary basins are of several main types: oval (Athabasca Basin in the Canadian Shield); wedge-shaped or more polygonal (Kilohigok Basin in northern Canada); linear (Pachelm Trough in east-central European Russia); or curvilinear (Midcontinent Rift in North America). Only in the Late Proterozoic Riphean did the cratonic platforms become stable enough to commence a slow, general subsidence and accumulate a predominantly sedimentary cover of stratal rocks which remained unmetamorphosed and largely undeformed to this day.

The relatively high-standing shields of ancient cratons experienced long-term predominant uplift due to upwarping. The shields are usually located in central parts of cratons, but can be found in off-center positions as well (e.g., Fennoscandian Shield in the East European craton or Aldan Shield in eastern Siberia); upwarping predominated in the cratonized crust in most of the African continent. In contrast, broad, continuous downwarping enabled the formation of vast cratonic platforms, with their long-lived, thick sedimentary cover, in most of the modern continents.

Fundamental restructuring of continental crust after cratonization occurred on all continents. The time interval between the end of cratonization and the formation of the sedimentary cover is not always well represented in the cratonic geologic record. A fundamental unconformity separates the cover from the crystalline basement. During that time, measured in many hundreds of millions of years, pre-existing rocks were affected by various forms of tectonism. New, cratonic, brittle structural patterns were imposed on the previous, ductile, orogenic ones. Rocks were altered, as evidence by geochemical studies. They were later eroded and beveled. Some earlier orogenic fault zones were partly reactivated by cratonic rearrangements of the cratonized crust. In most areas, new, brittle, polygenetic faults developed; new block-like structural patterns masked the original, orogenic structural style in the basement. The variable vertical movements of the crust and its blocks are reflected in the variations of composition and structure of the tectonic étages in the (volcano-)sedimentary cover. The Late Proterozoic to Recent cratonic platforms bear numerous structural-formational étages in their cover, corresponding to different cratonic crustal rearrangements during the cover's development.

From studies of fluid inclusions and vitrinite reflectance, epeirogenic vertical movements 4 to >7 km in amplitude have been estimated in the Appalachian Basin (in front of eponymous mountains) in the eastern North American craton (Friedman and Sanders, 1982; Friedman, 1987a-c). Large-amplitude tectonic movements of crustal blocks created deep Paleozoic sedimentary basins such as Illinois, Michigan, Williston and others. They appeared as a response to tectonic events rooted very deep in the cratonic crust and lithosphere. Sedimentary basins, which receive a lot of attention largely due to their hydrocarbon resources, are distinguished by their specific tectonic history, amplitude of subsidence, structural style, etc. On these criteria, the cratonic Midcontinent and Interior platforms of North America have more in common with, for example, the faraway Russian Platform than with younger platforms overlying the recently-inactive orogenic zones in the nearby Appalachian and Cordilleran provinces. The younger, non-cratonic platforms have more characteristics inherited from the antecedent orogens. They have a greater basement relief, and were affected by more-intense tectonism which produced more geologic diversity in the cover. As drilling in the U.S., Europe and Russia has shown, stratigraphic record in the cover of such young platforms contains evidence of vertical movements of 8 km or more.

Misleadingly, relief of the top of the crystalline basement is sometimes referred to as "basement structure". This is just confusing jargon, and it is best avoided because the modern relief of the basement bears the marks of the pre-cover erosional physiography as well as subsequent vertical tectonic movements. Paleogeomorphological features at the top of the basement might have resulted from differential erosion of contrasting lithologies (e.g., resistant quartzites or mafic intrusive plugs may be expressed as local basement highs); erosional sculpting of structural features (e.g., fluvial incision along faults zones); or tectonically induced scarps predating the cover. Pre-sedimentary tectonism affected the basement repeatedly after cratonization. Syn-sedimentary and post-sedimentary tectonism affected both the basement and the sedimentary cover. The bedded structural-formational étages under tectonic regimes specific to their corresponding tectonic stages of the craton.

As was noted above, many previous tectonic maps were intended to express the tectonic evolution of a region in terms of structural-formational étages and tectonic stages. However, most stages were defined on a geochronological basis obtained from the ages of main regional orogenies (e.g., *Tectonic Map of the U.S.S.R.*, 1956; *International Tectonic Map of Europe*, 1964; *Tectonic Map of Eurasia*, 1966; *Tectonic Map of North America*, 1969). New tectonic maps produced in Europe and North America (Ziegler, 1982; Nokleberg et al., 1994; Muehlenberger, 1996) related the timing of on-land orogenic activity to assumed tectonic events on the ocean floor. Oceanic-crust subduction and collisions of continents are mainly inferred based on data from oceanic magnetic anomaly lineations and presumed hot-spot tracks for Mesozoic to Recent time (e.g., Engebretson et al.,

1985; Stock and Molnar, 1988), but the interpretations of other geophysical anomalies such as gravity anomaly pairs (e.g., Gibb et al., 1980) and questionable paleomagnetic data (e.g., Symons, 1991; Beck et al., 1997), intended to fit these models, are often loose.

Hoffman (1988, 1990, 1991) used these types of data for his speculative mobilistic concept of the North American craton's origin. He attributed the birth of this craton to plate interactions in the Early Proterozoic, presuming that the Laurentian craton had been assembled from far-traveled pieces. Convergence and collision of continental-crust masses was proposed, but not proved, to be the leading process in the craton's formation. Hoffman's hypothesis is still very influential in North America, and it is applied to many parts of the cratonic platform basement in Canada and the U.S. (see, e.g., Reed et al., eds., 1993; Stott and Aitken, eds., 1993).

Only rocks and rock-made bodies can provide objective and reliable criteria for the evaluation of a geological hypothesis. Structural-formational étages in the basement and the sedimentary cover are a tool to check ideas about the craton's evolution. In the unmetamorphosed and usually little-deformed sedimentary cover, the étages are well preserved for geologic studies. In this book, we examine platformal étages on the North American craton to reconstruct some important aspects of this craton's evolution.

3   REGIONAL TECTONIC ANALYSIS OF THE WESTERN CANADA SEDIMENTARY PROVINCE IN ALBERTA, SASKATCHEWAN AND ADJACENT PARTS OF THE U.S.

Reasons for the choice of western Canada as an example for detailed study

The western Interior Platform of the North American craton is rich in oil, gas, coal, salt and other natural resources (see Figs. 6-41, Tables I-III). It is one of the best-studied platformal regions in the world. For a long time, especially since the 1940s, it has been a place of extensive exploration which involved drilling of hundreds of thousands of wells, whose cores and logs are available publicly.

The Canadian province of Alberta has for many decades been the lively center of the Canadian petroleum industry whose activity extends all across the Western Canada Sedimentary Province, including southern Saskatchewan and northeastern British Columbia, as well as some neighboring areas in the Yukon and Northwest Territories of Canada and in the U.S. This region is among the world's most densely covered with a wide variety of geological and geophysical surveys, as well as drillhole information, available in the public domain. It has been studied with a common-sense approach typical of practical exploration. The enormous volume of data has fed a vast literature on different aspects of regional geology and local peculiarities of the sedimentary cover and, to some degree, of the crystalline basement. This region has been a fertile ground for creation and application of the major techniques, models and concepts developed during the 20th century. Successes and failures in the use of various ideas may play a crucial role in the development of an adequate and practical tectonic theory.

The Western Canada Sedimentary Province comprises three main basins (Figs. 6-9): the Alberta Basin in the Interior Platform west of the Sweetgrass Arch and the Williston Basin of the Midcontinent Platform east of the Arch; north of the Alberta Basin, on the other side of the Tathlina Arch, lies the MacDonald Platform. This province is composed of sedimentary successions from the late Precambrian (in the north) and Cambrian to the Recent. Well-studied examples of a great many geologic situations, tectonic, eustatic and combined tectono-eustatic, of subsidence and uplift, broad warping and more-local block movements - all these are found in the Western Canada Sedimentary Province. They have been studied in great detail by professionals in the industry and academia (see, e.g., the many publications of the Canadian Society of Petroleum Geologists).

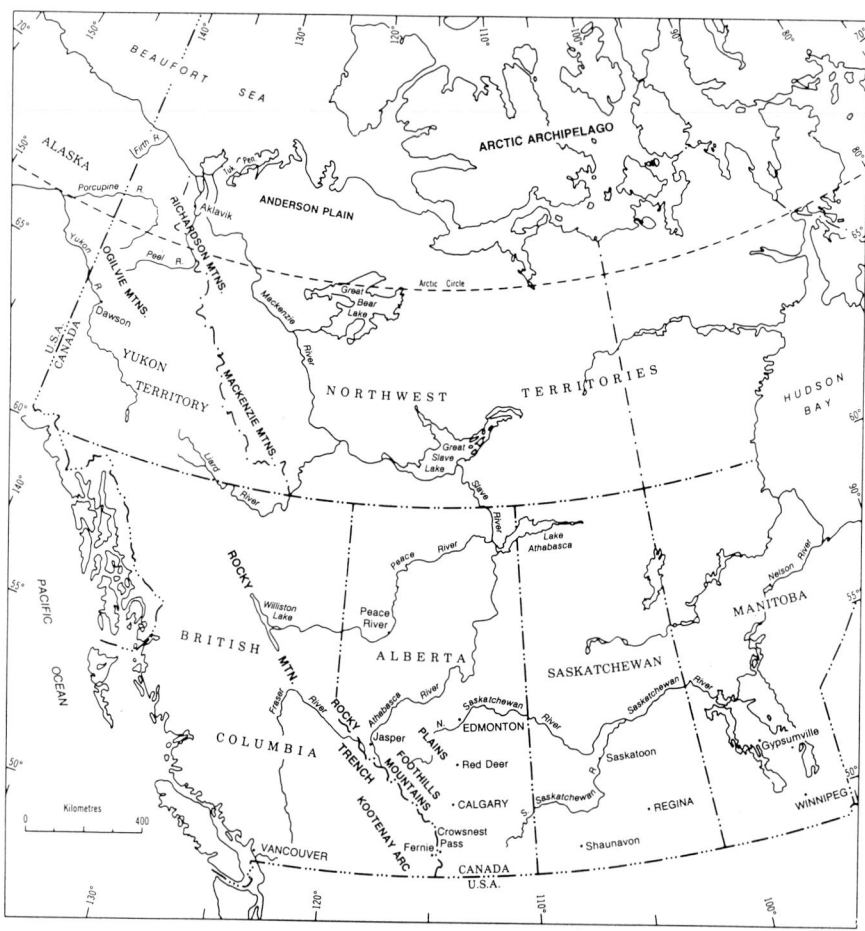

Figure 6a. General geographic index map of western Canada (modified from Poulton et al., 1993), showing major geographic features. To become familiar with the region of study discussed in this chapter, and particularly with the place names and localities mentioned in the text, the reader is encouraged to review this and subsequent diagrams before continuing with the reading.

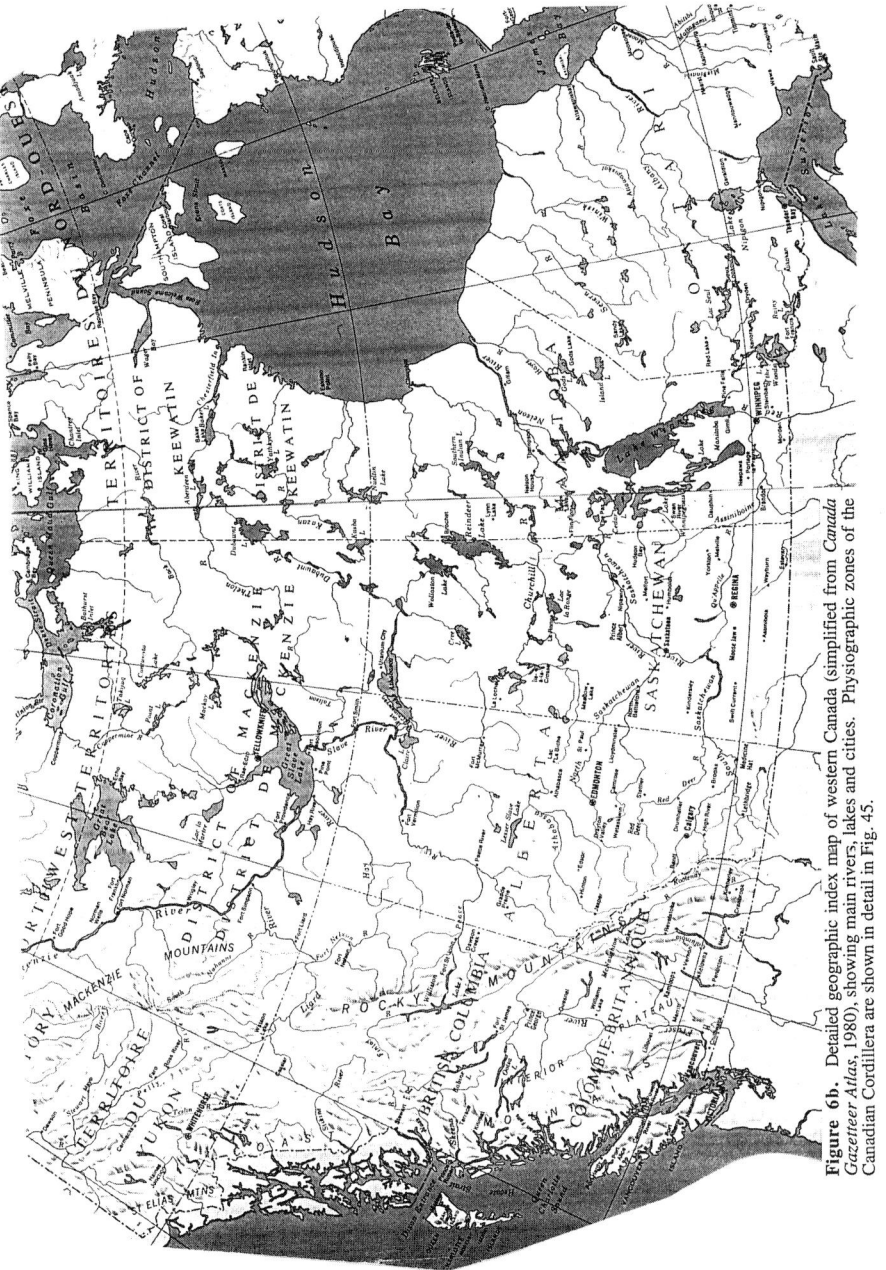

Figure 6b. Detailed geographic index map of western Canada (simplified from *Canada Gazetteer Atlas*, 1980), showing main rivers, lakes and cities. Physiographic zones of the Canadian Cordillera are shown in detail in Fig. 45.

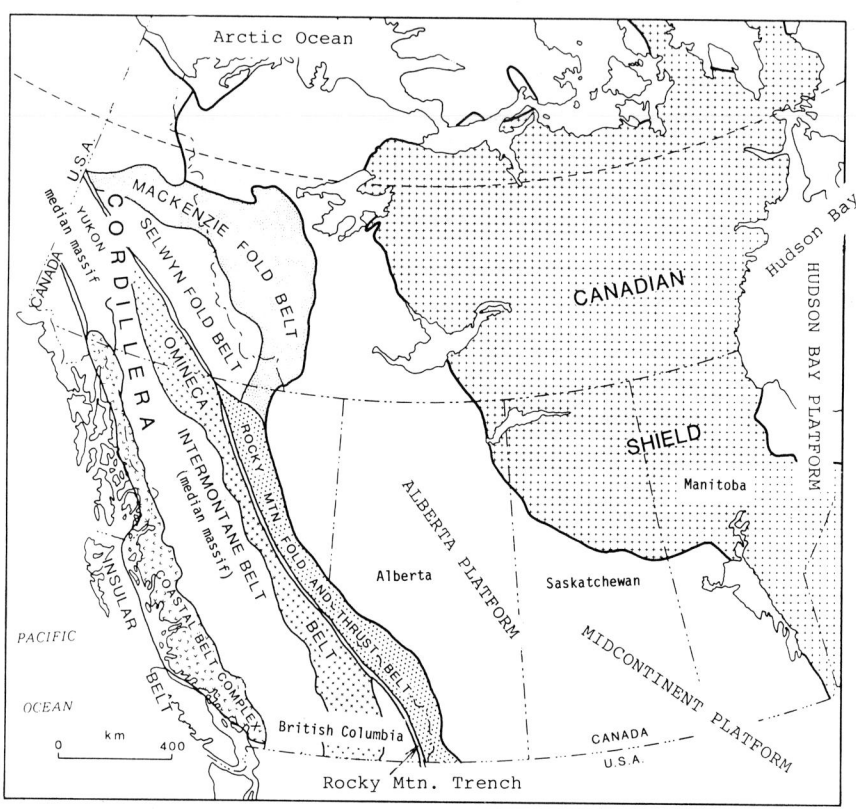

Figure 7. Principal present-day geological zoning of western Canada (modified from Douglas et al., 1970 and Richards, 1989). Additional details of the Rocky Mountain fold-and-thrust belt are shown in Figs. 35, 36, 44, 45.

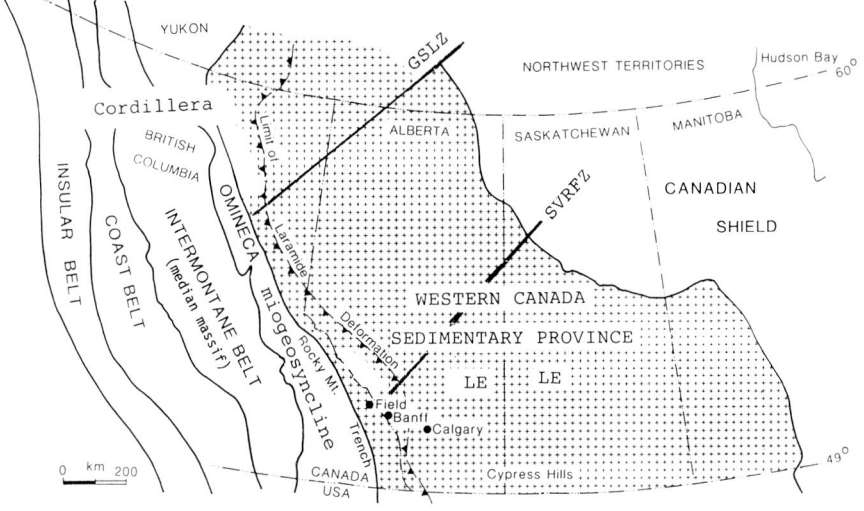

Figure 8. General geologic index map for Chapter 3 - Western Canada Sedimentary Province and vicinity (modified from Ricketts, 1989). GSLZ - Great Slave Lake fault zone; LE - Lloydminster embayment; SVRFZ - Snowbird-Virgin River fault zone.

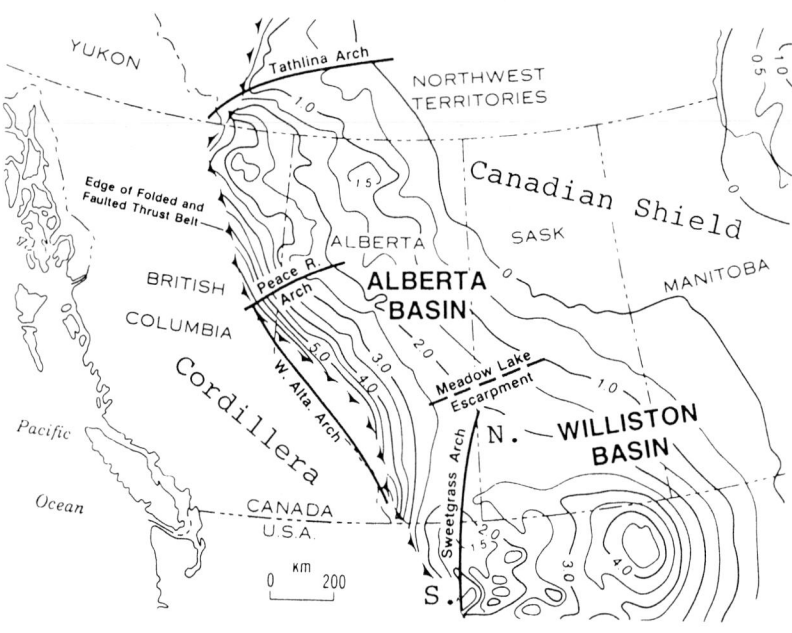

Figure 9a. Thickness of the sedimentary cover in the Western Canada Sedimentary Province (modified from Podruski et al., 1988). Contours in kilometers. The "Sweetgrass Arch" in fact consists of two separate arches with different trends: northern (N.) in Canada and southern (S.) in the U.S. Northern Sweetgrass Arch separates the Alberta and Williston basins, and the Interior and Midcontinent cratonic platforms.

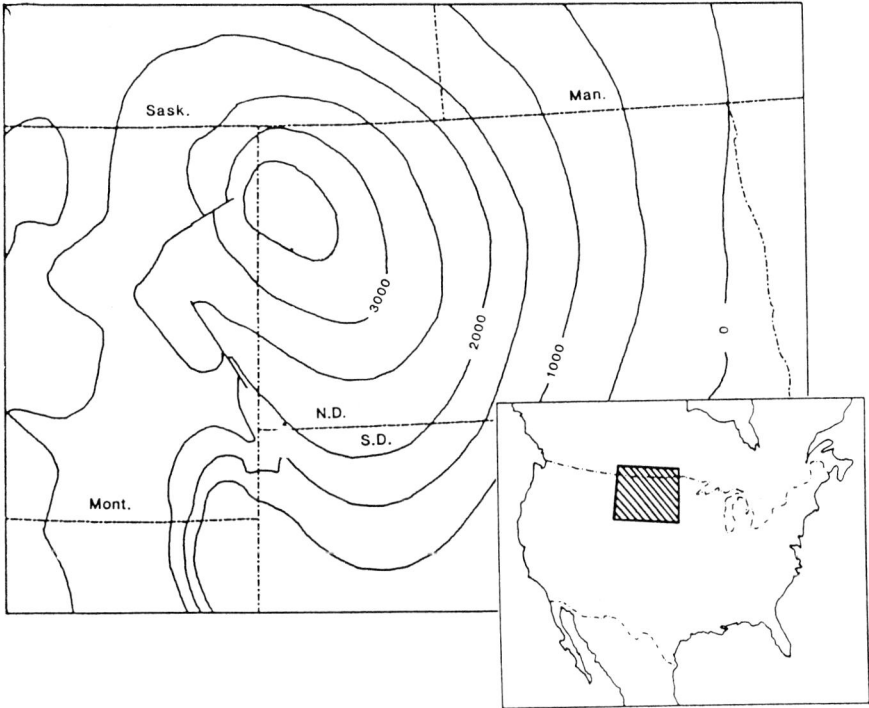

Figure 9b. Location and thickness (in meters) of the Williston Basin in the Midcontinent Platform. Simplified from Ahern and Ditmars (1985). Reprinted with permission from Elsevier Science.

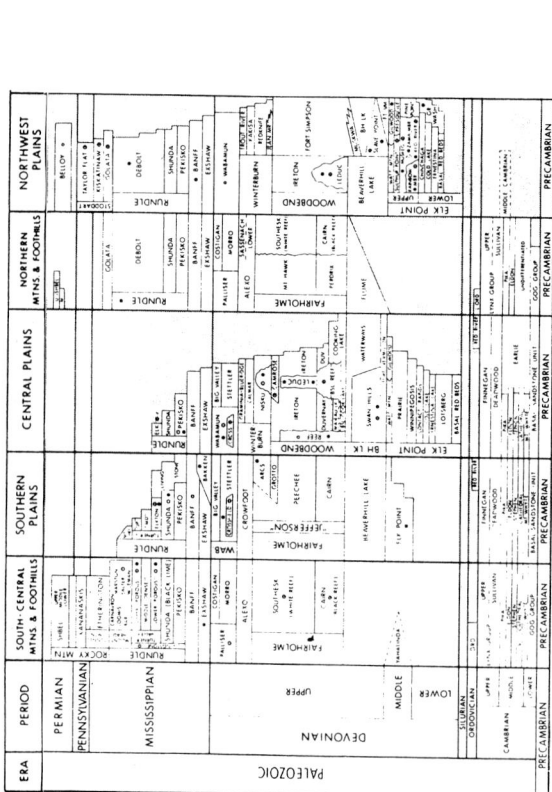

Table II. Paleozoic tables of formations of the Western Canada Sedimentary Province (sources: Energy Resources Conservation Board; and Saskatchewan Geological Survey, 1994).

Reliable geologic record, detailed cross-sections of the cover across Alberta from the Canadian Shield edge in the east into the Rocky Mountains in the west, and outstanding regional summaries compiled since the 1950s and 1960s have facilitated rock-based tectonic analysis of this region. The principal summaries have been published under the auspices of the Canadian Society of Petroleum Geologists and include two monumental atlases: *Geological History of Western Canada*; (McCrossan and Glaister, eds., 1964) and *Geological Atlas of the Western Canada Sedimentary Basin* (Mossop and Shetsen, comps., 1994).

These sources of geological information are used widely in the descriptions in this book. Extensive personal experience of the authors helped them make these descriptions and assessments critically.

Tectonic étage and stage between the Middle Cambrian and the Silurian

Accumulation of sediments that constitute the Western Canada Sedimentary Province began in Alberta and Saskatchewan in the Middle Cambrian. A peneplained cratonic region including parts of the former Archean Slave and Wyoming cratons and Trans-Hudson Orogen began its long-lived subsidence. Accumulation of sediments was initiated in a short time over a broad region. In what is now the eastern Cordillera, subsidence and gradual marine transgression began as early as in the latest Precambrian-Early Cambrian, but spread eastward only in the Middle Cambrian (Tables IB, II). Thin quartzose sandstones overlying the pronounced unconformity at the top of the crystalline basement mark the base of the new, sedimentary mega-étage. Coarser facies rimmed the Purcell and Peace River basement uplifts at that time.

In westernmost Alberta, the Lower Cambrian clastic deposits of the Gog Group restricted to the eastern Cordillera are thicker. In various localities, they are disconformable or conformable with the late Neoproterozoic Ediacaran/Vendian stratal rocks of the miogeosynclinal Windermere Supergroup (Aitken, 1989; Thompson, 1989; McMechan, 1990). But on the whole, the first platformal tectonic étage, from the latest Proterozoic/Early Cambrian to Ordovician and, in places, Silurian, comprises thin-bedded, fine terrigenous and carbonate rocks deposited in a very shallow epicontinental sea. As the sea transgressed eastward in Middle Cambrian time, the onset of sedimentation was diachronous from area to area. Still, platformal shelfal sedimentation spread rather rapidly over the Precambrian crystalline basement in a vast region. Much of the Interior and Midcontinent platforms was covered by a thin sheet of the initial sedimentary sequence in a short time. Only the irregularly scattered monadnocks, composed of resistive igneous rocks in plugs and sculpted by the pre-cover selective erosion, protruded slightly above sedimentary sheet at the shelf bottom. In time, shallow-marine carbonate and terrigenous clay to sand clastic sedimentation spread on both sides of the

N-S-oriented Western Alberta Arch. It extended as far as the Lloydminster Embayment and later farther east. Two prominent, NE-SW-trending fault zones, Snowbird/Virgin River and Great Slave Lake/Hay River, chiefly controlled the position of relatively subsided and elevated crustal blocks and the location of depositional centers.

Many elongated, narrow zones linked to then-active faults are marked by increased thickness of rocks, up to 2,500 m. Such pre-existing fault networks are recognized, for instance, in the Williston Basin. The Alberta, Central Montana-Coeur d'Alene and Eager troughs are related to such fault zones, too. It should be noted that these so-called troughs were no more than local disturbances in the overwhelmingly shallow shelfal region (e.g., Peterson, 1986; Figs. 10, 11). Presence in them of thick shale deposits does not constitute evidence for deep-water environments. From the embayment in central Alberta, the Cambrian succession thickens and grades into calcareous shales towards the Alberta trough, but farther west, in northeastern British Columbia, it again thins and grades into massive argillaceous and pure limestones and sandstones (for details, see Stott and Aitken, eds., 1993).

As the sea transgressed and the marine basin expanded, Cambrian sedimentation onlapped and then completely overstepped some areas of crystalline rocks that had at first escaped inundation. Only the broad, E-W-trending Athabasca-Peace River-Skeena basement promontory (arch) persisted during the entire Cambrian to Silurian time (Figs. 4a, 12). This prominent high-standing arch widened towards the shield, separating two regional sedimentary depocenters to the south and to the north (Pugh, 1973; Aitken, 1989). Some blocks in the arch subsided intermittently. History of the well-drilled southern depocenter, on the site of the future Alberta Basin, is known in more detail. It is part of a single, huge early Paleozoic sedimentary basin covering southern Alberta, southern Saskatchewan, Manitoba, Montana, North and South Dakota, Wyoming, Idaho, Colorado and Utah (Peterson, ed., 1986; Sloss, ed., 1988). This basin was surrounded by low-standing land areas of barren Precambrian terrain: the Athabasca-Peace River-Skeena basement arch in the north; the Transcontinental Arch (Sioux uplift) in the east and southeast; the so-called "Western [relative to the Middle Proterozoic Belt-Purcell Basin] craton" (with the Uncompahgre and Lemhi arches, the "Belt Island", and the North Cascade cores) in the west (Figs. 10a-b, 11, 12).

The usual 300-400 m thickness of the Cambrian succession is similar on both sides of the Alberta trough: on the so-called "Alberta shelf" to the east and in the Cordilleran miogeosyncline to the west (see maps in Slind et al., 1994). Tectono-sedimentologically, the cratonic Interior Platform and the Cordilleran miogeosyncline in the Middle to Late Cambrian do not seem well distinguishable in western Canada. With more confidence, the position of the Cambrian miogeosyncline may be inferred in areas south of the Wyoming craton (Sloss, 1988a). In Nevada and southern California, a rapidly subsiding west-facing embayment on the site of the miogeosyncline collected

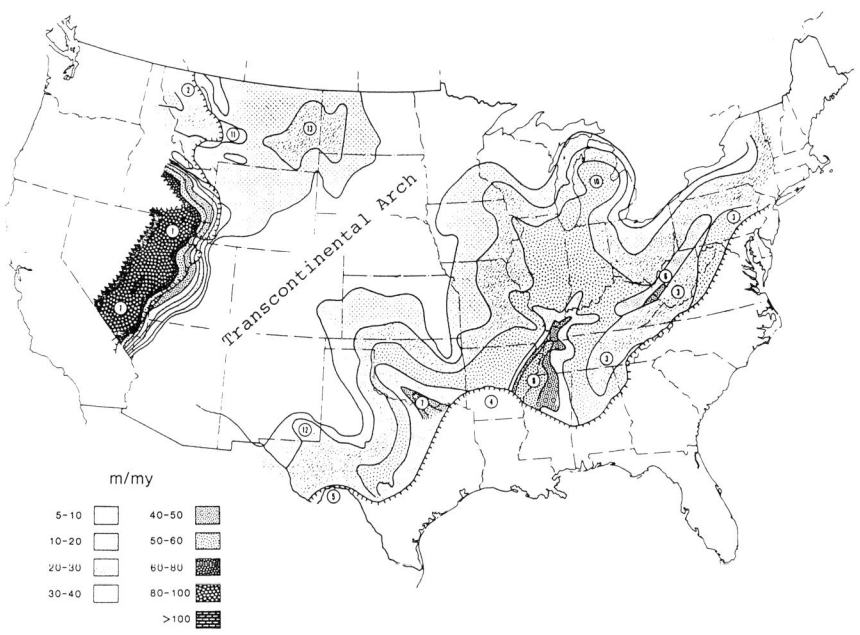

Figure 10a. Latest Proterozoic to early Ordovician net subsidence rates in the conterminous U.S. (simplified from Sloss, 1988a). 1 - Cordilleran Shelf; 2 - area of delayed subsidence; 3 - Appalachian Shelf; 4 - Ouachita province; 5 - Marathon province; 6 - Rome Trough; 7 - Southern Oklahoma Aulacogen; 8 - Mississippi River-Reelfoot Rift; 9 - Transcontinental Arch; 10 - Michigan Basin; 11 - Central Montana Trough; 12 - Tobosa Basin; 13 - Williston Basin.

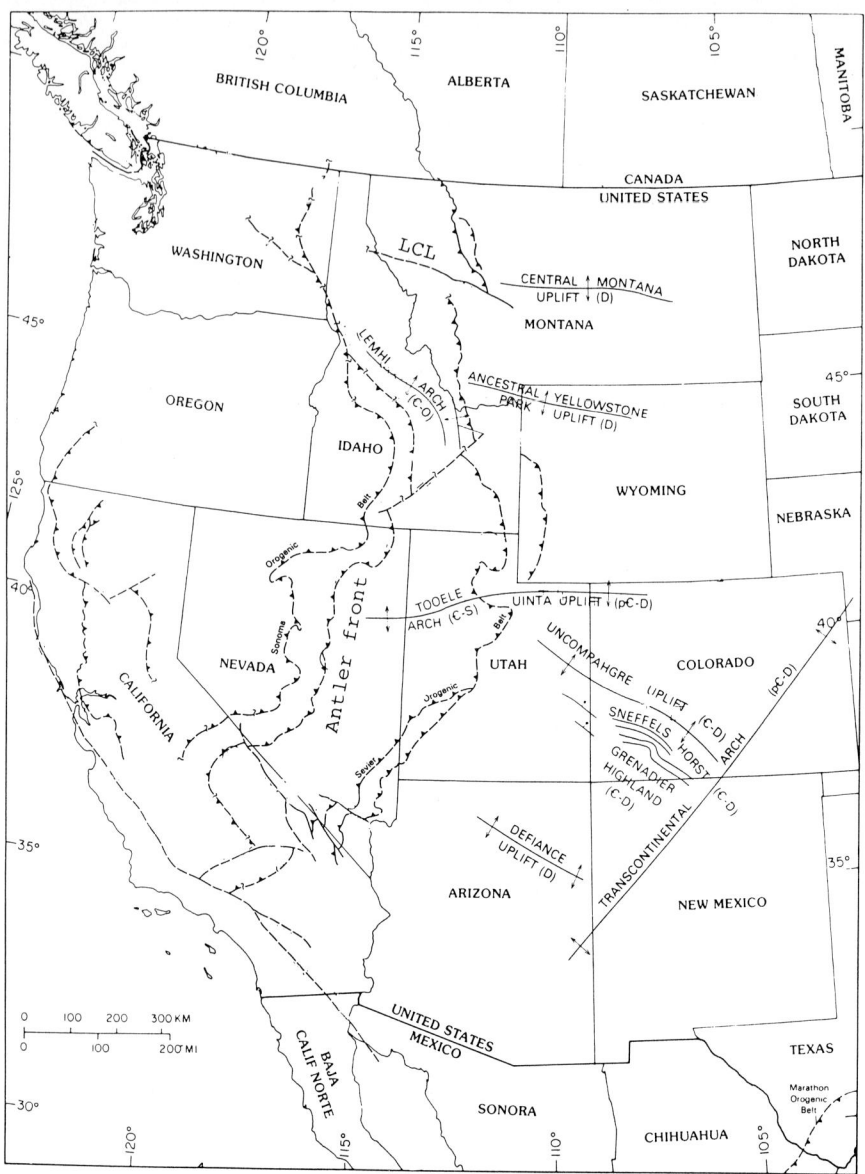

Figure 10b. Index map of major pre-mid-Paleozoic uplifts in western U.S. (modified from Poole et al., 1992). Younger thrust belts are added for reference. LCL - Lewis & Clark lineament.

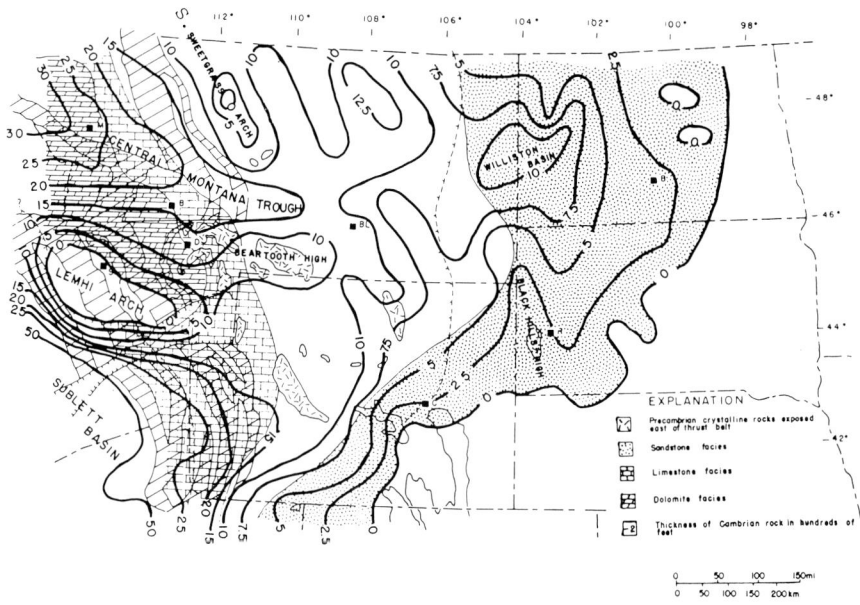

Figure 11. Cambrian isopach map of the Interior Platform and adjacent regions in the U.S. (modified from Peterson, 1985, 1988). Thickness in hundreds if feet (100 feet = 30.5 m). Eastern limit of Middle Cambrian rocks is shown by dashed barbed line. Unpatterned areas are shale, carbonate and sandstone. Cities in the area of study: B - Bismarck (in North Dakota) and Boise (in Idaho); BL - Billings (in Montana); M - Moscow (in Idaho).

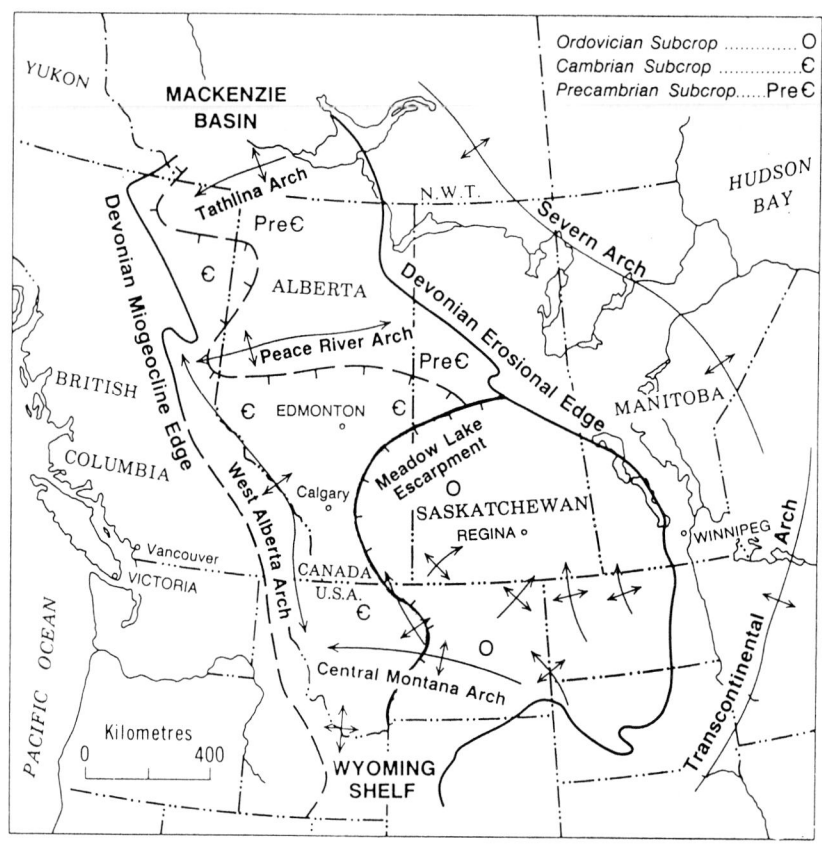

Figure 12. Major Paleozoic arches in the Western Canada Sedimentary Province (modified from Podruski et al., 1988).

several kilometers of sediments, an order of magnitude more than the Interior Platform to the east. As this embayment expanded, it was differentiated into distinct arches and basins (the Central Montana Trough was one of them). The existence at that time of "a simple, continuous miogeosyncline [along the Cordillera] is not evident" (Sloss, 1988a, p. 30).

It seems that cratonic rather than miogeosynclinal conditions of sedimentation prevailed in the early Paleozoic not just on the site of the North American craton but also in the eastern Cordillera. Although areas of rapid subsidence existed during that time in parts of the Cordillera in northern Canada (Aitken, 1993a-b) and southwestern U.S. (Sloss, ed., 1988), most sedimentary basins were relatively isometric, broad and shallow. In the Midcontinent, the Williston Basin and similar sedimentary basins east of the Transcontinental Arch were orthogonal to round. Yet, some elongated Cambrian grabens - Mississippi Valley or Robson Trough - were marked by sediment thicknesses in excess of 1 km. All of them were produced by slow subsidence.

Division of early Paleozoic North America into regions with cratonic and miogeosynclinal sedimentation and tectonic settings is based more on tradition (since Kay, 1951) than fact. No single miogeosynclinal belt has been traced along the eastern Cordillera: distinct troughs, grabens and rifts are recognized locally, separated by isometric basins or by block uplifts and arches (Fig. 13). North of the Wyoming-Montana state border, westward continuation of the Middle and Late Cambrian epicratonic shelf remained monotonous and lacking major changes. In the entire area of Cambrian-Ordovician rock exposures between the Shield and the Cordilleran Omineca Belt, there was no appreciable hinge. Variations of relatively restricted and open-marine, shallower and deeper parts in the same shelfal sea do not suggest the presence of a continental slope. In the Cambrian, shallow marine paleoenvironments predominated all over Montana and Alberta (Hein, 1987). There is no regular westward transition in sedimentary lithofacies that would suggest oceanic water depths in the west. This whole area was characterized by platformal tectonic conditions with epicontinental shelfal sedimentation and tectonic setting (possibly in contrast to the situation inferred in Nevada; cp. Wannamaker et al., 1997).

The western boundary of the cratonic environments has been proposed to lie near the famous Burgess Shale fossil locality in southeastern British Columbia. A big carbonate ramp was postulated there at the border with Alberta, because the carbonate platform lying to the east is depositionally juxtaposed against shales on the west (Aitken and McIlreath, 1982). But a gradational lithofacies transition was revealed there by detailed mapping: beds with similar ages of trilobite assemblages lie on both sides of this lateral paleoenvironmental break, without a structural ramp (Ludwigsen, 1989).

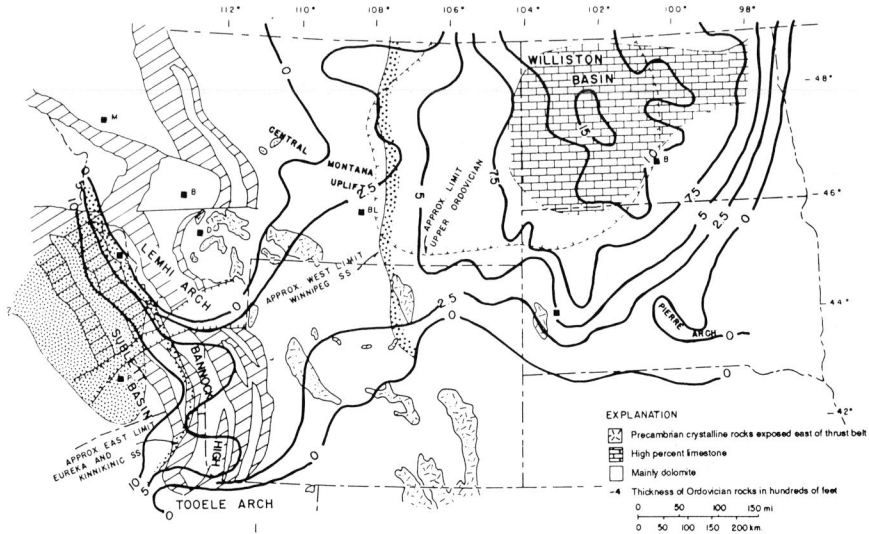

Figure 13. Ordovician isopach map of the Interior Platform and adjacent regions in the U.S., palinspastically reconstructed in the Laramian thrust belt (modified from Peterson, 1985, 1988). Thickness in hundreds of feet (100 feet = 30.5 m); ss - sandstone. Unpatterned area is primarily dolomite. City names as in Fig. 11.

An essentially similar, shelfal epicontinental paleogeography was probably maintained in Alberta throughout the Middle and Late Cambrian (Pugh, 1975). All formations show an increase in their clastic content towards the more-or-less stationary shield edge. Several disconformities and unconformities over shoal areas pass into conformable contacts in deeper troughs. The pre-Middle Ordovician unconformity is widespread, although the Cambrian usually passes into the Ordovician conformably in the Williston Basin. In Alberta, disconformities and non-depositional stratigraphic hiatuses are often found at the top of the Cambrian, while the Lower Ordovician is reduced in thickness or missing. However, the character of sedimentation during both these periods was essentially similar. No regional tectonic restructuring separated the Cambrian succession from the Ordovician one (Fig. 14).

Middle Ordovician clastic sedimentation in the Interior Platform reflected a widespread marine transgression from the south. In Alberta, shallow carbonate deposition took place in oxygenated water with little input of clastic sediments (Podruski et al., 1988, p. 15). Though Tonnsen (1986, p. 28) mentioned that "orogenic pulses in the Canadian Cordillera affected the terranes north of Wyoming". Except for the appearance of a few small carbonate subbasins in western Montana and northern Wyoming, tectonic quiescence in the Interior Platform at that time is evident from the lack of sharp local changes in thicknesses and facies, and from the limited extent of unconformities.

On these criteria, the pre-Middle Ordovician unconformity does not mark a new tectonic étage in the Western Canada Sedimentary Province. This interpretation differs from that of Sloss (1963, 1988a), who put the Cambrian and Lower Ordovician into his Sauk Sequence while placing the Middle Ordovician to Lower Devonian into another Sequence - Tippecanoe. That division probably holds for some areas in the southern Interior Platform and the Midcontinent Platform to the east, but even in the latter region it is not always obvious (e.g., in the Michigan Basin). In Alberta, the distinction between the proposed Sauk and Tippecanoe sequences encounters strong reservations (Podruski et al., 1988). In Saskatchewan, that division is ambiguously justified on paleontological grounds and is not reliably supported based on lithological characteristics (Slind et al., 1994; Binda et al., 1996).

The lack of evidence for a considerable regional tectonic restructuring before the Middle Ordovician is tectonically more important than the presence of this or that local or even regional unconformity, disconformity or stratigraphic gap. Thus, our tectonic division of the Western Canada Sedimentary Province cover into structural-formational tectonic étages is based on different criteria than those used by Sloss for his Sequences.

The Ordovician through Silurian formations contain sedimentary rocks much like those in the Cambrian, dominated by shales and argillaceous carbonates. They too accumulated in a warm, shallow sea. Marine subtidal depressions were the loci of restricted

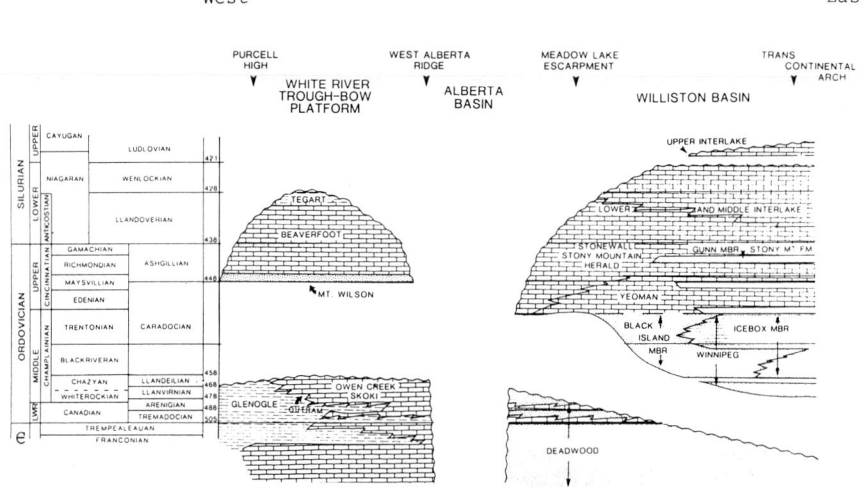

West   East

Figure 14. Lithofacies scheme for Ordovician and Silurian successions in the Western Canada Sedimentary Province (modified from Osadetz and Haidl, 1989).

environments, where salt was precipitated locally. A change to deeper-water conditions is reported at the Alberta-British Columbia boundary, where the presence of graptolite shales (Cecile and Norford, 1993) suggests the Cordilleran miogeosyncline was differentiated there as a separate tectonic entity.

The Ordovician ocean water inundated broad lowlands all across Canada and the U.S., where for a brief time the Ordovician seas, connected by narrow seaways north of Lake and Winnipeg and near the junction of Wyoming, Utah and Colorado, covered a large part of the North American craton (Fig. 15). Slight fluctuations of the sea level could have been induced by epeirogenic warping of the crust, small vertical movements of the crustal blocks, and eustasy. Where preserved, the Middle Ordovician-Silurian sedimentary succession is up to 500 m thick. In most parts of this vast transcontinental shelfal basin, rocks were later thinned or removed completely by erosion due to general uplift that occurred prior to the next period of continuous regional sedimentation in the Devonian. Locally, however, subsidence persisted steadily in the Williston Basin, and probably on the northern and western flanks of an elongated uplift sometimes referred to as Montania.

The formation of an extensive sedimentary cover in the Interior Platform thus began with the deposition of transgressive, thin Cambrian sedimentary strata south of the Athabasca-Peace River-Skeena Arch, including large parts of Alberta, Saskatchewan, Montana and Wyoming (Sloss, ed., 1988; Stott and Aitken, eds., 1993), where the peneplained and subsided western part of the Precambrian craton acquired a rather simple broad embayment. Even gentle crustal warping could easily shift the shoreline, changing the outline of sea and land areas. Hence, importantly for regional tectonics, the distribution of arches and depocenters remained fairly steady. Structural complications in the form of grabens/troughs only slightly disturbed the general pattern of the quiescent tectonic regime.

The Cambrian to Silurian deposits were more extensive than their remnants preserved today, and isolated outliers of Ordovician and Silurian rocks are found even in the modern Canadian Shield areas. Low content of terrigenous clasts in the rocks indicates the relief was low (Norford et al., 1994). Outlines of paleogeographic provinces assigned to the cratonic Canadian Shield, Interior and Midcontinent cratonic platforms and the Cordilleran miogeosyncline were different than today, and their modern shape was established much later (Fig. 16).

The base of the platformal sedimentary cover of the North American craton (Fig. 17) is sometimes treated as an early Paleozoic "breakup unconformity [which] recorded separation of continents moving apart" (Aitken, 1993a-b). A Paleozoic continental slope of an ancestral North American continent has been hypothesized to lie in the eastern Cordillera (Bond and Kominz, 1984) and equated with the eastern Cordilleran miogeosyn-

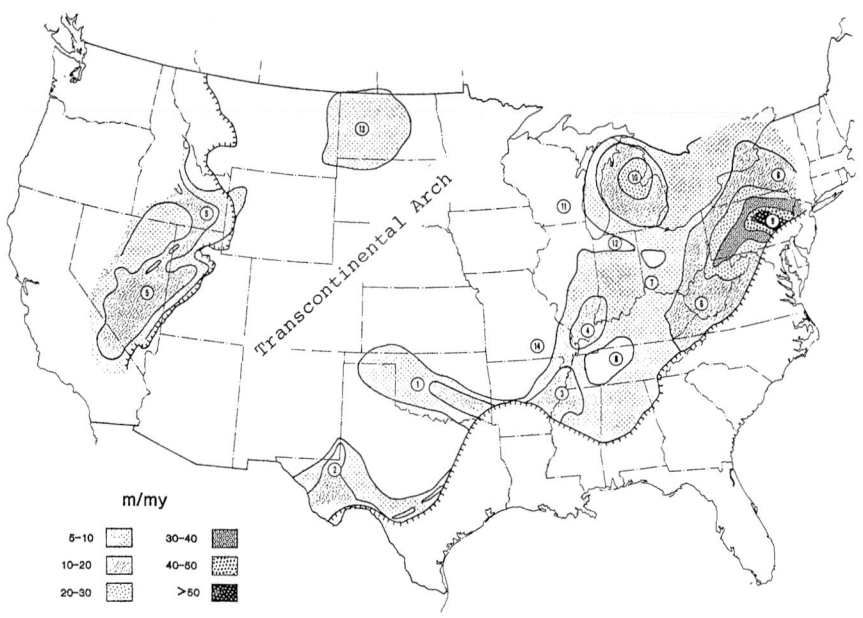

Figure 15. Middle Ordovician to Early Devonian net subsidence rates in the conterminous U.S. (simplified from Sloss, 1988a). 1 - Anadarko Basin; 2 - Tobosa Basin; 3 - Mississippi Valley Graben; 4 - Illinois Basin; 5 - Cordilleran Shelf; 6 - Appalachian Shelf; 7 - Cincinnati Arch; 8 - Nashville Dome; 9 - northern Appalachian Basin; 10 - Michigan Basin; 11 - Wisconsin Arch; 12 - Kankakee Arch; 13 - Williston Basin; 14 - Ozark Dome.

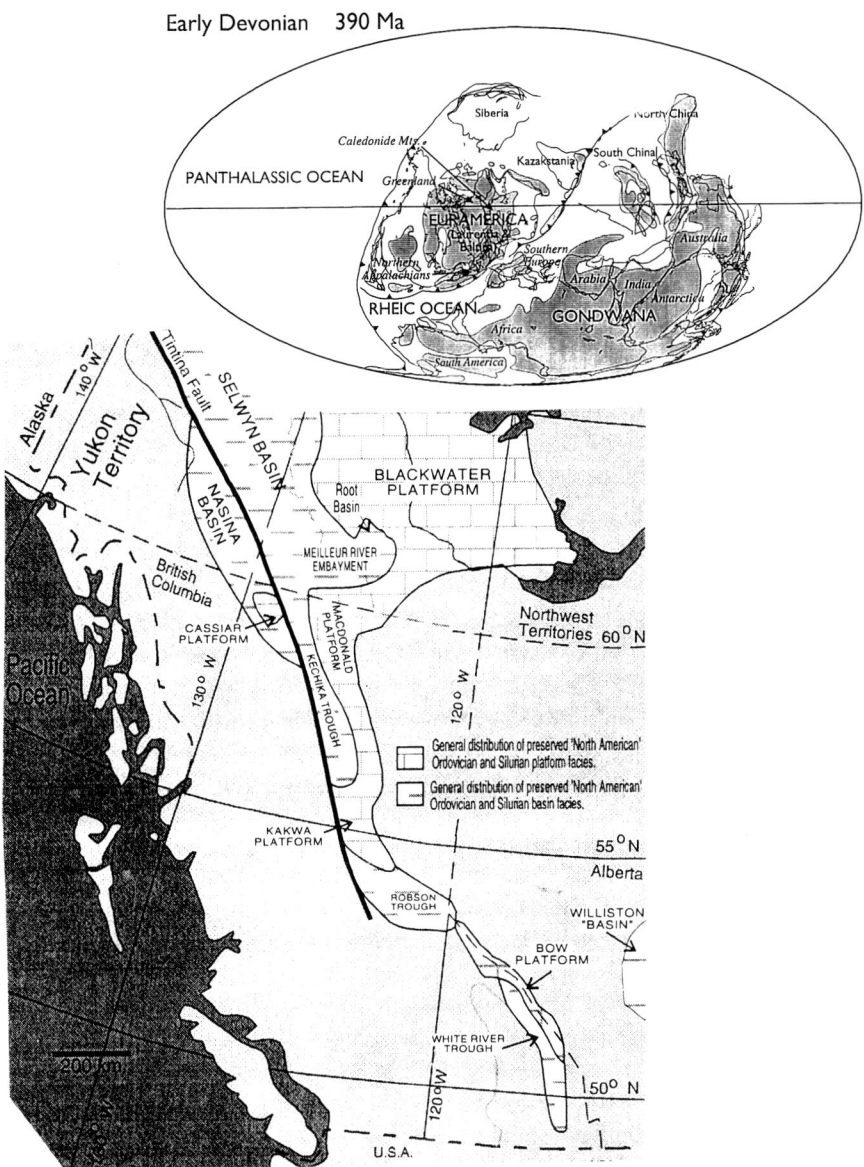

Figure 16. Maps of early Paleozoic carbonate platforms and deeper-water basins in the Cordilleran miogeosyncline and adjacent regions (A), and occurrences of alkalic and potassic volcanic rocks (B) of that age (modified from Cecile et al., 1997). Growth of carbonate platforms and reef complexes at that time was possible as North America lay in warm climatic zones, presumably at low latitudes (oval inset, simplified from Scotese, 1997; as explained in text, however, reliability and exactness of such reconstructions are uncertain).

West   East

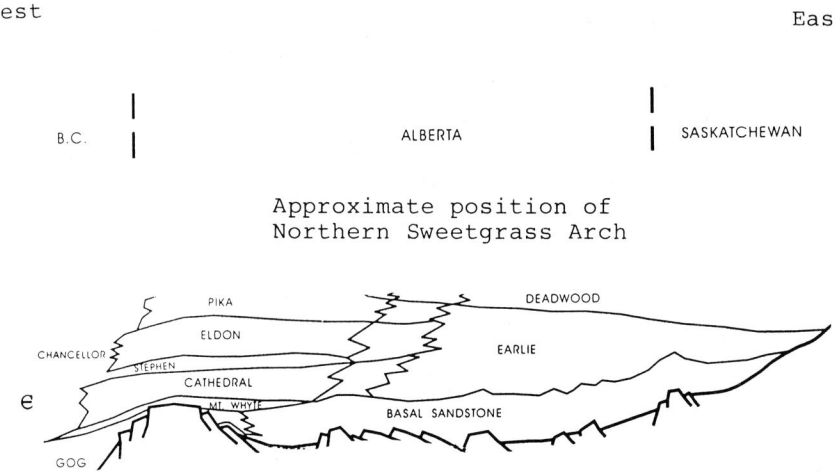

Figure 17. Geometric relationship of the lower part of the Cambrian to Silurian etage with the Precambrian cratonic crystalline basement below, Western Canada Sedimentary Province (modified from Hein et al., 1995). Influence of the pre-cover relief of the basement on the Phanerozoic sedimentary cover is restricted mostly to the lower horizons of this etage (usually Cambrian and Ordovician), leaving the overlying younger rocks unaffected.

cline (e.g., Bally, 1989). Characteristics of the lowest tectonic étage in the cover of the Interior Platform does not support these ideas. Evidence indicates that the eastern Cordilleran miogeosyncline was not expressed as a stationary paleogeographic feature during the Cambrian to Silurian tectonic stage. Instead, the western continuation of quiet, uniform, shallow shelfal and cratonic tectonic and sedimentation settings points to intermittent similarity of tectonic regime all across the western craton and miogeosyncline. No clear causal connections can be established for that time between the evolution of platformal sedimentary basins and the Cordilleran miogeosynclinal regime (cp. Osadetz and Haidl, 1989). Within the craton, the Cambrian Lloydminster Embayment appeared in Alberta, but the Williston Basin was not yet differentiated (van Hees, 1964). It became a distinct entity in the Western Canada Sedimentary Province only in the Middle Ordovician (e.g., Slind et al., 1994).

No rocks of Cambrian, Ordovician or Silurian age that could be associated with oceanic-crust magmatism are found in the eastern Canadian Cordillera. Even in Ordovician-Silurian time, the assumed eastern Cordilleran miogeosynclinal basins are estimated to have been only a few hundred meters deep in some localities (Cecile and Norford, 1993). The eastern Cordilleran miogeosyncline, which is well recognized in later Paleozoic tectonic stages, was only weakly discernible in the Ordovician and Silurian. Even for that time, a hinge line which would demarcate the deeper-water miogeosyncline containing graptolite shales and chert is not located clearly. On structural characteristics, the Middle Cambrian to Silurian tectonic étage extended all across western Alberta. But the general tectonic tilt varied: westward in the Cambrian, southward in the late early Paleozoic.

Tectonic étage and stage between the Devonian and the Early Mississippian

By the end of Silurian and in earliest Devonian time, the sea was in retreat from most of the Interior Platform, and subaerial erosion of bedded rocks of the Cambrian-Silurian étage was prominent in many regions (Table II). It was especially deep in Alberta. In many areas of the platform, warping of the crust and differential movements of crustal blocks prior to the start of Devonian sedimentation rearranged the positions of depocenters and arches radically.

The new tectonic étage began in central Alberta with the deposition of Lower Devonian Gedinnian marine sediments in the Elk Point Basin. Sedimentation spread mostly to the south, far and wide, in the Middle Devonian. Platformal carbonates of that epicontinental marine basin are recognized in many parts of the Cordilleran miogeosyncline along the Rocky Mountain and Omineca belts, where they have been mapped both east (Bow, Kakwa and MacDonald platforms) and west (Cassiar platform) of the NNW-SSE-trending Rocky Mountain Trench. The lack of differentiation in the tectonic conditions of sedimentation over the cratonic Interior Platform and the eastern Cordilleran miogeo-syncline at that time is consistent with the commonality in these regions of a network of active faults which determined the distribution of basins and arches of that age (e.g., Mountjoy, 1980; Meijer Drees, 1986; Moore, 1989; Halbertsma, 1994). The NE-SW fault zones inherited from the Early Proterozoic connected structures in the crystalline basement with those in the cover (Lyatsky et al., 1992; Burwash, 1993; Edwards et al., 1996).

The sharp difference in the patterns of basins and arches in the Early-Middle Devonian to Early Mississippian étage in Alberta and Saskatchewan from the underlying Cambrian to Silurian étage has nevertheless been recognized for decades (Grayston et al., 1964; Moore, 1993). A considerable difference in the tectonics of the Interior Platform north and south of the Canada-U.S. border since Late Devonian time has also been known for years (Figs. 18a-b, 19; Peterson, 1986). In the south, the Central Montana Arch mani-fested itself in an area straddling the Canada-U.S. border, while to the north, the Alberta Trough became inverted and rose. The western part of the former Athabasca-Peace River-Skeena Arch was mainly inverted and subsided, though its eastern part remained elevated. A wedge-shaped westward protrusion of the Canadian Shield embraced north-ern Alberta. To the southeast, the Transcontinental Arch began to grow anew, and the Severn-Sioux Arch rose in the east. In British Columbia and Idaho, subsidence occurred in the middle of the previously broad and shallow Cambrian-(?)Early Devonian basin on the site of the eastern Cordillera. Rapidly expanding, this subsidence created a vast sedimentary basin between the Yukon-Tanana median massif in the north and the Cas-cade-Lemhi uplift in the south. An oval, sink-like marginal sea in that region was al-most completely separated from the Pacific Ocean. East of this basin, the reduced shel-fal sea accumulated sediments under cratonic platformal conditions, while the shield edge retreated gradually eastward (Sloss, 1988a; Osadetz and Haidl, 1989; Moore, 1993).

The pre-Devonian regional emergence and erosion produced a big unconformity, so the new étage in the Interior Platform rests on different older rocks, eroded in different areas variably but deeply. On the site of the Williston Basin of the Midcontinent Platform, subsidence was more or less continuous, and although erosion was widespread, it was not deep. A very gentle regional westward tilt of the Interior Platform at the beginning of the Devonian sedimentation was similar to that during the initial, Cambrian episode of cover formation.

The inception of Devonian deposition was marked by the appearance of thin basal red-beds over older rocks, including in places crystalline rocks of the Precambrian basement. But eluvial facies are encountered rarely, having been removed by marine and river cur-rents. Thin quartz-rich clastic beds are found in abundance on the rim of the Peace River Arch and along the edge of the shield. While basal redbeds continued to accumulate at the lowlands denoted as the Tathlina and Peace River highs, the Early Devonian sea

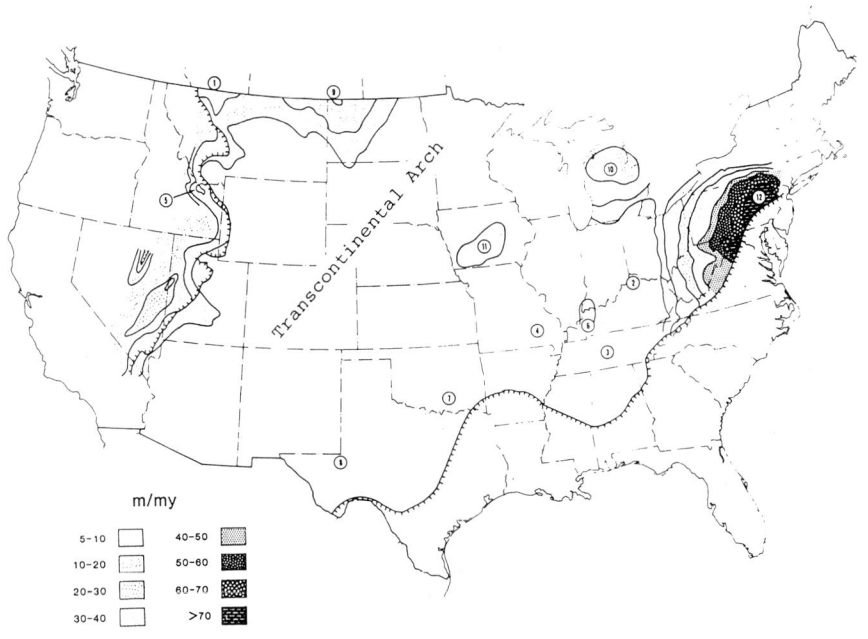

Figure 18a. Mid-Early Devonian to Late Devonian net subsidence rates in the conterminous U.S. (simplified from Sloss, 1988a). 1 - Southern Sweetgrass Arch; 2 - Cincinnati Arch; 3 - Nashville Dome; 4 - Ozark Dome; 5 - Lemhi Arch; 6 - Illinois Basin; 7 - Anadarko Basin; 8 - Tobosa Basin; 9 - Williston Basin (during Early-Middle Devonian Elk Point time); 10 - Michigan Basin; 11 - Iowa Basin; 12 - Catskill clastic wedge (fan-delta complex; cp. Friedman, 1988a) in the northern Appalachian Basin.

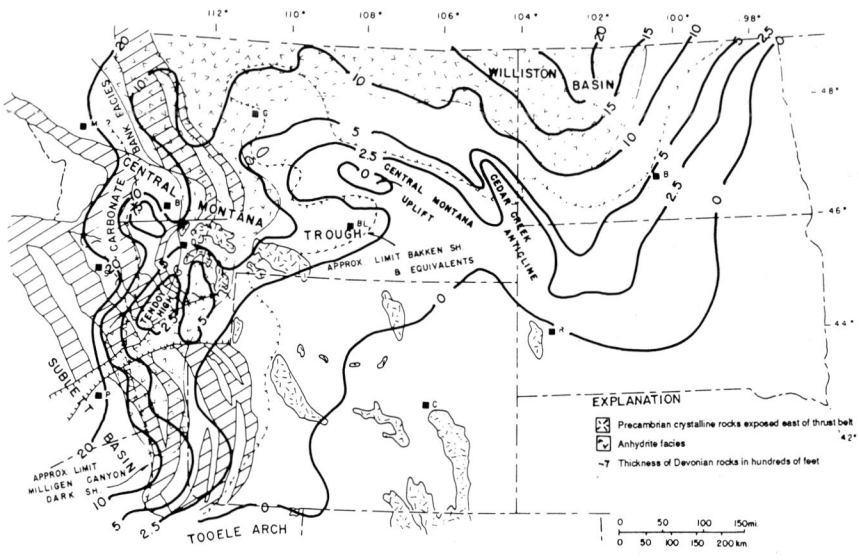

Figure 18b. Devonian isopach map of the Interior Platform and adjacent regions in the U.S., palinspastically reconstructed in the thrust belt (modified from Peterson, 1985, 1988). Thickness in hundreds of feet (100 feet = 30.5 m). Unpatterned areas are primarily carbonate rocks. City names as in Fig. 11.

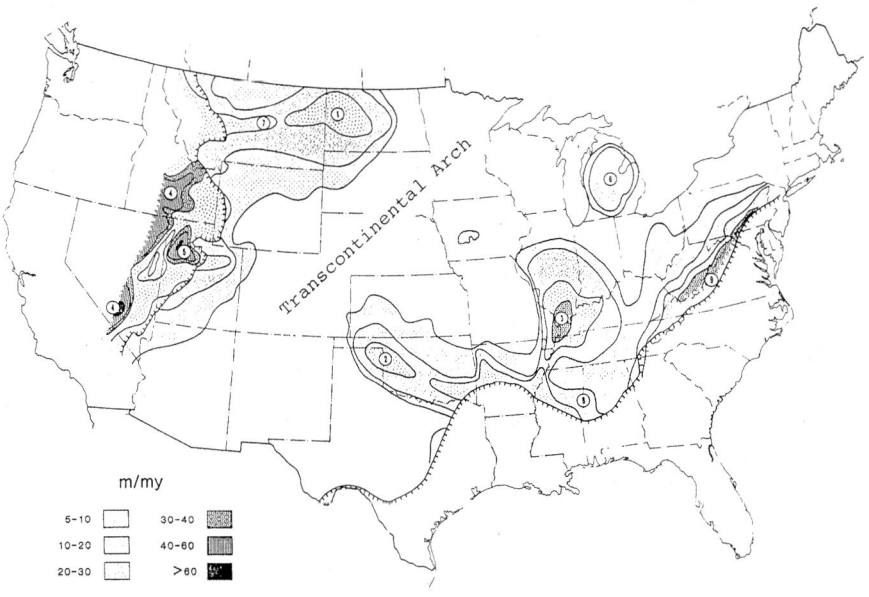

Figure 19. Latest Devonian to Late Mississippian net subsidence rates in the conterminous U.S. (simplified from Sloss, 1988a). 1 - Williston Basin; 2 - Anadarko Basin; 3 - Illinois Basin; 4 - Cordilleran foreland basin; 5 - Oquirrh Basin; 6 - Michigan Basin; 7 - Central Montana Trough; 8 - Black Warrior Basin; 9 - Appalachian foreland basin.

covered central Alberta, mostly since the Gedinnian but locally as late as the Emsian. Thin, clay-rich, silty and dolomitic sediments with beds of halite indicate shallow and restricted shelfal environments in the Elk Point Basin, which gradually expanded (Meijer Drees, 1986, 1994). Only in mid-Eifelian time did a probable short interruption occur, but it was soon followed by renewed sedimentation. In contrast, the Williston Basin experienced some reverse upward block movements accompanied by erosion. Since that time, the northern part of the Interior Platform, from northern Montana to the North-west Territories, evolved separately from its parts to the south.

Increasing differentiation in tectonic activity across the Interior Platform took place in the Devonian. To the south from central Montana, timing of sedimentation was differ-ent, tectonic movements more intense, regional westward tilt greater. The part of the Interior Platform embracing Alberta and northern Montana has been denoted as the Al-berta Platform for this and subsequent geologic time. This platform experienced a pre-dominant gentle regional tilt to the north. Sedimentation there occurred under more stable tectonic conditions than in the south. The following description focuses mainly on this platform.

The distribution of Devonian basins and arches in Alberta and Saskatchewan indicates that the regional paleotopography was determined to a large degree by penecontempora-neous movements of crustal blocks bounded by then-active faults. Those block move-ments influenced the distribution of lithofacies and thicknesses in Early Devonian ba-sins which have strikingly dissimilar stratigraphic records. Nonetheless, sedimentation was by and large durable and fairly stable, with only slight stratigraphic gaps probably caused by low-amplitude local uplifts and fluctuations of the sea level.

Since Eifelian time, the Devonian sea covered most of the Alberta Platform, leaving exposed only some islands caused by high-standing blocks. The epicontinental sea spread from the Arctic areas far to the south and east, into Saskatchewan and beyond. In southwestern Alberta, the NNW-oriented, high-standing Western Alberta Ridge, sepa-rated elongated seaways. In the Midcontinent Platform, the Williston Basin began to subside again.

During Middle Devonian time, many spectacular coral and algal reefs grew in Alberta (Figs. 20a-b). Wide, though not thick, bodies of laminated carbonates (carbonate plat-forms) served as a base for barrier reefs and more-complex reefal buildups. Isolated pin-nacle and patch reefs readily grew where the sea bottom was uneven. The biggest barrier reefs between the Southern Alberta Barrier Complex and the Peace River Arch have dis-tinct NE-SW trends. Palinspastic restorations in the modern Rocky Mountain fold-and-thrust belt have allowed to trace these reefs far to the southwest, into the eastern Cordil-lera (Klovan, 1974; Davies, ed., 1975a-b; Geldsetzer et al., 1982).

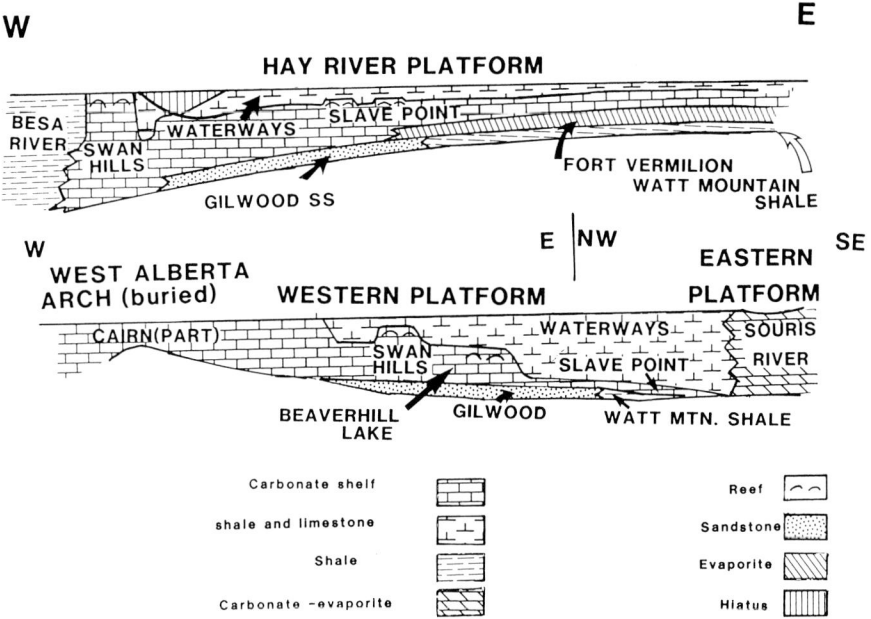

Figure 20a. Schematic stratigraphic profile of the Middle Devonian succession (modified from Moore, 1989), showing carbonate platforms and reefs in the Beaverhill Lake Group. Note the lack of westward thickening of this sedimentary succession.

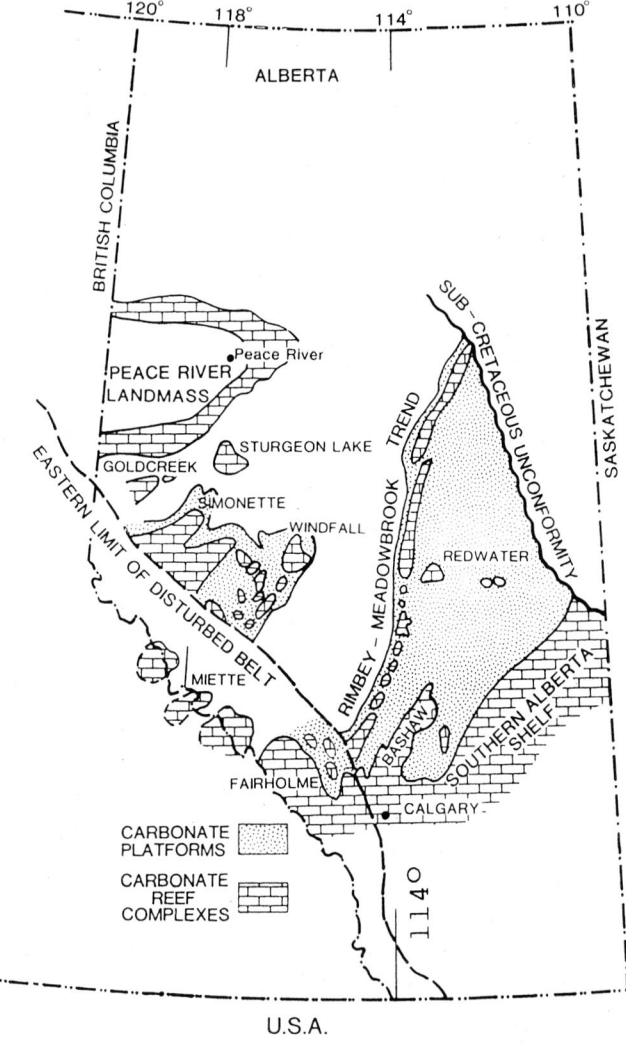

Figure 20b. Paleogeographic maps of the Upper Devonian Woodbend Group, showing main reef trends in the Alberta Basin (modified from Stoakes and Wendte, 1987; Osadetz, 1989). The Rimbey-Meadowbrook chain of reefs is often also referred to in the geologic literature as Rimbey-Leduc or just Leduc.

Small basins (subbasins) lying between elevated basement blocks, platformal arches and carbonate buildups were filled with organic-rich muds, in places grading laterally into and interbedded with coarser clastic or carbonate and evaporite deposits (Fig. 21). Total thickness of these packages reaches 600 m. Extensive lagoonal environments created good conditions for deposition of oil-prone shales, from which oil was later expelled into very porous reef reservoirs. These hydrocarbon systems resulted in the formation and preservation of a huge amount of oil and gas (mostly oil), making Alberta and Saskatchewan one of the most prolific petroleum provinces in North America.

Sensitivity of reef-building organisms (corals, stromatoporoids) to sea-water and sunlight conditions caused many reefs to grow on top of bathymetric highs and on the flanks of islands. Many ramps, even those of small amplitude, were controlled by faults. The largest and longest-lived chains of barrier reefs in Alberta have NNE-SSW trends, correlatable with NE-SW basement faults. Other reef complexes are linked to other fault systems whose control on reef distribution is in places indirect and subtle. Builders of coral chains, in their search for the best physical conditions and nutrient supply, tend to expand outward, away from the straight linearity of faults. Festoon shapes are common in modern barrier reefs, as well as in ancient ones. But whatever their shape at maturity, they tended to nucleate initially at sea-floor scarps which are often fault-induced.

Repeatedly, after interruptions in growth, new reefs again nucleated along the same old trends. The interruptions could have been caused by eustatic changes in sea level, local transgressions and regressions due to tectonic block movements, increased flux of clastic sediments that choked the reef growth, fluctuations in nutrient supply, etc. Possibly, in the Western Canada Sedimentary Province epeirogenic crustal movements at times had "a profound effect" on reef growth (Podruski et al., 1988, p. 20). But fault control on the general regional distribution of many reef complexes and chains remained apparent. Knowledge of reef trends is often used to predict the distribution of reef traps in petroliferous regions worldwide. The largest of Alberta barrier reefs (Slave Point, Swan Hills, Leduc; Figs. 20-22) stood 100-250 m high above their carbonate-platform foundations. Some of the prominent barrier reefs, up to tens of kilometers across, are stacked vertically.

In Alberta, the Devonian marine transgression reached its peak in the early Late Devonian Frasnian, when shelfal seas spread over most of the Interior Platform in the north and south. Narrow near-shore zones rimmed the shelf in the east along the elevated shield lands. In the west, this shelfal sea extended far into the eastern Cordillera. No deep-water deposits have been found in the Alberta Platform: the sea bottom in the west and east, excepting the noted complications, was at a similar depth of ~200 m, much like in shelfal seas at present.

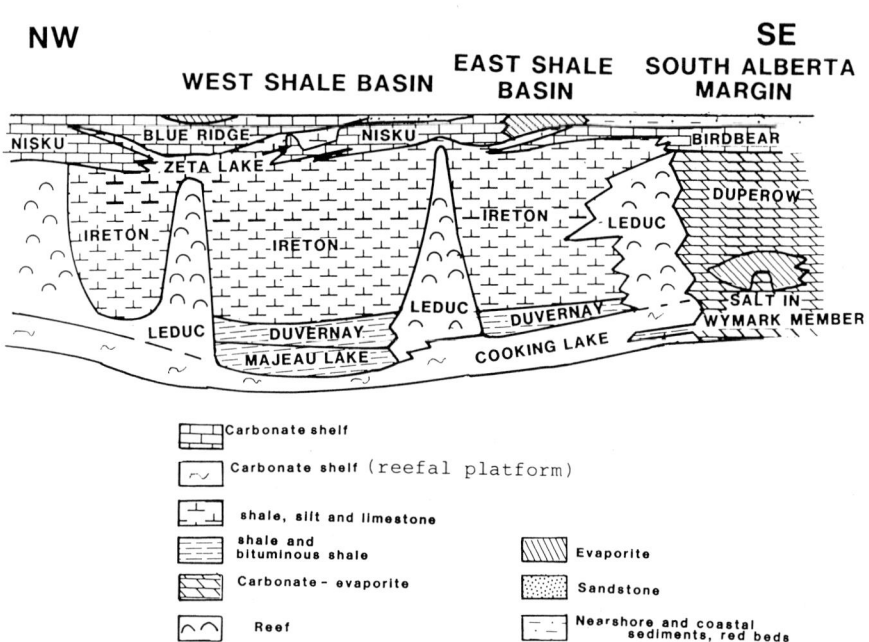

Figure 21. Schematic stratigraphic profile of a Devonian reef-bearing succession (modified from Moore, 1989), showing Leduc reefs encased in Ireton shale. This encasement of porous reefs in impermeable shale creates excellent hydrocarbon source-trap systems, making these reefs an attractive exploration target. Note the lack of westward thickening of this sedimentary succession.

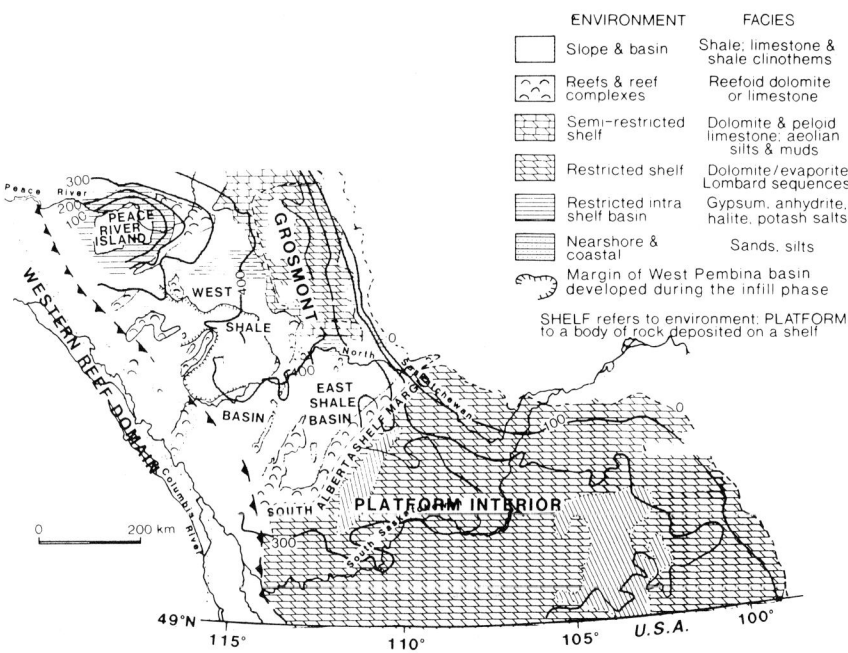

Figure 22. Isopach map of the Upper Devonian Woodbend Group and its equivalents, also showing the main reef trends (modified from Moore, 1989). Thickness contours in hundreds of meters. No westward thickening of this sedimentary succession is observed.

Remnants of Devonian carbonate platforms recognized in many places in the eastern Canadian Cordillera are made up of thick Devonian carbonate beds similar to those in the Alberta Platform. Significantly, however, in the Cassiar platform in the Cordilleran Omineca miogeosynclinal belt (Fig. 4a), quartzose sandstones are interbedded with carbonate strata. They were derived not from shield terrains in the east, but from land sources in the west - presumably, from the uplifted Yukon-Tanana or Intermontane median massifs (see Douglas et al., 1970).

Distinct sedimentary basins which had previously developed north and south of the Peace River Arch gradually expanded all over the Alberta Platform. Restricted shallow marine conditions, common in the Mackenzie River area and in northern Alberta, produced carbonate and evaporite marine successions of intercalated thin beds of halite, anhydrite, dolomite and shale, with a total thickness reaching 150 to 400 m in depocenters. In the southern Interior Platform in the U.S., such rocks are also common as far south as the Central Wyoming Basin. This evaporite-carbonate package in southwestern Montana and Wyoming is conventionally assigned to the so-called "Wyoming shelf", distinguished from the "Alberta shelf" to the north (e.g., Peterson, 1986). Tectonically, these two distinct shelves were assigned to different crustal blocks rejuvenated in the Devonian (Tonnsen, 1986). Recently, this idea was developed further (Cecile et al., 1997).

A local uplift occurred in the Devonian in eastern Montana along the Miles City Arch, while the Central Montana arches bounded the Williston Basin on the southwest and west. In northeastern Montana and southeastern Alberta, the northwestern side of that basin was delineated by the Southern (in Montana) and Northern (in Alberta) Sweetgrass arches which meet at an obtuse angle. Late Devonian time was also marked by the reactivation of the Transcontinental Arch in the east and of the Uncompahgre Uplift in Utah in the west. In northern Alberta, Northwest Territories and southeastern Yukon, prominent fault networks are seen in the sedimentation patterns of the Devonian Mackenzie Basin. In the Mackenzie and Selwyn Mountains in northeastern Cordillera, Ziegler (1969) and Douglas et al. (1970) noted three subparallel, NNW-trending zones: from east to west, Mackenzie Trough, Mackenzie Arch, and Selwyn Basin. Other authors (Law, 1971) combined the first two into the Western Mackenzie Basin, which was initiated in the Late Silurian. In more detail, during the Devonian there were several positive features linked with the original Mackenzie Arch or with the NE-SW-trending Tathlina Arch. Elongated depressions were localized between them. This gives a distinct horst-and-graben structure to this part of the northern Interior Platform (e.g., Morrow, 1991). Elongated relative depressions in the extended platform are reminiscent of those in the lower part of the Cambrian-Silurian étage (Fig. 23).

In the northern part of eastern Cordillera, thin, dark Lower Devonian shales conformably overlie the Silurian shales in the Peel and Selwyn basins. These shales are sometimes

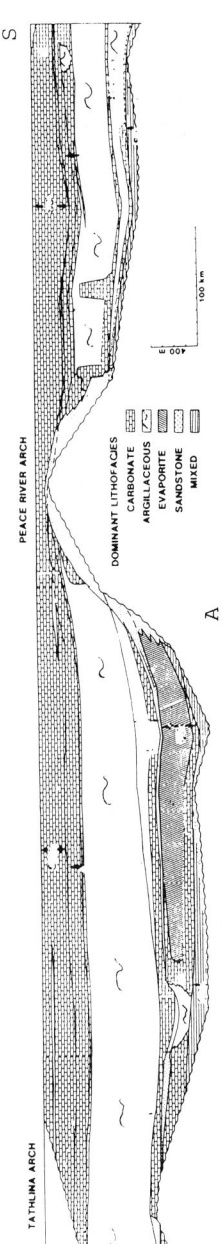

Figure 23a. Position of the Devonian shale basins, reefs and carbonate platforms, and of the Peace River and Tathlina arches, illustrated in a N-S stratigraphic cross-section through northern Alberta (modified from Moore, 1989).

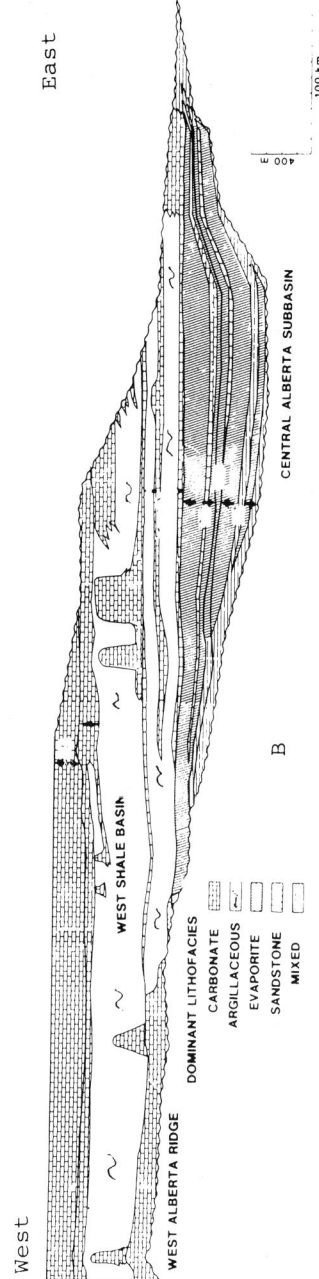

Figure 23b. Position of the Devonian shale basins, Leduc reefs and carbonate platforms, and of the Western Alberta Ridge, illustrated in an E-W stratigraphic cross-section through central Alberta (modified from Moore, 1989).

assigned to a miogeosynclinal environment of sedimentation, although they were probably laid down under intracratonic platformal conditions similar to those proven in the east. In the Selwyn Basin, shales grade into carbonates and carbonate breccias. Although the breccias at carbonate-platform edges may be caused by non-tectonic factors (cp. Cecile et al., 1997), such a facies change to the NE and SW may be another indication of motion of underlying cratonic crustal blocks.

Selwyn Basin shales apparently thicken to the south. Together with carbonate beds, they make up the thick Lower Devonian package in the narrow Kechika Trough east of the Cassiar platform. This package contains sandstones rich in "quartz presumably derived from an emergent western part [of this area]" (Douglas et al., 1970, p. 394). Similar sandstones have also been noted in the MacDonald platform, whose clasts could have been sourced from local Precambrian exposures in uplifted blocks (Meijer Drees, 1994).

An otherwise very informative description of the Mackenzie Mountains by Morrow (1991, p. 5, 6) contains examples of confusion between words whose meaning is different in different contexts: geologic or geomorphologic, paleogeographic or tectonic. He stated, for example, that "in the study area, Silurian-Devonian strata lie almost entirely within the Mackenzie Shelf or miogeosyncline... Most modern continental areas are surrounded by topographic shelves of greater or lesser breadth... The shelves are also commonly covered by a thick sequence of sediments... The shelf sequence examined in this study is part of the Cordilleran Miogeocline, a sedimentary prism of shelf deposits... [An assumed hinge line separates] uniformly thin, cratonic, shelf platform sequences of shallow water sediments from shallow water shelf sequences that thicken rapidly basinward". With similar confusion, Cecile et al. (1997, p. 56) wrote: "The term miogeocline is used to mean ancestral continental shelf"; and "uppermost Proterozoic to Lower Devonian rocks in the Cordillera of Western Canada accumulated within two ancestral continental margin systems - the Canadian Cordilleran and southwestern Franklinian miogeoclines" (op. cit., p. 54). While these authors regarded the Interior Platform east of the Cordilleran deformation front and the eastern Cordilleran miogeosyncline (miogeocline, as they preferred to call it) as separate tectonic entities, they suggested that both of them share several underlying long-lived large crustal blocks: from north to south, Mackenzie, Blackwater, Macdonald, Alberta, and Montana, with boundaries at the long-recognized by NE-SW-trending fault zones inherited from the ancient Precambrian. This division is compatible with Peterson's (1986) "Alberta and Wyoming shelves" (or blocks).

A major feature in the Middle Devonian in northern Alberta and northeastern British Columbia was the Athabasca-Peace River-Skeena arch system, composed of variously elevated basement blocks. At that time, the Peace River Arch (as it is conventionally denoted) extended from Lake Athabasca in the east to the future Rocky Mountains. Intermittently, it was probably linked with the Skeena Arch farther west in the Cordilleran interior, in the Intermontane median massif. This WSW-ENE structural zone cut the general NNW-SSE structural grain of the Cordilleran mobile megabelt at a right angle.

In the Early Devonian, the Alberta Trough was inverted to form the Western Alberta Ridge, which merged with the Peace River Arch (see Oldale and Munday, 1994; Switzer et al., 1994). This big landmass shrank gradually during later Devonian time. In the early Late Devonian, it was divided in two parts and only slightly elevated above the sea level. The Peace River Arch finally disappeared beneath the sea in the Famennian, though it still formed a buried high in the subsurface. The Tathlina Arch disappeared by the early Frasnian, while the Western Alberta Ridge during that time was inverted back down to again become a trough. An unconformity (actually, a disconformity) at the base of the Frasnian and upper Givetian rocks marks this shift in paleoenvironments (Fig. 23).

A uniform, very shallow sea covered the northern and southern parts of the Interior Platform by the end of the Devonian. In the Alberta Basin, carbonate, evaporite and clastic sedimentation kept up with a very slow subsidence. These fine-clastic and limestone rocks change gradually to dolomite and evaporite to the southeast, in the Williston Basin. Salt bodies tens of meters thick were formed in southeastern Alberta and southern Saskatchewan. In Montana and the Dakotas, coeval strata contain abundant redbeds. The distribution of lithofacies and thicknesses established in Late Devonian time generally persisted in the Western Canada Sedimentary Province into the Early Mississippian. The structural changes that affected this tectonic étage were slow and happened gradually over a long period of some 40-50 m.y. All the up and down reversals occurred in generally steady tectonic regime, and by the end of this stage in the Early Mississippian the region achieved quiescence. The Devonian-Early Mississippian étage in western Canada bears a lot of internal compositional and structural complexities. But it is regarded as a single étage because these complexities were a product of a continuous tectonic regime characterized by differential and inverted warping and block movements.

Organic-rich mudstones were widespread, but Late Devonian lithofacies still varied. Colonial organisms still created various buildups, commonly with high vuggy porosity. The 1,000-m-thick Frasnian stratigraphic interval is one of the most productive petroleum horizons in Alberta and Saskatchewan. To the south, the shale-carbonate sequences grade into less petroliferous dolomitic lithofacies. Restricted paleoenvironments there caused the accumulation of halite.

Many Devonian salt bodies in Saskatchewan are more than 60 m thick (Meijer Drees, 1986), and they are mined for potash. Evaporitic facies are also widespread in the Upper Devonian in Montana and Wyoming, although thinner Upper Devonian deposits are recorded in the area of the Central Montana Trough. Their thickness increases to 600 m

to the west, suggesting a westward regional tilt. Farther south, this stratigraphic horizon continues, with its normal regional thickness, into the Central Wyoming Basin.

In many areas of the Western Canada Sedimentary Province, Mississippian rocks were
truncated deeply by erosion before the Late Mississippian to Triassic tectonic stage.
The zero edge of the continuous Mississippian strata in subcrops is erosional. Yet,
many isolated remnants of Mississippian rocks all over the Alberta Platform suggest a
wide original areal span. Similar lithofacies characterize both the Famennian and the
Lower Mississippian rocks; most of them are argillaceous carbonates, with thin, organic-rich shales, intercalated with evaporites including anhydrite and salt. Slight fluctuations in a very shallow sea locally produced cyclic repetition of calcareous shale,
massive crinoidal limestone, dolomite and anhydrite. Wide shoals were associated with
lagoons and sabkhas.

The best-recognized Carboniferous depocenters developed in the Wyoming and Williston
basins and the Peace River Embayment (formed over the inverted Arch), and in the
Prophet Trough parallel to and over the eastern Cordilleran miogeosyncline in Canada.
This trough is now conventionally regarded as a direct continuation from the U.S. into
western Canada of the Early Mississippian Antler foreland or foredeep (cp. Sloss, ed.,
1988; Richards, 1989; Richards et al., 1993, 1994).

Upper Devonian-Lower Mississippian clastic deposits (a probable Antler molasse complex) have been mapped along the eastern Canadian Cordillera in a series of separate
grabens (Gordey et al., 1987; Gordey, 1991). They contain pebble and cobble conglomerates whose total thickness reaches 500-700 m. Syndepositional graben-bounding
high-angle fault might have served as conduits for felsic volcanism and exhalite mineralization (see also Cecile et al., 1997). The probable equivalents of these molassic
rocks in the U.S. differ by the absence of volcanics.

In the Cordillera, especially in Nevada, the appearance of elongated highlands called the
Antler High marked a new stage in the miogeosyncline evolution. As a result of the
Antler orogeny (Late Devonian to Early Mississippian at its tectonotype), the Lemhi
Arch, which had long been a source of sediments, became inverted into the Muldoon
Trough, never again to be a positive feature (Fig. 24). Together with the Sublett Basin
farther south, this trough formed an elongated foreland along the Antler orogenic high-
lands, where a thick Lower Devonian-Upper Mississippian flyschoid sequence was laid
down. This foreland basin was not structurally uniform and contained a series of isolated subbasins separated by several broad NE-trending arches (cp. Tonnsen, 1986;
Sloss, 1988a). In the north, in Canada, NNW-SSE structural trends were active, besides
the NE-SW trends linked to Precambrian fault zones (Great Slave Shear Zone, Liard
Line, Macdonald fault; e.g., Aitken, 1993a-b; Burwash et al., 1993). The near-E-W

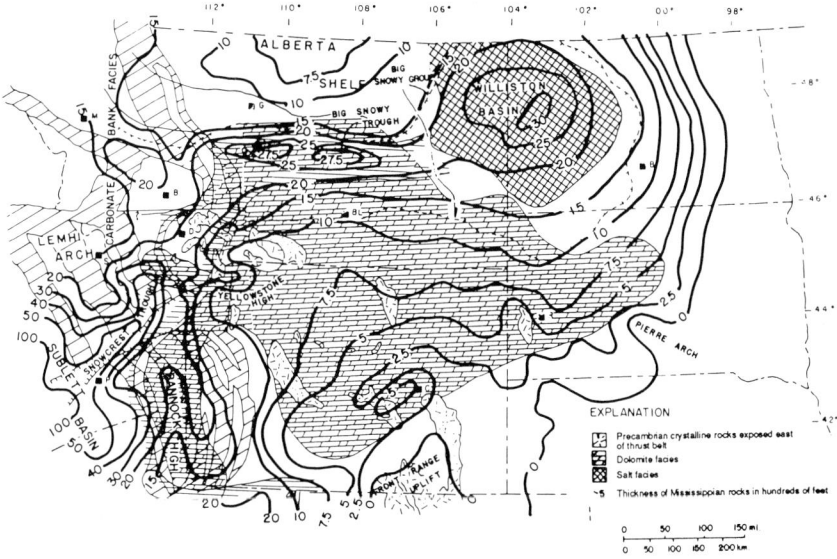

Figure 24. Mississippian isopach map of the Interior Platform and adjacent regions in the U.S., palinspastically reconstructed in the Laramian thrust belt (modified from Peterson, 1985, 1988). Thickness in hundreds of feet (100 feet = 30.5 m). Limit of the Upper Mississippian Big Snowy Group is shown by barbed line. Unpatterned areas are primarily carbonate rocks in the Lower Mississippian, and fine siliciclastics, carbonates and evaporites in the Upper Mississippian. City names as in Fig. 11.

trends, inherited from the ancient Athabasca-Peace River-Skeena zone of crustal weakness, were active, too.

The pre-Frasnian unconformity was identified in many parts of the Western Canada Sedimentary Province, including some locations in the Williston Basin. It was less strongly developed in the Canadian Rockies (Tempelman-Kluit, 1979; Workum and Hedinger, 1987). Across Alberta, the lack of extensive unconformity breaks or abrupt changes in structural style puts the entire Devonian to Lower Mississippian lithostratigraphic interval into a single, distinct tectonic étage. On the whole, the tectonic regime in the entire Devonian through Early Mississippian tectonic stage was much the same. A persistent general structural network determined the rather stationary position of basin depocenters, despite the tendency of some basins to expand. Even though some of the arches shrank due to partial to full, and repeated, inversion, their axial zones mostly maintained their position. All across Alberta, against the background of continuous regional subsidence, relatively intense differential block movements took place. The relatively high differential block activity set distinguishes this tectonic stage from the previous, Cambrian to Silurian one.

Platformal carbonate sedimentation on the site of the eastern Cordillera was interrupted frequently by increased crustal dynamics (e.g., along the previous Windermere rift and the future Southern Rocky Mountain Trench). Yet, the geology of the entire eastern Cordilleran miogeosyncline north and south of latitude 56°N shows considerable differences in lithostratigraphy as well as in structural and magmatic styles (Pell, 1987; Norford, 1990; O'Connell et al., 1990; Cecile and Norford, 1993). The first ophiolitic formations (in the Cache Creek Group) appeared in the central Cordillera west of the miogeosynclinal Omineca Belt only in the Permo-Carboniferous (Monger, 1989). Struik (1987) attributed the appearance of this distinctive rocks assemblage to a final separation of an ancestral North America from undefined continental masses to the west. Monger et al. (1982) linked it with accretion of a Pacific oceanic-crust terrane.

A still-influential hypothesis relates the Cordilleran miogeosyncline to a continental slope of ancestral North America, which in the Paleozoic supposedly lay on the site of the Omineca or Rocky Mountain belts of the Canadian Cordillera (see Aitken, 1993b-c for details). Several orogenic episodes affected the eastern Cordilleran miogeosyncline in Canada in the late Precambrian (McMechan and Price, 1982; McMechan, 1991) and after (Gabrielse and Yorath, eds., 1991). Episodes of rifting along the eastern Cordilleran miogeosyncline have been reported in the Early Cambrian, Middle Ordovician and Late Silurian (Cecile and Norford, 1993; Lickorish and Simony, 1995). The last two episodes were accompanied by sporadic magmatism (Cecile et al., 1997).

It has been shown that the tectonic rock assemblages, i.e. structural-formational étages, in the Canadian Cordilleran miogeosyncline vary from area to area (Gabrielse and

Yorath, eds., 1991; Wheeler and McFeely, comps., 1991). Some of them are correlative with étages in the Alberta Platform, whereas others are not (cp. also Stott and Aitken, eds., 1993). Platformal regime in Paleozoic time extended repeatedly into the site of the Cordilleran miogeosyncline. This makes it possible to distinguish in the western marginal zone of the North American craton transitional tectonic characteristics recorded throughout the western Interior Platform and the Cordilleran miogeosyncline (cp. the miogeosyncline-craton domain of Price, 1994).

Late Devonian conglomerates, found sporadically all along the Canadian Cordilleran miogeosyncline, indicate the appearance of mountainous provenance areas nearby, in the Omineca and/or Intermontane belts. This is now conventionally linked with the Late Devonian to Early Mississippian Antler orogen (Fig. 25), which extended into Canada from the south (Richards, 1989). The Alberta Platform at that time experienced intermittent but prevailing expansion of subsidence. The reef complexes gradually waned. The semi-restricted basins merged, overwhelming the arches between them. Even the former Peace River Arch drowned and vanished beneath the overlapping sedimentary cover, making room for a long-lived late Paleozoic and early Mesozoic depocenter. The full volume and extent of the Devonian to Mississippian structural-formational étage is uncertain because of the subsequent profound unconformity.

Tectonic étage and stage between the Late Mississippian and the Triassic

The Antler orogeny in its tectonotype in Nevada took place in the Late Devonian-Early Mississippian. It was manifested by the formation of a huge allochthon of thick Middle Cambrian through Lower Mississippian eugeosynclinal rocks, translated to the east (Roberts, 1949, 1964; Gilluly and Gates, 1965). This thrusting occurred on the low-angle Roberts Mountains fault, which is now mapped from southeastern California to central Idaho. No direct continuation of this thrust into the Canadian Cordillera is apparent (Read and Wheeler, 1976). Only some steep west-dipping faults in the Kootenay Arc area may represent its northern splay (Smith and Gehrels, 1991).

Where it exists, the Roberts Mountains allochthonous thrust sheet is composed mainly of early Paleozoic polymictic sandstones, argillites and cherts. They are usually interpreted as turbidites, but this interpretation is poorly justified; the clastic beds are intercalated with abundant pillow-basalt flows and breccias. These rocks have been attributed by some authors (e.g., Burchfiel and Davis, 1975) to an oceanic crust, but this assumption has been questioned all along. Many of the allochthon's sandstones are arkosic, suggesting that pre-existing granitic rocks were exposed nearby; and mineralogical and paleocurrent studies indicate that the source area probably lay to the west (Turner et al., 1989). Presence in some, mostly Devonian, intervals of argillaceous black shales and radiolarian cherts may be linked with truly deep-water environments, but it does not

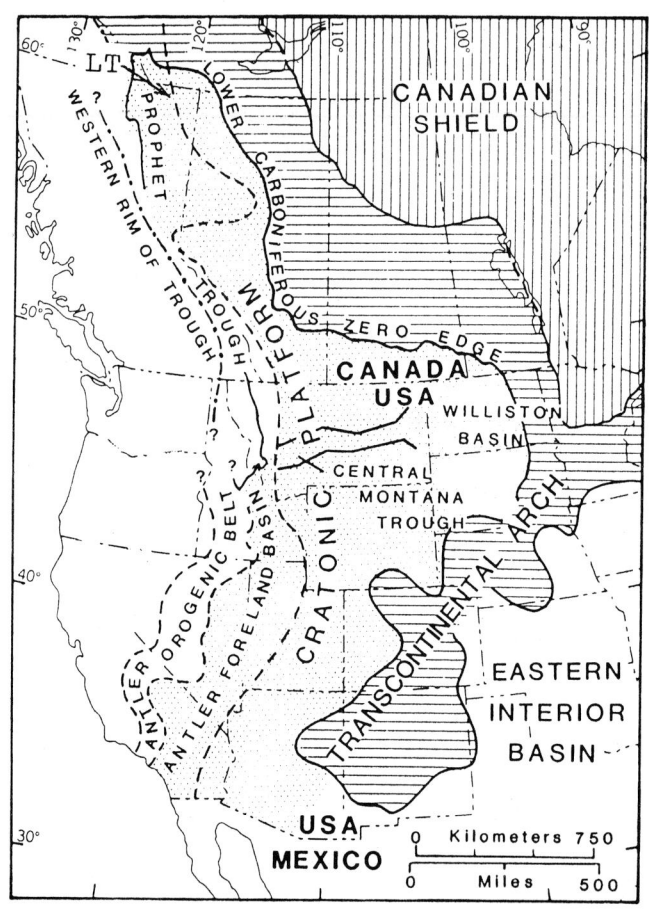

Figure 25. Position of the cratonic foreland basin and its Prophet Trough in Canada in relation to the Antler orogen (modified from Richards, 1989).

necessarily mean an oceanic-crust setting (cf. Speed and Sleep, 1982). Basalts in the Roberts Mountains allochthon are alkalic and REE-enriched, indicating a probable intra-continental-rift tectonic setting (Turner et al., 1989). Although Turner et al. (1989, p. 343) supposed rocks of this allochthon could have originally been part of "the outer continental margin" of North America in the Paleozoic, later geochemical and sedimentological studies confirmed the presence of a continental crystalline basement even to the west from the location of the allochthon's assumed roots (also Armstrong et al., 1991). Significantly, the formation of this allochthon was "not accompanied by any plutonism or severe metamorphism" (Nilsen and Stewart, 1980).

Nilsen and Stewart (1980, p. 298) argued for "the persistence of a positive belt along the [Antler] orogenic zone through the remainder of Paleozoic time, and even later". This Antler belt seems to have been coupled with an associated sedimentary basin to the east. To the north, in the Kootenay Arc straddling the U.S.-Canada border, the Cambrian-Ordovician succession is considered to be parautochthonous but belonging to a western extension of the craton of that time (Klepacki and Wheeler, 1985; Aitken, 1993a-b). To the east of the Kootenay Arc, in Canada, the Upper Proterozoic and younger rocks have always been considered autochthonous (Douglas, ed., 1970; Devlin and Bond, 1988; Gabrielse and Yorath, eds., 1991). The radiometrically dated Early Proterozoic rocks exposed there in metamorphic core complexes have been linked to a cratonic basement (Parrish, 1991, 1995). The Upper Cambrian-Ordovician rocks west of the Kootenay Arc, assigned to the Cordilleran eugeosyncline (Smith and Gehrels, 1991, p. 1271), were deformed and metamorphosed before the Early Mississippian: they supplied clasts for Mississippian conglomerates (Read and Wheeler, 1976). In the eastern Cordilleran miogeosyncline in Canada, discontinuous sedimentary basins separated by highs have been recognized for the time through the Devonian and Early Mississippian. Carboniferous stratal rocks in the miogeosynclinal Omineca Belt rest on the late Precambrian and early Paleozoic rocks with a strong regional unconformity and with "boulders of granitic and high-grade metamorphic rocks in conglomerates" (Douglas et al., 1970, p. 411).

Peterson (1986) noted a major change in tectonic conditions all over western North America around the Middle-Late Mississippian boundary. In the U.S., the relatively stable subsidence in the Interior Platform continued, but carbonate sedimentation of the previous tectonic stage gave way to unstable clastic sedimentation that occurred locally (e.g., in the E-W Big Snowy Trough in central Montana; Figs. 25, 26). Shorelines pulsated back and forth, producing many episodes of subaerial exposure and erosion.

In the Canadian part of the Interior Platform, the distribution of late Paleozoic rocks is also restricted. There are many unconformities between individual stratigraphic units, which are eroded deeply. Late Carboniferous sedimentary rocks are thought to have been laid down over a broad region in Alberta and Saskatchewan. However, the Pennsylva-

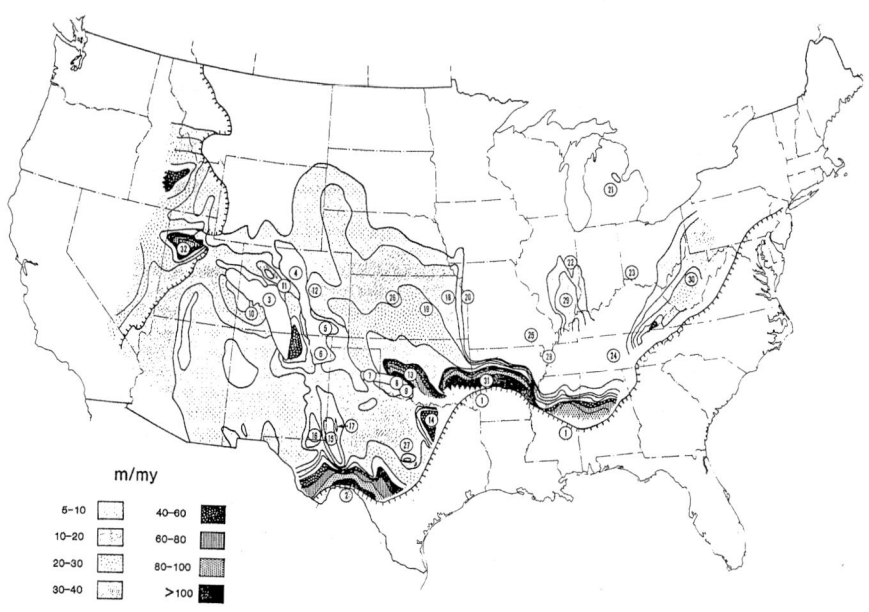

Figure 26a. Latest Mississippian to Early Permian net subsidence rates in the conterminous U.S. (simplified from Sloss, 1988a). 1 - Ouachita province; 2 - Marathon province; 3 - Uncompahgre uplift; 4 - Front Range uplift; 5 - Apishapa uplift; 6 - Sierra Grande uplift; 7 - Amarillo uplift; 8 - Wichita uplift; 9 - Arbuckle uplift; 10 - Paradox Basin; 11 - Eagle Basin; 12 - Denver Basin; 13 - Anadarko Basin; 14 - Fort Worth Basin; 15 - Central Basin Platform; 16 - Delaware Basin; 17 - Midland Basin; 18 - Nemaha uplift; 19 - Salina Basin; 20 - Forest City Basin; 21 - Michigan Basin; 22 - La Salle Anticline; 23 - Cincinnati Arch; 24 - Nashville dome; 25 - Ozark dome; 26 - Central Kansas uplift; 27 - Llano uplift; 28 - Pascola Arch; 29 - Illinois Basin; 30 - Appalachian Basin; 31 - Arkoma Basin; 32 - Oquirrh Basin.

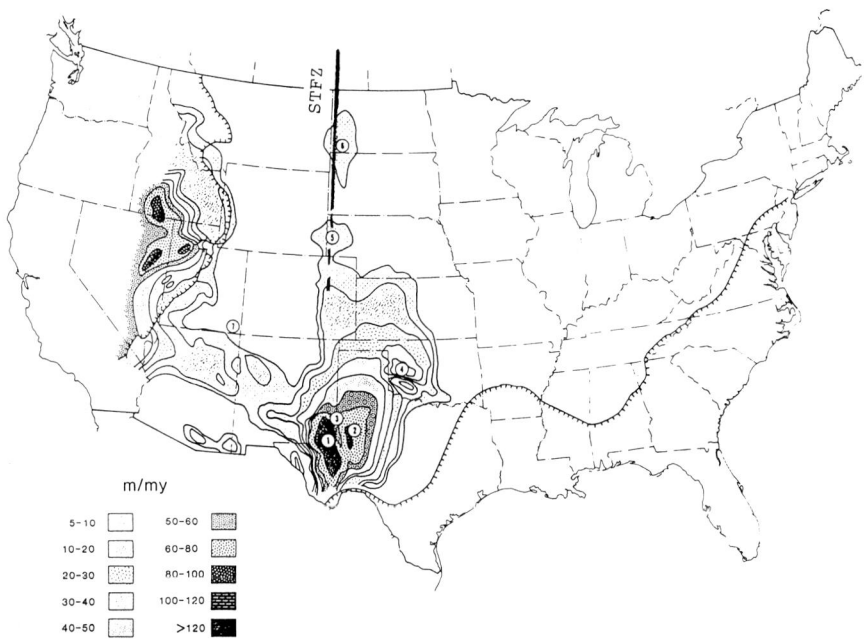

Figure 26b. Mid-Permian and Late Permian net subsidence rates in the conterminous U.S. (simplified from Sloss, 1988a). 1 - Delaware Basin; 2 - Midland Basin; 3 - Central Basin Platform; 4 - Anadarko Basin; 5 - Alliance Basin; 6 - Williston Basin; 7 - Colorado Plateau uplift. STFZ - southward continuation from Canada of the N-S-oriented Saskatoon-Tabbernor fault system; along its trend lies a N-S boundary between regions of subsidence and uplift, continuing as far south as the Gulf of Mexico coast in western Texas. This suggests the Precambrian Saskatoon-Tabbernor fault system represents a continental-scale zone of weakness, which was reactivated in the Permian.

nian is regionally mostly missing due to non-deposition and deep erosion, whereas the Mississippian is preserved in the eastern parts of the platform in separate remnants in subcrops.

The most continuous belt of the Late Mississippian to Triassic rocks is preserved in the Foothills and Rocky Mountains. The ancient West Alberta Trough and the two newly-formed troughs, Liard and Sylvester, gradually merged into a single elongated depression denoted Prophet Trough in the Carboniferous and Ishbel Trough in the Permian (Bamber et al., 1991; Henderson et al., 1993; Richards et al., 1993; Fig. 25). Rocks from the shield to the east provided sediments for these troughs, but importantly, clastic influx has also been reported from the west, from the gradually rising Cordilleran interior.

Podruski et al. (1988, p. 15) described the tectonic regime in the Western Canadian Sedimentary Province during that time in terms of a "general instability". In their description, "the stratigraphic record is interrupted by a series of minor unconformities. The unconformity surfaces merge eastward into a zone where most of the stratigraphic record is missing... [At] no time during deposition of this sequence is there evidence of prolonged or extensive transgression of the seas over the interior of the craton... A gradual transition from dominantly carbonate deposition to clastic sedimentation is evident... Subtle tectonic features and structures in the continental crust were accentuated during this period of minor continental instability. In particular, uplift of the Sweetgrass Arch and subsidence in the Peace River and Williston Basin regions influenced the depositional framework. Williston Basin was characterized by terrestrial and shallow marine deposits whereas north of Sweetgrass Arch sedimentation occurred under the influence of open marine conditions".

Many of the tectonic movements in Carboniferous-Permian time in the Alberta Platform were the reverse of those in the Devonian. In the Devonian, major NE-SW structural trends were expressed in the shapes of the Tathlina Arch, Peace River Arch, Meadow Lake Hinge, rectangular MacDonald platform, broad Elk Point/Central Alberta Basin, massive Western Alberta Ridge (Morrow and Geldsetzer, 1988; Moore, 1988, 1993). The subsequent, late Paleozoic pattern of the Alberta Platform (Fig. 27) was marked by very different structural trends. Subsidence occurred in the almost E-W-elongated Peace River Embayment and the NNW-SSE Prophet/Ishbel Trough. Uplift took place along the NNE-trending Northern Sweetgrass Arch running from the Montana-Alberta border into Saskatchewan. The Williston Basin was connected with the Antler foreland south of the U.S.-Canada border by the elongated Central Montana Trough.

The so-called Cassiar "terrane" in the eastern Cordillera was rimmed on the west by a continental magmatic belt. Still farther west lay a subduction zone which induced thrusting in the Kootenay Arc. Thrust-sheet loading might have caused the formation

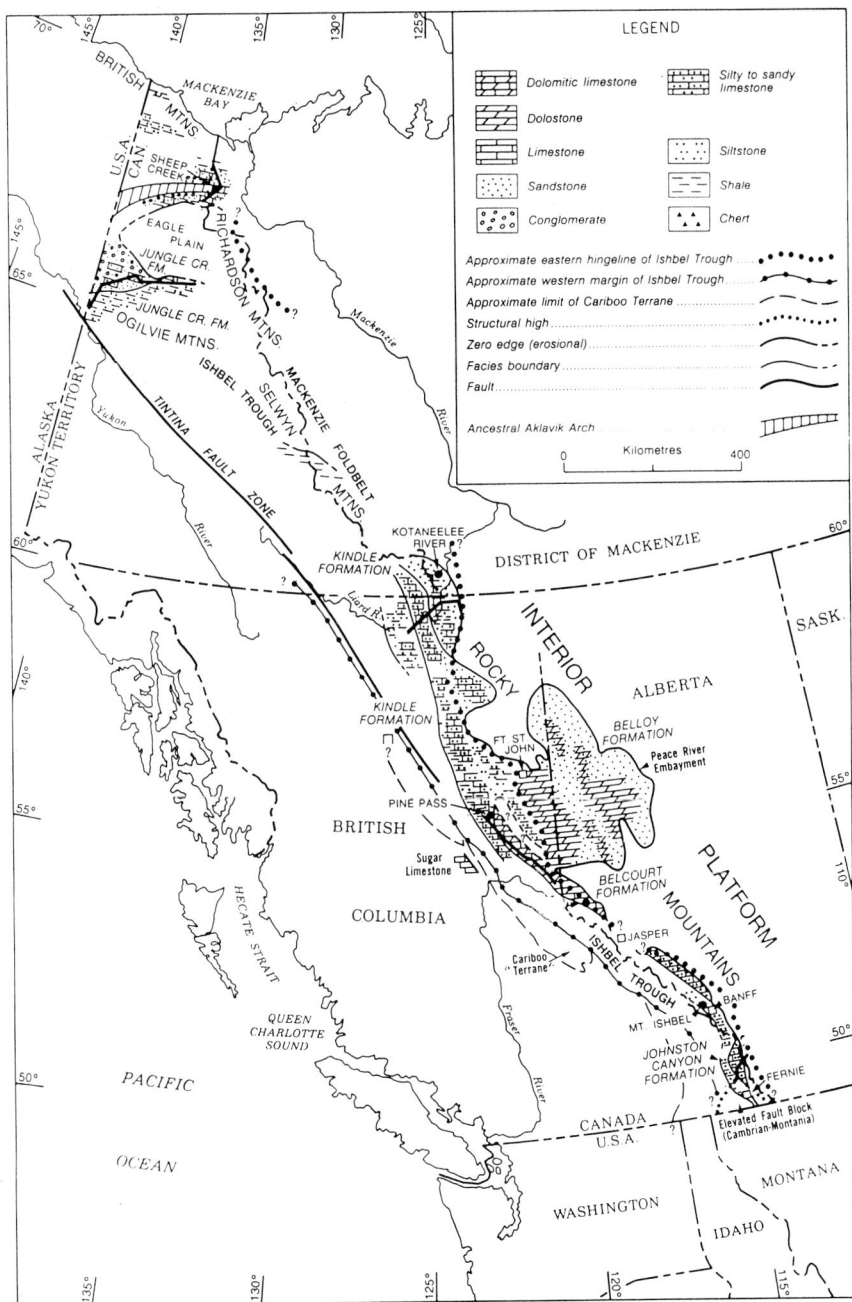

Figure 27. Position of the Early Permian Ishbel Trough and Peace River Embayment in western Canada (modified from Henderson, 1989). Location of rocks in the Rocky Mountains is not palinspastically restored.

of the Antler/Prophet/Ishbel foreland basin. In time, this area of subsidence shrank: the Permian Ishbel Trough was only half as wide as its predecessor the Prophet Trough, and is disappeared by the Triassic.

The Montania landmass that had existed since the late Proterozoic in an area straddling the Canada-U.S. border finally subsided as it was suppressed by the Prophet Trough. In its widest part, this depression formed a single, complexly shaped feature joined with the Peace River Embayment. Movements of many small blocks occurred during Late Carboniferous and Permian time, with a rather new E-W/N-S and NNW/ENE fault network in the Peace River area. Many small blocks moved differently (Cant, 1988; O'Connell et al., 1990), and fault offsets in places reached many hundreds of meters (Barclay et al., 1990). In the Athabasca polymetamorphic domain in the crystalline basement of the Alberta Platform, horst-graben differentiation had vertical amplitudes up to 150 m. The strong influence of faults on thicknesses and lithofacies of different local formations in the Mississippian-Triassic étage is recognized from the extensive drilling in this area (O'Connell and Bell, eds., 1990; in particular, Dix, 1990).

Long-term tectonic instability of the Peace River region has produced many kinds of hydrocarbon traps, which have attracted a lot of petroleum-exploration drilling. As a result, the subsurface geology of this area is known well, but the explanations of tectonics there have remained speculative. Extensional, isostatic and rifting mechanisms have been proposed, but because these discussions are merely ill-founded speculations, we will not consider them further. The Lithoprobe deep seismic refraction data show no disturbances in the Moho surface and in the upper mantle beneath the Peace River Arch/Embayment area (Halchuk and Mereu, 1990). Stephenson et al. (1989) pointed out that the Peace River Arch/Embayment is not clearly expressed in the pattern of magnetic anomalies. O'Connell et al. (1990) stressed that the nearly E-W axis of this zone of weakness does not match the regional Proterozoic fault pattern of the Canadian Shield. Important points to note are that the rise of the Paleozoic Peace River Arch and its subsequent transformations into an embayment, depression or horst-graben system (a) occurred without an evident involvement of the subcrustal upper mantle, and (b) was essentially restricted to the Alberta Platform, involving neither the shield nor the Cordilleran interior. The block movements could reflect an intracratonic thermotectonic instability in the long-lived Athabasca-Peace River-Skeena zone of crustal weakness. Heat-flow values are elevated in this area even at present, compared to other parts of the North American craton (Drury, 1988).

Besides the Peace River Arch/Embayment, other major structural peculiarities in the Alberta Platform were the prominent E-W-trending Tathlina Arch in the north and the NNE-SSW-oriented Northern Sweetgrass Arch in the south. The Tathlina Arch separated the distinct Mackenzie Platform farther north. The Northern Sweetgrass Arch (sometimes called Bow River Arch) completed the separation of the Alberta Basin from the Williston Basin (Porter et al., 1982; Podruski et al., 1988). The NNE trend of the Northern Sweetgrass Arch is also rather unusual in the Precambrian crystalline basement in western Canada. It makes an obtuse angle with the Southern Sweetgrass Arch in Montana, which is oriented NNW. Together, they clearly demarcate a big cratonic block, which manifested itself in the subsequent tectonic stages as well, and helped the Williston Basin to acquire its familiar round shape.

During Mississippian-Triassic time, both the Peace River Embayment and the Williston Basin gradually diminished in size, while a broad, elongated shelf on the Alberta Platform provided favorable conditions for carbonate sedimentation till at least the Middle Pennsylvanian Bashkirian (Barclay et al., 1990). In the later Pennsylvanian Moscovian and in the Permian, the Peace River subsidence was localized mainly in the west. A general elevation of the Alberta Platform caused the spread of very shallow marine environments sensitive to even small transgressions and regressions of the sea. Later, erosion stripped away much of the bedrock in a broad area. Late Carboniferous Pennsylvanian rocks are preserved only in grabens on the western side of the Alberta Platform, indicating a general westward tilt of the platform at that time.

Carbonate sedimentation was subdued in the Permian, and Permian through Triassic rocks in western Alberta are mainly clastic. Their original eastward extent along the eastern Rocky Mountains is hard to restore because they were truncated by several major regional unconformities (Henderson, 1989). Many of the so-called "deep-water" lithofacies in the Permian are actually shallow marine, deposited below the storm wave base. A deeper waterway and depocenter existed in the Ishbel Trough farther west. Provenance areas for sediments lay on both sides of this trough, in the east (the shield) as well as in the west (the Cordillera). This situation was similar to that reported for the Carboniferous trough, which was also two-sided (Richards, 1989; Barclay et al., 1990; Henderson et al., 1993). In the Triassic, rejuvenated horst-graben movements provided small but identifiable depocenters. Sands and shale with some carbonate predominate in them, locally reaching a thickness up to 1,200 m (O'Connell et al., 1990).

Tectonic étage and stage between the Jurassic and the Early Cretaceous

A sharp restructuring occurred prior to Jurassic time in the entire Western Canada Sedimentary Province. It was followed by the development of a new structural-formational étage (Fig. 7, Table III). This tectonic stage largely coincided with one of the most dramatic global sea-level rises in geologic history, and to determine how much of the Jurassic sedimentation was caused by tectonic subsidence of the Alberta Platform and how much by eustatic marine transgression is not always easy, but it is not very important for our descriptions.

106

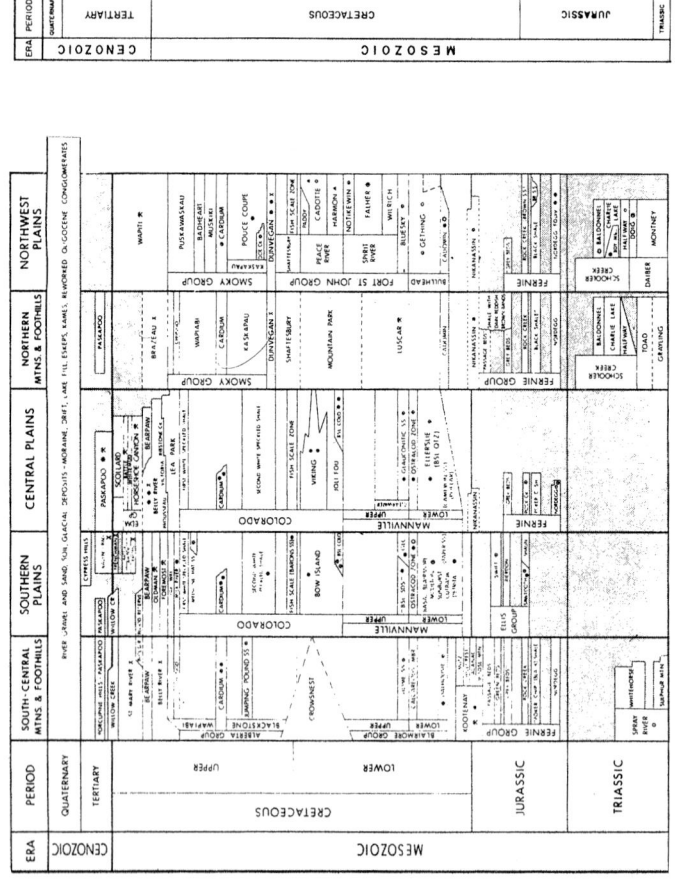

Table III. Mesozoic and Cenozoic tables of formations of the Western Canada Sedimentary Province (sources: Energy Resources Conservation Board; and Saskatchewan Geological Survey, 1994).

In Early Jurassic time, the sea covered most of southern and central Alberta. Middle Jurassic marine sediments overlapped the subsided Sweetgrass arches and reached the Williston Basin (Poulton, 1989). Despite a common effect of the eustatic sea-level rise, distinct tectonic patterns characterized the newly formed Fernie Basin near the Alberta-British Columbia border, the pericratonic Alberta Basin in the foreland, and the intracratonic Williston Basin.

In the Fernie Basin in the southern part of the Canadian Cordilleran miogeosyncline, tectonic subsidence began in the Middle Jurassic Bajocian, ca. 180 Ma. This subsidence extended eastward from the narrow former foreland basin which had existed there from the Late Carboniferous to the Permian. Influx of clasts from the growing Cordilleran mountains in the west became appreciable in latest Jurassic time (Poulton et al., 1993, 1994). Before then, it was just one of the basins that appeared from time to time in the miogeosyncline-platform tectonic domain. As the Canadian Cordillera grew, compensatory subsidence to the east from and along its main NNW-SSE structural trend created the Alberta foredeep. This was another one of the successive foredeeps that developed since the late Paleozoic on the western side of the Western Canadian Sedimentary Province. Each new generation of these foredeeps was shifted cratonward from the previous ones. On the other, eastern side of the Western Canadian Sedimentary Province, since the late Middle Jurassic Callovian, the Williston Basin again became a distinct depression due to differential tectonic subsidence in the northwestern Midcontinent Platform.

A profound change from predominantly carbonate successions with secondary and mainly shield-derived clastic sediments to almost entirely clastic successions with west-derived debris occurred across the Alberta Platform in the Late Jurassic (Figs. 28, 29). This change in sedimentation setting was rather transitional in the miogeosynclinal Omineca Belt. Farther east, early Mesozoic, Triassic and Jurassic paleogeographic environments were also transitional: the seas shrank, and non-marine conditions broadened since the Pennsylvanian. Structural rather than compositional peculiarities compel us to separate the Triassic and Jurassic successions into distinct unconformity-bounded units. As the Cordilleran mountains grew intermittently, three major orogen-sourced clastic influxes spread over the Alberta Platform in the Mesozoic (Stott et al., 1991, 1993). The early Oxfordian (Late Jurassic) influx strongly affected the sedimentation patterns in the Alberta Basin. In the Early Cretaceous Berriasian to Aptian, Cordilleran-derived detritus reached the Williston Basin.

In terms of regional sedimentological processes, this reorganization was fundamental. For petroleum exploration, this change was of great importance: previous predominantly carbonate sedimentation changed to predominantly clastic, and shield-derived clasts were replaced with orogen-derived ones. Former petroleum systems ended and new ones were formed, requiring a different exploration strategy to locate the traps at this level. But tectonically, this was not a fundamental regional break.

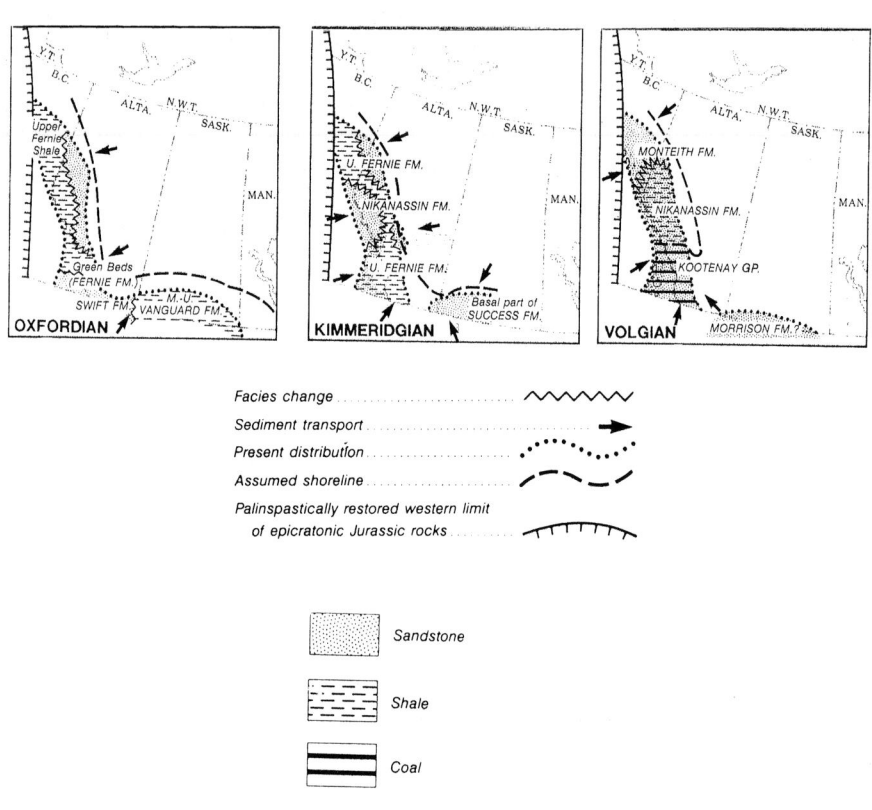

Figure 28. Distribution of Jurassic lithofacies in the Western Canada Sedimentary Province (modified from Poulton, 1989).

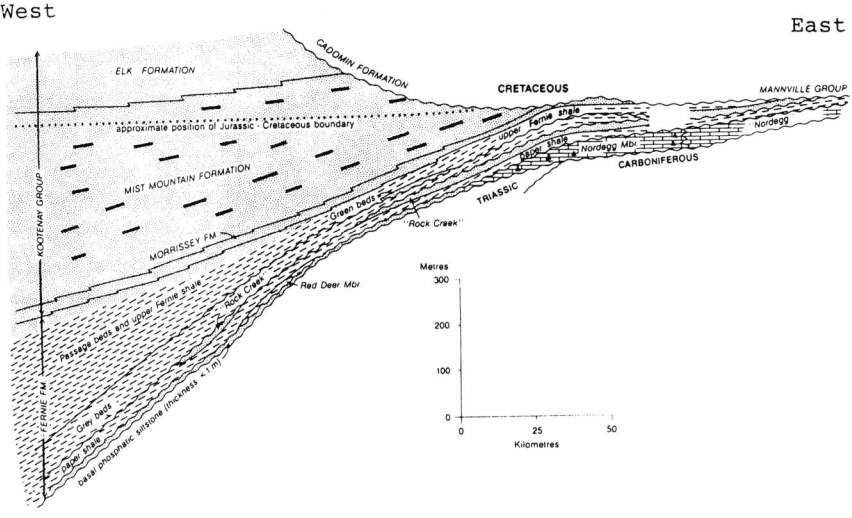

Figure 29. Westward-thickening Jurassic sedimentary wedge in the Western Canada Sedimentary Province, illustrated in a schematic E-W lithostratigraphic cross-section (modified from Poulton, 1989). Bars indicate coal; triangles indicate chert.

Pre-Jurassic relief of the Alberta Platform was leveled, and regoliths are common at the base of the Jurassic rocks. A very thin, <1 m, layer of marine phosphatic siltstone covered the beveled relief even in the Omineca miogeosynclinal belt (see figures in Poulton, 1989; Poulton et al., 1994). The initial, five or six small marine incursions in the eastern Cordillera produced a rather thin, 200-400 m, clastic succession in the Jurassic Fernie Basin. Farther east, on the Alberta Platform, they left several uncon-formity-bounded packages of strata, each probably no more than a few tens of meters thick. This indicates the Jurassic vertical tectonic movements in the miogeosyncline-platform domain were moderate, though they seem to have been slightly more distinct in the miogeosyncline than on the cratonic platform (Fig. 29).

Because the Jurassic epicontinental sea was still shallow, even minor sea-level fluctua-tions had a big effect on the depositional paleoenvironments on the platform. Detailed studies of numerous Jurassic coal seams indicate a combination tectonic and eustatic influences on sedimentation in the Alberta Basin. The structural unconformity over the older, truncated stratal rocks developed at the base of the Lower Jurassic Sinemurian across most of the Alberta Platform. This permits to pinpoint the beginning of the Jurassic structural-formational étage quite reliably.

The base of the Jurassic tectonic étage should be put at the base of the Sinemurian strata, even though this stratigraphic hiatus was of short duration. Several other Juras-sic unconformities, mentioned often in the literature, are of secondary importance be-cause they did not correspond to major regional restructuring.

By the Early Jurassic, the locus of deposition in the Alberta Platform was in the south. The arcuate Fernie Basin in front of the southern Omineca Belt expanded. The Peace River Embayment finally disappeared as an identifiable bathymetric feature. Increased mobility is noted in the Northern and Southern Sweetgrass arches. Sedimentation was mostly marine in southwestern Alberta and Montana, but in the rest of the Interior Plat-form farther south, the prevailing conditions were non-marine (Peterson, ed., 1986).

Sloss (1963) included Lower Jurassic strata in the U.S. into his Absaroka Sequence, which begins in the Late Carboniferous Pennsylvanian. However, in the Interior Plat-form in the U.S. and Canada, the succession of structural-formational étages varied from region to region and is likely different than in the Midcontinent. In the large region of so-called Ancestral Rockies, in New Mexico, Colorado, Utah and southwestern Mon-tana, gradual uplift reduced the areas of sedimentation, and deposition in the early Meso-zoic continued, in subaerial conditions, only near the Cordillera (Westermann, ed., 1984).

A regional restructuring between the Hettangian and the Sinemurian (~200 Ma) marked a cessation of subsidence in northeastern British Columbia. Areas of subsidence shifted from Peace River to Bow River in southwestern Alberta. During the entire Middle Jurassic, for over 20 m.y., the tectonic depression expanded. In the Oxfordian, it extended northward to its former locus (Poulton, 1989), and in the latest Jurassic, the sea spread over most of the platform. Deposition of conglomerates marked the appearance of high mountains in the Intermontane and Omineca belts (McMechan, 1989, 1997). Westerly-derived, Cordilleran-sourced sediments gradually overtook the sediment supply from the shield in the east. With the increase of sediment influx, the Jurassic depression was filled, and the sea retreated. A new rock succession, rich in coal seams, marked a new tectonic setting. Transformation of the entire region into a domain of continental subaerial sedimentation continued from the Late Jurassic into the Early Cretaceous smoothly, without appreciable stratigraphic gaps (Hayes et al., 1994; Poulton et al., 1994).

The inversion and rise of the Omineca geanticline in the eastern Cordilleran miogeosynclinal zone also involved elevation of the western part of the Alberta Platform, including a big pre-existing downwarp. In the Foothills and Plains, the Jurassic rocks extended more widely (though possibly not by much; Poulton, 1989) than their present-day erosional edge. Remnants of Late Jurassic sediments are preserved sporadically in the grabens that now lies in the Main and Front Ranges of the Rocky Mountain Belt (Wheeler and McFeely, comps., 1991).

In most of the platform, many fluvial valleys cut into gently elevated areas. Sea incursions along some of these broad valleys happened during occasional rises in sea level (Hopkins, 1981). A Late Jurassic-Early Cretaceous drainage system developed even over the temporarily subsided Northern Sweetgrass Arch. In many areas, detritus from the Cordilleran mountains was channeled along river valleys into the Alberta Basin. Patterns of lithofacies and thicknesses reveal several principal structural trends active at that time: N-S and E-W, which are inherently cratonic; and especially NW-SE/NE-SW, parallel to the structural trends in the Jurassic Cordillera.

Several pulses of regional uplift caused unconformities in broad areas of the Alberta and Williston basins. In the Williston Basin, sediment supply was abundant enough to keep pace with the Jurassic subsidence. In the Early Cretaceous, the structural pattern of the Western Canada Sedimentary Province was dominated by the again-elevated Northern Sweetgrass Arch; by the broad Swift Current Platform which rose on the northwestern shoulder of the reduced Williston Basin; and by the NNW-oriented Pembina High induced by forebulging on the eastern rim of the Jurassic-Early Cretaceous foredeep. The resulting subaerial topographic relief of 100-150 m was dissected by river valleys. A general southward regional tilt (with a probable axis near the Snowbird fault zone) caused the remnants of Jurassic rocks to be preserved only in southern Alberta.

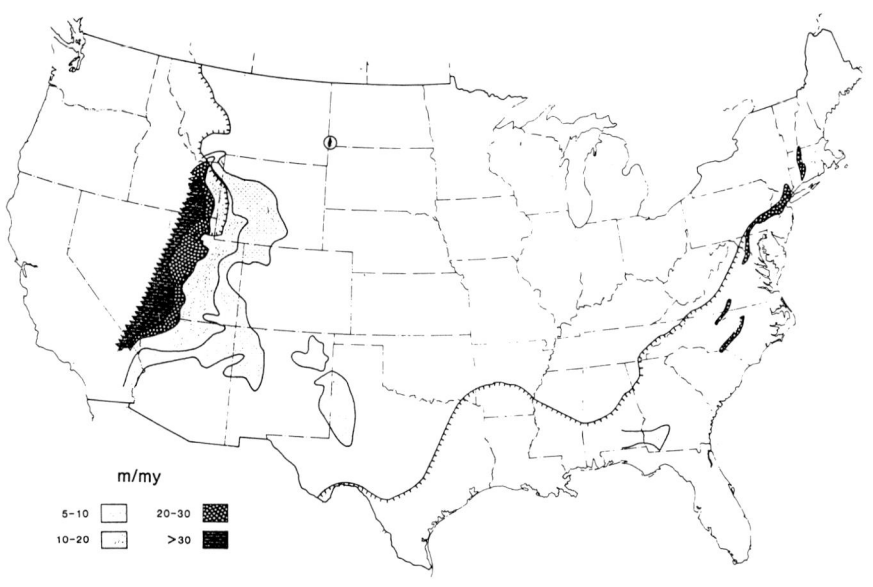

Figure 30. Triassic to Early Jurassic net subsidence rates in the conterminous U.S. (simplified from Sloss, 1988a). 1 - Williston Basin.

Sloss (1988a, p. 42-43) pointed out that in the western Midcontinent region of the U.S. (Fig. 30), the Late Jurassic-Early Cretaceous structural features "...are more the inheritance of syn-Absaroka tectonics (particularly Pennsylvanian-Early Permian events) than of differential vertical movements immediately prior to Zuni deposition". This is apparently different from the Jurassic-Early Cretaceous tectonic regime in the Alberta Platform, where block movements, albeit low in amplitude, had a crucial significance in the platform evolution during the Jurassic to Early Cretaceous tectonic stage. Upper Jurassic rocks pass conformably upsection into the thick Lower Cretaceous lithostratigraphic interval. The top of this structural-formational étage is marked by another big regional unconformity, in the Early Cretaceous Barremian.

Tectonic étage and stage between the late Neocomian and the middle Late Cretaceous

A pronounced regional Early Cretaceous Neocomian unconformity is widespread over much of the Alberta Platform, excepting the foredeep where sedimentation was more continuous. Transgression from the north began in the late Neocomian Barremian and eventually reached the Williston Basin (Hayes et al., 1994). The resulting thick Cretaceous structural-formational étage is one of the best-studied in the region due to its shallow depth and a high oil and gas potential (Deroo et al., 1977; Podruski et al., 1988).

In Wyoming, Colorado and Utah, an approximate counterpart of this étage began earlier, in the early Neocomian, and was confined to a foredeep in front of the late Nevadan Sevier orogen (Burchfiel et al., eds., 1992). Seaways (Fig. 31) progressed from the Arctic Ocean and the Gulf of Mexico, eventually overflowing the uplift in the southern Ancestral Rocky Mountain province only later (Jackson, 1984; Sloss, 1988a). In the Alberta Platform, a broad but simple basin dominated by general subsidence formed in the Cretaceous. This structural-formational étage is colloquially (if loosely) referred to as Mannville, after its main lithostratigraphic unit which has been explored extensively for decades (e.g., Hayes et al., 1994).

Up to 700 m of Barremian through mid-Albian sedimentary strata compose the narrow foredeep basin along the Rockies from the Alberta-Montana border to the Peace River area. In the rest of the Alberta Platform, thickness of equivalent units is much less and rather uniform (Fig. 32), in the tens to a few hundreds of meters. Slow subsidence and abundant supply of clastics changed the initial shallow marine paleoenvironment to subaerial (Fig. 33). In the inner part of the Alberta Platform, numerous fluvial channels continued to be incised into the underlying rocks. In many places, newer channels cut into older ones, creating complex distribution patterns of channel bodies that hamper petroleum exploration (Fig. 34). Delta complexes were built where streams reached epicratonic seas. Though the major streams, some with a "Cordilleran" NNW-SSE orientation, were relatively stationary, their tributary and distributary valleys migrated

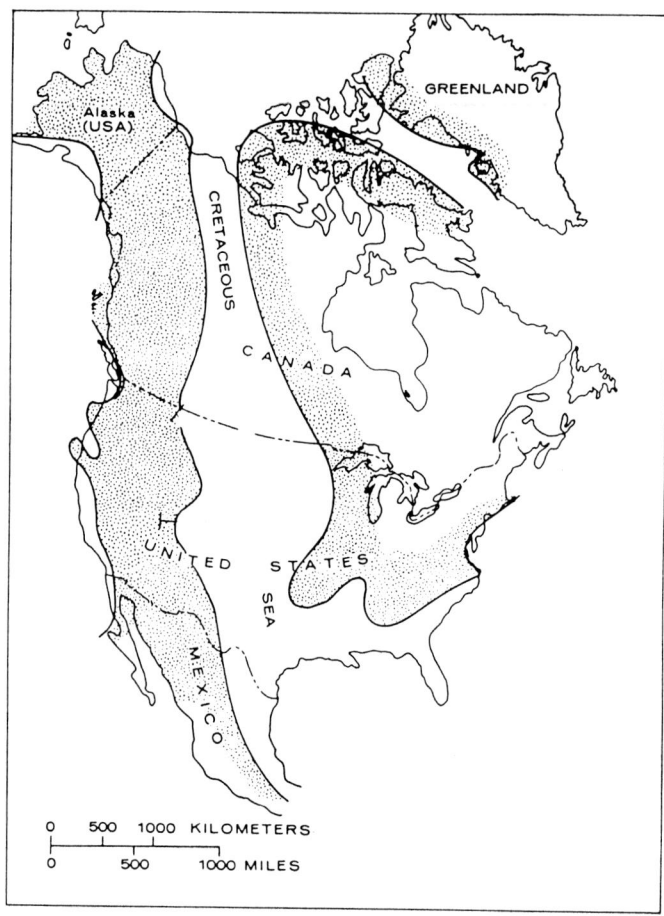

Figure 31. Position of the epicontinental Western Interior Seaway during Late Cretaceous Campanian time (modified from Molenaar and Rice, 1988). Shading indicates land regions.

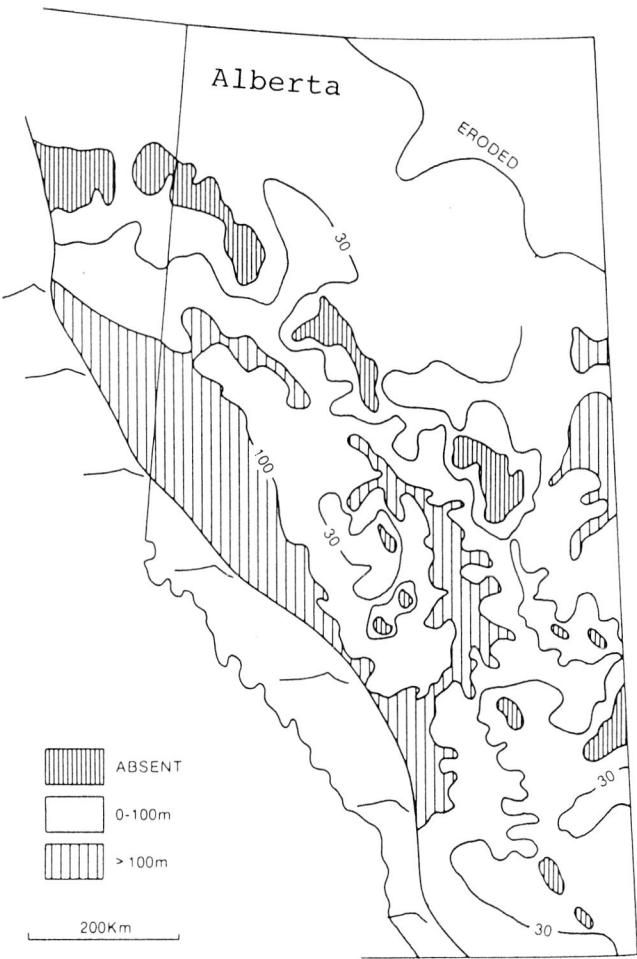

Figure 32a. Lower Mannville isopach map in Alberta (modified from Cant, 1989). Thickness contours in meters. Note the NNW-SSE and NE-SW alignment of thickness changes.

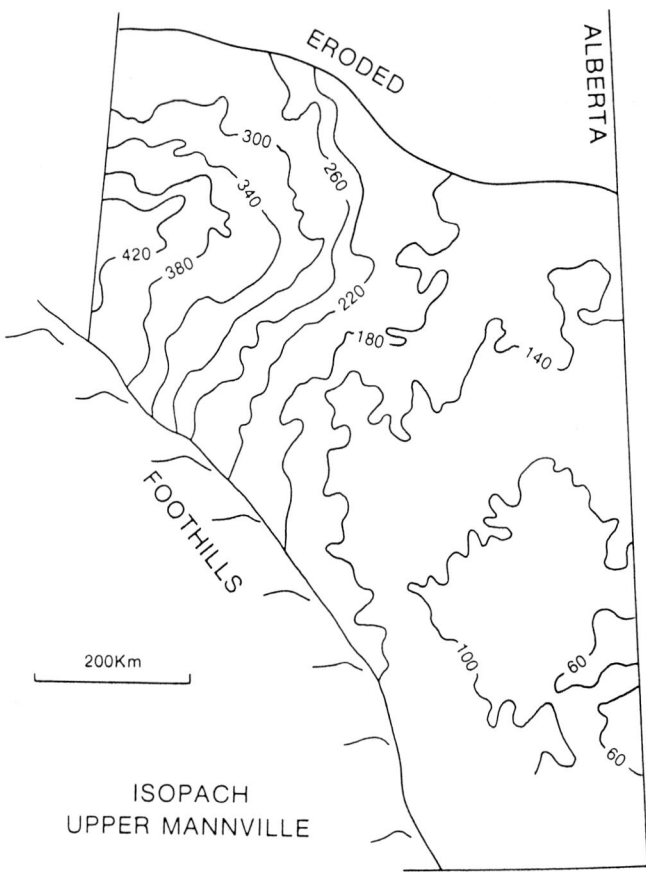

Figure 32b. Upper Mannville isopach map in Alberta (modified from Cant, 1989). Thickness contours in meters. Note the box-shaped thickness contours with NW-SE and NE-SW orientations.

117

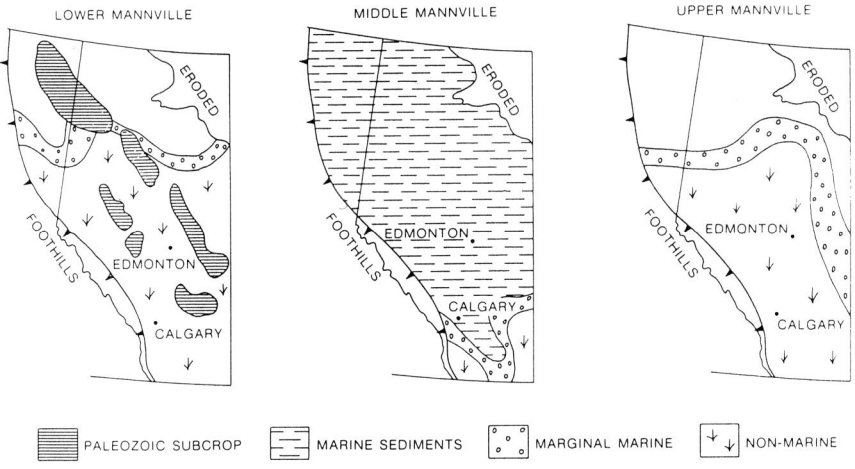

Figure 33. Generalized facies distribution at different stratigraphic levels in the Mannville etage (modified from Cant, 1989).

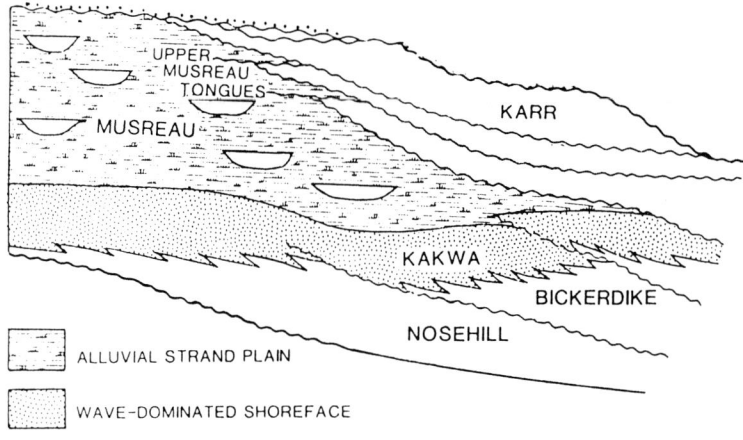

Figure 34. Laterally discontinuous lithofacies in the Upper Cretaceous Turonian Cardium Formation in Alberta, illustrated in a schematic cross-section (modified from Osadetz, 1989). Absence of reliable regional markers complicates correlation.

often. The resulting sand deposits, providing high-quality reservoirs for oil and gas, are commonly elongated NW or NE; they are encased in fluvial and off-channel shale and mudstone (e.g., Robson, 1980; Hopkins, 1987; Cant and Abrahamson, 1996). These bodies may be hundreds of meters or some kilometers long, but only a few meters or tens of meters thick and tens of meters wide, which makes it difficult to detect them seismically. The complex configuration of channels and the sandstone bodies in them make their search notoriously hard and risky.

Besides, stratigraphic correlations of rocks in this structural-formational étage are complicated by the absence of reliable biostratigraphic and lithostratigraphic markers of regional extent. Continuous lithofacies are rare, and the boundaries of many formation are very diachronous. This reduces the reliability of Mannville paleogeographic maps (Hayes et al., 1994; also Flach, 1984).

The Cretaceous structural-formational étage bears traces of two types of tectonic movement: vertical, inherently cratonic, on newly created or reactivated steep faults all over the Alberta Platform; and extracratonically induced, orogenic, Cordilleran (Columbian or Nevadan), producing a gradually deepening foredeep along the platform's western margin. A picture of the fault networks is provided by potential-field linear anomalies (as will be discussed in the following chapters) and maps of the distribution and thicknesses of lithofacies. Several sets of faults were obviously active, although their displacement amplitudes were small. Reactivated old NE-SW-trending faults transected the entire platform including the foredeep. A NNW-trending hinge line marked the inboard limit of the foredeep basin some 200 km east of the Alberta-British Columbia boundary and 50-80 km east of the Mesozoic front of orogenic deformation. A small escarpment south of Calgary, a bigger one northwest of Edmonton, and the Liard Basin in the northwest, were conspicuously centered on the sites of the Paleozoic Alberta Syncline and Peace River Embayment. The most pronounced subsidence occurred at the intersections of the foredeep axis with the major NE-SW-trending fault zones and especially the E-W-oriented Athabasca-Peace River-Skeena zone of crustal weakness (e.g., Cant, 1988, 1989; O'Connell et al., 1990). Burwash et al. (1994, p. 53) rightly stated that fault frameworks established in the Proterozoic "played a critical role in the basement" of the Western Canada Sedimentary Province. Box-shaped Cretaceous isopachs record the renewed block motions in the Peace River area (Hayes et al., 1994). The rest of the Alberta Platform also felt the influence of crustal block movements. Newer fault patterns were no less important; though their influence was often subtle and they are often missed in seismic images, they can be detected with gravity and magnetic data.

Faults controlled the positions of Mannville-age channels oriented NW-SE, but NE-SW trends were active as well. Both these fault trends, reflected in potential-field anomalies, are correlatable with trends of many sandstone bodies containing oil and gas. Dissolution of Devonian salt, which was considerable in Cretaceous time, often occurred along fronts that followed these fault trends (e.g., Anderson and Brown, 1991; Edwards et al., 1998); some fronts could also correspond to open fractures that conducted salt-dissolving water. Salt-dissolution trends are widely known to have affected the position of some of the petroliferous Mannville sand-bearing channels (e.g., Hopkins, 1987). The so-called Punichy Arch on the northeastern edge of the Williston Basin is thought to be a product of salt dissolution (Christopher, 1984), although direct fault control is also possible.

Near the Aptian-Albian boundary, much of Alberta and southern Saskatchewan was flooded by a prograding shallow sea, and the associated fairly uniform swamp-like paleogeographic conditions were common in Alberta at that time. This paleogeographic change was probably not induced by a considerable tectonic restructuring, and most topographic lineaments, nearshore sand barriers, prograding deltas, shorelines, etc. continued to match the same fault-block patterns as before.

On a regional scale, tectonic activity of two crustal blocks separated by the N-S Saskatoon fault in Saskatchewan (cp. Fig. 26) was of great importance. The eastern block defined the contours of the Cretaceous Williston Basin; it subsided moderately but continually during the Cretaceous. The western block raised the broad Northern Sweetgrass Arch, although by Middle Albian time it was completely overlapped by sediments. The Alberta Platform north of the Snowbird lineament was also affected by movements of two distinct blocks, one in western and the other in east-central Alberta. Their boundary induced changes in thickness of many stratigraphic units in the Alberta Basin in Cretaceous time.

The Pembina upland bounding the foredeep basin still persisted at that time. A fault-induced Red Belt-Kindersley zone of plateaus between the Assiniboia and Edmonton valleys corresponded to a median horst zone. The Medicine Hat-Swift Current plateau corresponds to the rise of a major basement block straddling the Alberta-Montana border.

The foredeep in front of the Early Cretaceous Columbian (Nevadan) Cordilleran mountains developed to the east of its Antler predecessor, and was deeper. From the analysis of abundant coal deposits, Langenberg and Kalkreuth (1991) concluded that maturation of vitrinite material largely preceded the deformation of these strata. Burial-related heating produced vitrinite-reflectance values from ~0.85% to 2.0%, increasing consistently along the Foothills and Rocky Mountains to the SSE (also Kalkreuth and McMechan, 1984, 1988). The value of 1.97% was obtained in a sample from a present-day 2,780-m depth in the outer Foothills. But, depending on the assumed paleo-geothermal gradient, the local semi-anthracite occurrences may indicate past burial to 5 km or more.

Most of the Alberta Platform remained subaerially exposed after the mid-Albian due to moderate but broad crustal warping. Shallow marine sedimentation continued only north of the Peace River Arch area, while a regional unconformity developed at the base of the late Middle and Upper Albian. On the whole, the regional tectonic regime remained essentially unchanged. The upwarp was mild and occurred without a regional restructuring. Since the Cretaceous, the region's platformal settings were reduced in width. To the west grew the mountains, induced by orogenic processes in the eastern Cordilleran miogeosynclinal zone.

Tectonic étage and stage between the Late Cretaceous Campanian and the Early Tertiary

The development of the Late Cretaceous-Early Tertiary tectonic étage in the Alberta Platform was strongly influenced by the Laramian orogeny, which was one of the most intensive in the North American Cordillera (Fig. 35). Spectacular thick sheets bounded by thrust faults were mechanically displaced on major low-angle fault planes from the eastern Cordilleran miogeosyncline westward over the Intermontane median massif and eastward over the Alberta Platform. Some rocks of the pre-existing platformal cover were also involved in these displacements (Fig. 36). In many areas of the Rocky Mountains, imbricated thrust sheets consisting of older rocks lie on top of undeformed younger strata. In the Rocky Mountain fold-and-thrust belt, some typically miogeosynclinal successions are now mechanically interlayered with western parts of the platformal succession. Hidden blind thrust faults and gentle folds affect the platformal sedimentary cover along the Foothills (Thompson, 1989; Lawton et al., 1994; Lebel et al., 1996).

The intensity of Laramian thrusting and folding decreased cratonward, so in Alberta the Outer Foothills are less deformed than the Inner Foothills, and the latter are less deformed than the Main Ranges of the Rocky Mountains. As the Late Cretaceous-early Tertiary Laramian orogeny progressed, the cratonward-verging fold-and-thrust belt and its associated mountains prograded eastward (Mack and Jerzykiewicz, 1989). The continental drainage at present lies in the Main Ranges of the Rockies (Holland, 1964; Mathews, comp., 1986).

An orogen-induced general uplift affected the disturbed and undisturbed parts of the Alberta Platform alike. The Late Tertiary-Quaternary uplift brought the Alberta Plains to their present elevation of about 1 km above sea level. But the uplift on the site of the Rocky Mountains could have been two to three times greater, as inferred from coal-rank and other evidence (Nurkowski, 1984; Issler et al., 1990).

Conservative estimates suggest the depth of erosion of stratal rocks of the elevated foredeep basin after the Paleocene and Early Eocene was 1-2 km (Kalkreuth and McMechan, 1984). These estimates change abruptly along the Rockies from one block to another, across big NE-SW transcurrent faults. Vertical movements of fault-bounded blocks

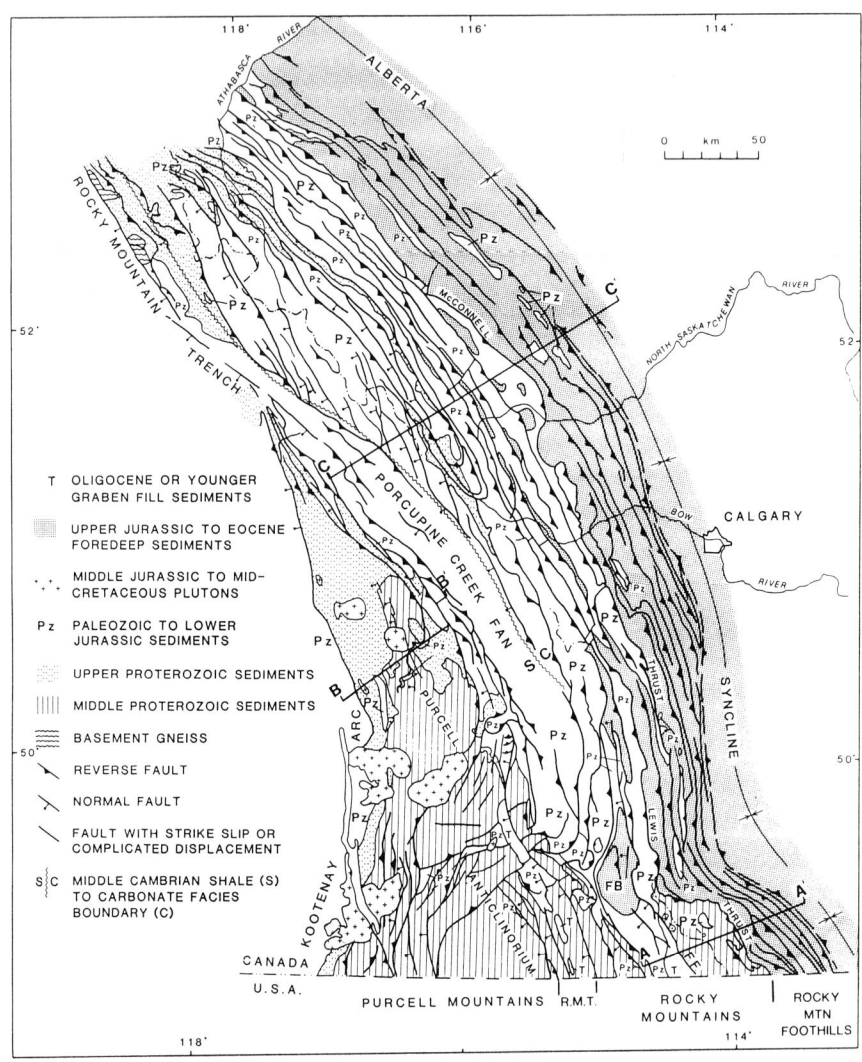

Figure 35. Generalized structural map of the Laramian Rocky Mountain fold-and-thrust belt in the southern Canadian Cordillera, Alberta and British Columbia (modified from McMechan and Thompson, 1989). Cross-section C-C' is shown in Fig. 36.

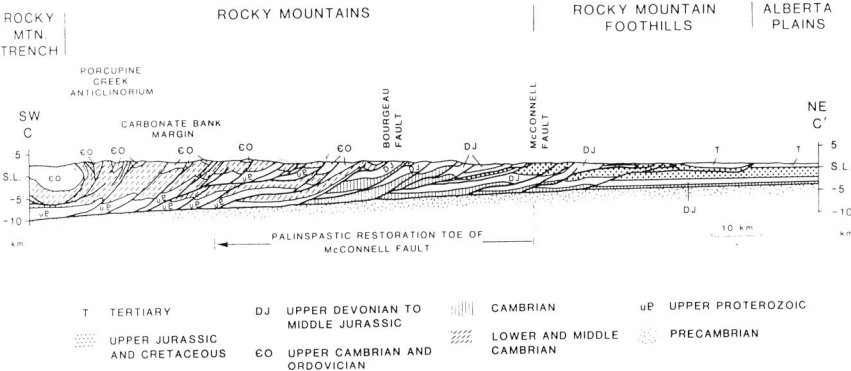

Figure 36. Structural cross-section C-C' of the Rocky Mountain fold-and-thrust belt (modified from McMechan and Thompson, 1989). Location of the cross-section is given in Fig. 35.

affected the distribution of basins and arches as well as the deformation style in the Rocky Mountain Belt, in both Canada and the U.S. (e.g., Hoppin, 1974; Maughan and Perry, 1986; McMechan and Thompson, 1993).

The magnitude of Laramide shortening varies considerably along the Rocky Mountain Belt. A general decrease of the shortening from Montana through Alberta to the Northwest Territories has been reported, along with changes from the predominance of thrusts in the south to the predominance of folds in the north (McMechan and Thompson, 1993). The Laramian foredeeps, where present, vary in age and dimensions. In the north, the foredeep is mostly Albian, and shallow (Jeletzky, 1975). In west-central and southern Alberta, the foredeep is Late Cretaceous-Early Tertiary and larger; it is known as the Deep Basin (Masters, ed., 1984).

From the Albian to the Paleocene-Early Eocene, enough detrital material was shed from the mountains to keep pace with tectonic subsidence and even to overcompensate it (Fig. 37). Up to 3,500 m of sedimentary rocks accumulated in the narrow Laramian foredeep in Alberta. On the whole, they are represented mostly by a coarse continental molasse. Lobate clastic sediment fans overlying the mid-Campanian Pakowki marine horizon have developed since 80 Ma (Jerzykiewicz and Norris, 1994). Deposits in the foredeep contained numerous peat, mudstone and volcanic ash horizons laid down in swamps (Brown, 1962; Gibson, 1977; Hughes, 1984). Thickness of the Upper Cretaceous rocks reach some 1,500 m. Thickness of the Lower Tertiary rocks, reduced by erosion, is now only ~500 m. From studies of coal ranks (Kalkreuth and McMechan, 1984; Bustin, 1991), maturation of hydrocarbons (Snowdon, 1989) and fission-track analysis (Issler et al., 1990), thickness of the Cenozoic succession exceeded 2 km in the Inner Foothills. In places, the molassic pile on the site of the Foothills could have been as thick as ~3,500 m (Jerzykiewicz, 1992). This pile could be a distinct structural-formational étage, developed locally in the foredeep basin, although no simultaneous significant regional restructuring is apparent in the foreland farther east.

An important change in the evolution of the Alberta Platform occurred at the end of the Maastrichtian. Inundations of the Campanian sea were short-lived, and the orogen-derived clastic succession prograded rapidly (Mack and Jerzykiewicz, 1989). Because pulses of thrusting and mountain rise were separated by tectonically quiescent intermissions, the Late Cretaceous-Tertiary foredeep succession contains several separate wedges. Laramian thrusting in the southern Canadian Rockies took place between ~83 Ma (earliest Campanian) and 60-55 Ma (latest Paleocene to Early Eocene; Oldow et al., 1989; Jerzykiewicz and Norris, 1992, 1994). Apatite fission-track data, which help determine when the rocks cooled below 115°C, showed that burial-related heating ended and uplift-related cooling began at ~60 Ma, in the late Paleocene Selandian.

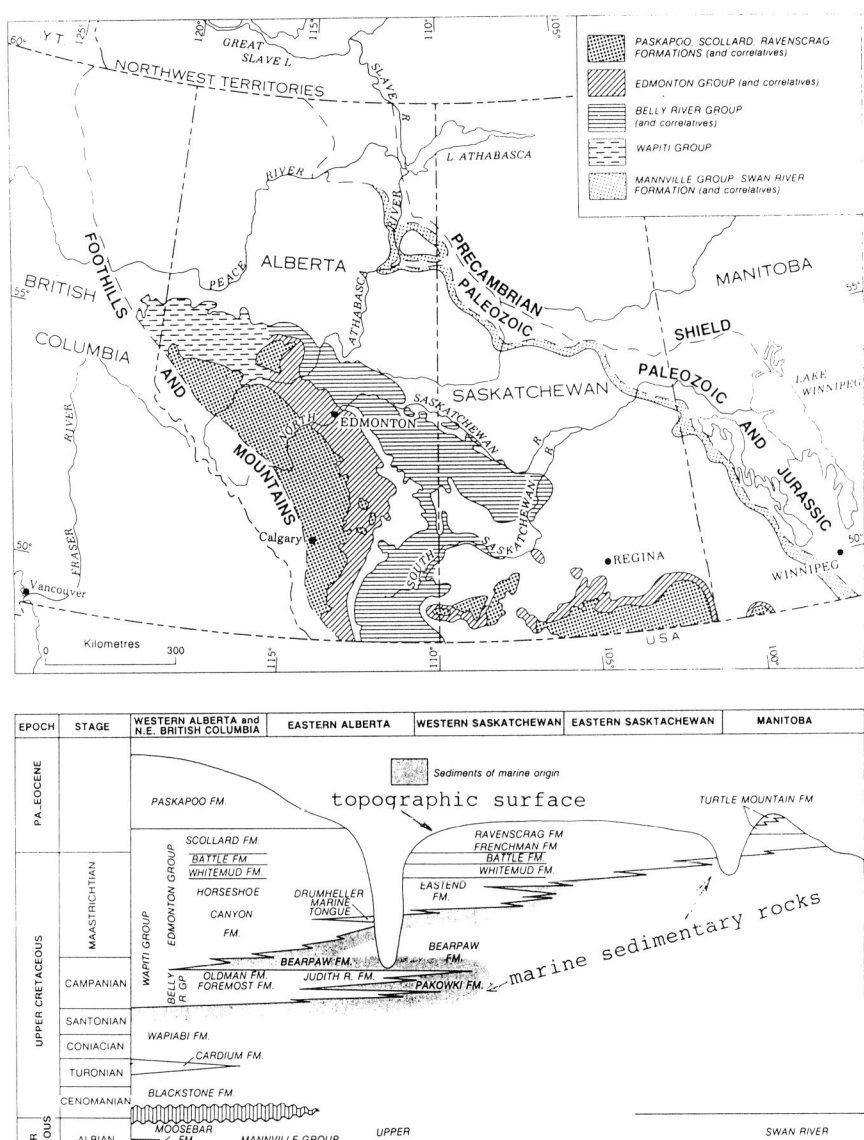

Figure 37. Distribution of Upper Cretaceous and Paleocene coal-bearing formation outcrops in the Western Canada Sedimentary Province (modified from Smith, 1989). Note the WNW-ESE, NW-SE and NE-SW alignments of these outcrop belts in Alberta. Due to similar lithologies and lack of reliable regional stratigraphic markers, the nomenclature of stratigraphic units varies from area to area.

In the Alberta Syncline, the best reference horizon to study Late Cretaceous and Early Tertiary tectonic movements is the Pakowki Formation (the Claggett Formation in Montana) of shale created by the last marine inundation of the Alberta Platform. These strata were later uplifted in several folded thrust sheets. At the latitude of Calgary, spectacular thrusts with eastward vergence form the 70-km-wide Front and Main Ranges of the Rocky Mountains. Where this rugged terrain is maintained by resistant, cliff-forming Paleozoic carbonates, some summits along the continental divide reach 4,000 m high. But even the strongest rocks could not resist erosion and sustain such high elevations for tens of millions of years. This suggests continuous upward movements of the crust. In the Foothills, thrust sheets exposed at the surface contain the more-recessive Mesozoic rocks, and the thrusts are shallower. The Outer Foothills, closer to the platform, are lower than the Inner Foothills just to the west. If up to 2 km of rocks were indeed eroded in the Foothills during the Cenozoic (Issler et al., 1990), even more rocks must have been removed by erosion in the Rocky Mountain ranges. Farther west into the Cordillera, the boundary of the Foothills with the fully developed Rocky Mountains often follows surface traces of big thrust faults (Lewis, McConnell, and others). The thrust stack stands 8-9 km above the cratonic basement at the highest topographic peaks (e.g., Price, 1994, his Fig. 2.7).

At the latitude of Calgary, in the eastern part of the thrust belt, a 20-km-wide zone contains a complex set of stacked thrust slices with rock ages from the Cambrian to the Tertiary. West of the Bull River thrust, the sedimentary rocks are mostly Cambrian to Devonian. The Bourgeau thrust was previously thought to have developed over a hinge zone demarcating the outer side of a late Paleozoic foredeep, but as was noted above, the Prophet/Ishbel Trough was two-sided (Henderson et al., 1993; Richards et al., 1993). A Laramian hinge zone and foreland bulge was not strongly developed in Alberta, either. Only a small and inconsistent upland areas seem to appear in the modern topography of the Prairies in Alberta.

Since the Campanian-Maastrichtian or Paleocene, Jerzykiewicz and Norris (1992, 1994) located the Laramian foredeep axis along the axis of the Alberta Syncline. They estimated the total eastward displacement on the major Lewis thrust fault in the western part of the foredeep to be up to 70 km. The latest Cretaceous marine incursions continued as far as 70 km west of the position of the Alberta Syncline. From geologic mapping, the Alberta Syncline is cut by several transcurrent faults. A box-like, keyboard-like pattern of crustal blocks is apparent from the shapes of gravity and magnetic anomalies over the Alberta foredeep basin. The Laramian orogenic processes did not significantly affect the deep crustal or subcrustal levels in the zone of thin-skinned thrust deformation (cp. Bally et al., 1966).

A general rise of the North American craton is thought to have proceeded for a period of 10-20 m.y. from the mid-Campanian (Weimer, 1984) to the early Paleocene. A contrast between the tectonic settings in the Paleocene and Eocene in the U.S. Interior Plat-from has been related to rapid vertical block movements (King, 1977; Weimer, 1984). These movements created a number of depressions, filled with detritus from local up-lifts. Laramian uplifts and basins defined the structure of the foreland until the Middle Eocene (the end of Sloss' Tejas I Sequence), and Sloss (1963) stressed the dissimilarity in the position and trends of many of these structural elements from older ones (Fig. 38). Sloss (1988a) noted also that, in comparison with the U.S., the Canadian Plains region was "devoid" of significant Laramian orogenic influences. Maughan and Perry (1986) put a boundary between the southern and northern Interior Platform at the WNW-trending Lewis and Clark Line and the Big Snowy Trough. Indeed, north of the South-ern Sweetgrass Arch, the Alberta Platform was more structurally coherent during the Late Cretaceous to Paleogene tectonic stage than the Interior Platfrom in the south. To determine if basin sedimentation ended due to tectonic rise or because the depression was filled is difficult, but block movements at that time did influence the distribution of certain facies and thicknesses in the Laramian foredeep in Alberta and elsewhere, and in the foreland farther east. Post-Paleocene uplift was common for the foredeep and the inner foreland alike (Langenberg et al., 1989). Generally, outside the foredeep, low maturation levels of organic matter indicate that no strong heating occurred in the fore-land basin of the Alberta Platform in the Cretaceous and Cenozoic, and the maximum burial depth of the corresponding rocks was just in the hundreds of meters. Only in the Peace River area do apatite fission-track studies suggest heating was intense enough to reset the apatite clocks (Issler et al., 1990).

Stratal rocks of the Campanian to Paleocene étage are commonly preserved in block-like areas delineated by parallel faults transverse to the general NNW-SSE trend of the Cor-dillera. One such block affected the area of the Northern Sweetgrass Arch south of the Bow River; another one, north of the Athabasca River, underlay the fault-related south-ern flank of the rejuvenated buried Peace River Arch. Two N-S-clongated blocks were active in central Alberta: one parallel to the Laramian deformation front, another just to the east. The NE-trending faults controlled stratigraphy and structure in front of the Rockies (cp. Stott, 1984; Smith et al., 1994). Crustal blocks in the western Peace River area, in the Bow River area, and in the area between them had moved up and down episodically and differently since the Jurassic through the Cretaceous and Early Tertiary, with inversions (also Jerzykiewicz, 1992). In the rest of the Alberta Platform, positive block movements created low-relief physiographic uplands. These highs were mostly bypassed by sediments, except for those around the Sweetgrass arches and the Snowbird structural zone (Leckie et al., 1994, their Figs. 17.8-17.12).

During some 50-60 m.y. in the Cenozoic, 1.0-1.3 km of sedimentary rocks might have been eroded in Alberta (e.g., Issler et al., 1990). This suggests a rate of Cenozoic ero-sion of about 20 m/m.y. A similar or higher rate could have existed in Alberta during

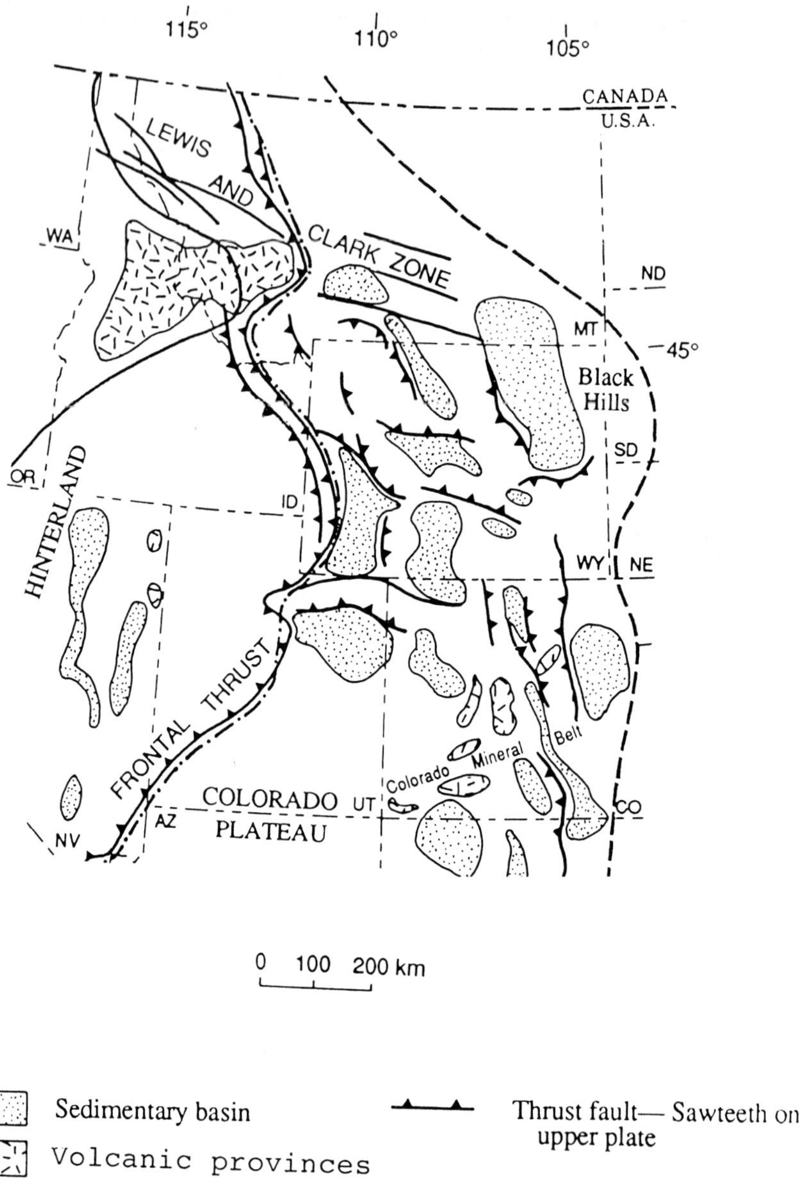

Figure 38. Distribution of major Late Cretaceous to Early Eocene sedimentary basins and igneous-rock complexes in the U.S. Cordillera (modified from Miller et al., 1992). Dashed line marks the edge of the Plains.

the "pre-Cretaceous" unconformity time, and during most of the lifespan of the Interior Platform in the U.S. (Sloss, 1988a).

From the Early Eocene to Early Oligocene, in different parts of the eastern Cordillera crustal extension took place with variable magnitude of stretching (Parrish et al., 1988; Beratan, ed., 1996). Most grabens (Fig. 39) are filled with undeformed Miocene clastic deposits (Gabrielse and Yorath, eds., 1991; Constenius, 1996). Early Oligocene fan-glomerates in a graben in southeastern British Columbia near the Lewis thrust may be more than 2 km thick, and they are not significantly folded (Price, 1994, his Fig. 2.9). But in the Alberta Platform beyond the Foothills, the Late Cretaceous-Early Tertiary structural-formational étage does not fit the timing of crustal restructuring in the Cordillera.

In most of the Western Canada Sedimentary Province, abundant lacustrine, swampy and fluvial settings indicated a continuing tectonic quiescence; even the foredeep in the west subsided rather gradually. An economically important belt of Late Cretaceous-Early Tertiary coal-bearing rocks (the "Great Lignite"; see Brown, 1962) stretches parallel to the Laramian deformation front of the Cordillera from Canada (Fig. 37) to Mexico. Early Campanian to Paleocene coal deposits are widespread in the southern Alberta Plains (e.g., Gibson, 1977; Hughes, 1984), where late Campanian and Maastrichtian coal deposits are typical in the area between Calgary and approximately the Snowbird zone (Smith et al., 1994, their Fig. 33.2). Tertiary coal fields are especially common in the Williston Basin, though some are also present in Alberta. At least two Early Tertiary coal zones are aligned along the NE-trending Snowbird lineament, suggesting that despite overall regional tectonic quiescence, slight block movements intermittently affected sedimentation in Alberta and Saskatchewan. In the still-subsided Williston Basin and along the rejuvenated Snowbird zone in western Alberta, block movements on transcurrent faults were stronger but seemingly short-lived.

Coal seams more than 20 m thick have been traced in outcrop and in the subsurface across central Alberta for several hundred kilometers usually N-S or NW-SE (e.g., Lyatsky and Lawton, 1988, 1989; Smith et al., 1994). In the Rocky Mountain Belt, the coal-bearing rocks were folded and faulted. Syn-thrusting and post-thrusting coalification has been reported in the southern Alberta Foothills (England and Bustin, 1986; Langenberg et al., 1989). In the eastern part of the foreland, differential syn-sedimentary subsidence is often indicated by splitting of some coal seams. Lateral variations in thermal maturity of coal and oil also indicate dissimilar block movements in the Tertiary: for example, blocks south and north of a basement fault that runs south of Calgary apparently had different subsidence and uplift histories after the end of sedimentation (e.g., Bustin, 1991). Coal-rank contours the well-studied central Alberta region have deflections along some NE-oriented basement faults (Smith et al., 1994, their Fig. 33.14).

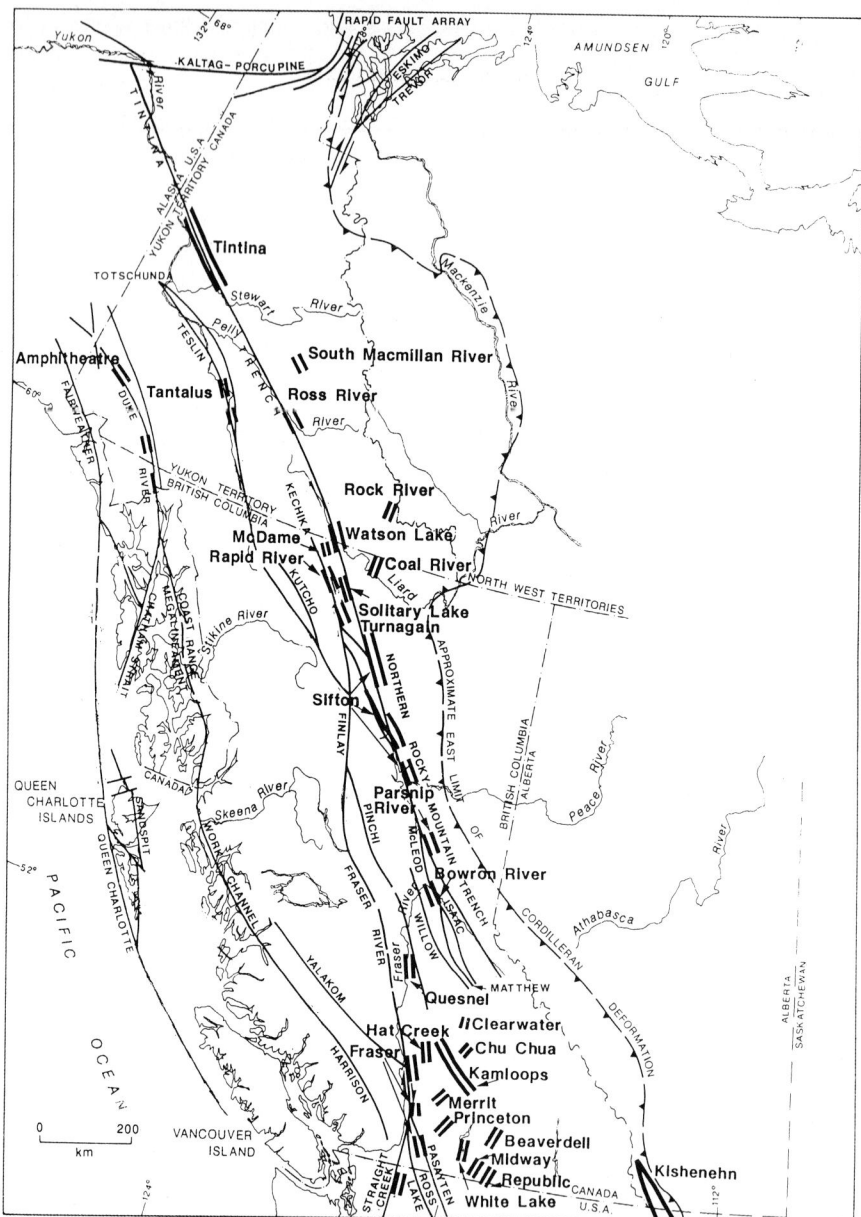

Figure 39. Major crustal faults and fault-controlled Tertiary sedimentary basins in the Canadian Cordillera (modified from Gabrielse, 1991). Heavy double lines mark the grabens.

There were volcanic eruptions and dike emplacement in southern Alberta and Montana in the Eocene, usually attributed to regional crustal extension. The common Late Cretaceous and Early Tertiary age of kimberlite pipes in various parts of the Western Canada Sedimentary Province suggest that some very complex and deep-seated lithospheric processes were under way in the craton. In Alberta, the Eocene, Oligocene and Miocene left no continuous rock record of events during some 50 m.y. But there is no evidence that this period of time was tectonically different from the preceding time interval. For this reason, this unrecorded interval spanning most of the Tertiary is included into the same tectonic stage that created the Late Cretaceous to Early Tertiary étage.

Tectonic étage and stage between the Pliocene and the Recent

In the Alberta Platform, Early Tertiary and older formations are the bedrock for unconsolidated sediments of mostly Quaternary age, and for Recent soils. The lower part of this structural-formational étage may be Pliocene, but in Alberta it mostly contains deposits genetically related to the Pleistocene continental glaciers. These glaciers, thousands of meters thick, advanced and retreated several times. They modified the ground surface, variously flattening or accentuating older topographic features (for instance, river valleys). When they melted, they left behind moraines and outwash deposits. Influences on the cratonic lithosphere from this enormous and variable ice load must have been large, but they are still not well understood.

At the onset of glaciation, most of the topography in western Canada was gentle, with broad valleys separated by low uplands (Fenton et al., 1994). Precambrian rocks in the Canadian Shield had been peneplained long before. Because uplift in the Tertiary was large, some pre-glacial channels were incised 100-200 m deep, mostly into fairly soft rocks of the Cretaceous and Tertiary strata in the bedrock in the Plains. The first glaciation in Canada occurred around 2 Ma, in the late Pliocene (Klassen, 1989). Advances and retreats of several heavy Pleistocene continental glaciers reactivated some of the old basement faults. Carving by glaciers accentuated these fault-induced valleys, and new deposits filled them as the glaciers melted. These valleys follow a distinct rectangular pattern of the Pleistocene and Quaternary fluvial drainage system all over the Plains (Figs. 6b, 40).

The distribution of Quaternary channels and modern topographic lineaments in the western Canadian part of the craton reveals two principal directions, NE-SW and NW-SE, with a subsidiary E-W/N-S pairset. These lineaments are apparent in the modern airphoto and remote-sensing images (Misra et al., 1991; Penner and Mollard, 1991), and they have been mapped in fracture systems on the ground (Babcock, 1973, 1974; Stauffer and Gendzwill, 1987). Despite some local deviations, the NW-SE topographic trends have been found to be predominant in Saskatchewan, NW-SE and NE-SW trends in southern Alberta, NW-SE and E-W trends in central Alberta, and NE-SW/NW-SE and

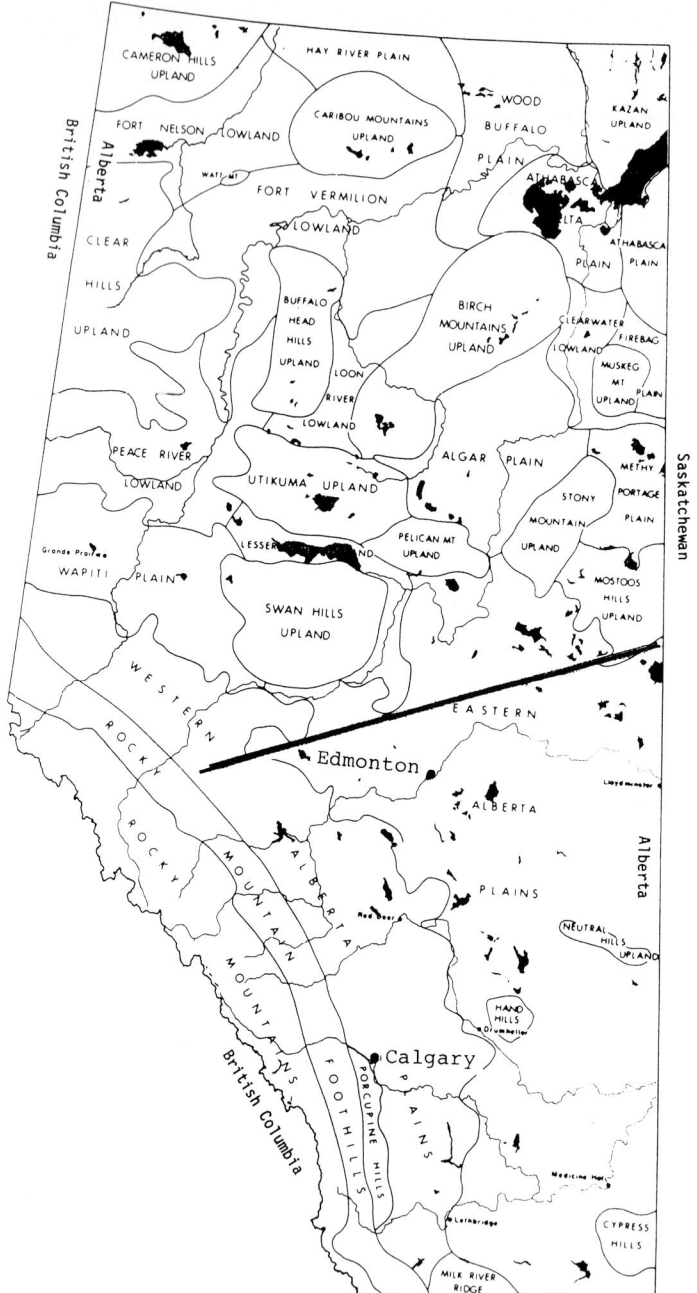

Figure 40. Principal physiographic domains of Alberta (modified from Jackson, 1981). Note the rectangular, NE-SW/NW-SE pattern of rivers and lakes. Heavy solid line through central Alberta, trending NE-SW, marks the approximate modern position of the Arctic-Atlantic continental divide of drainage. The Pacific-Atlantic continental divide in the Cordillera coincides with the Alberta-British Columbia provincial border in the southern part of the map.

E-W/N-S trends in northern Alberta. With regards to the NE/NW and E-W/N-S pair-sets, Misra et al. (1991, p. 149) concluded that "the orthogonal symmetry of the two lineament sets suggests they originated... in the same stress field". As was shown above, this tectonic stress field was in large part inherited from the Tertiary, and the glacial loading failed to obliterate it.

On global experience, the density of topographic lineaments tends to be the highest along major fault zones bounding large blocks in the crystalline basement (e.g., Gol'braykh et al., 1966; cp. also Piskarev and Tchernyshev, 1997). Many such linea-ments in the western cratonic areas in Canada have been linked to zones of extensive fault reactivation (Babcock and Sheldon, 1979; Langenberg, 1983) such as the Peace River (Sikabonyi and Rodgers, 1959) and Sweetgrass (Arnott et al., 1995) arches, which have many reactivated blocks. A high density of topographic lineaments is noted also on the northeastern side of the Williston Basin, east of the Saskatoon fault.

Another zone of topographic lineaments, identified recently from remote-sensing data, runs from Edmonton northward along a boundary between two major crustal blocks (Misra et al., 1991; Fig. 41). This NNW-trending lineament belt of dense fractures, some 150 km wide, is parallel to the structural trend of the Cordillera. The mapped fracture zone lies just east of the Cordilleran deformation front (Fig. 42), possibly mark-ing the boundary of two major tectonic domains in the Alberta Platform: miogeosyn-cline-platform and platform-shield.

Joints found on the surface in the Interior Platform are often correlative with major structures in the Precambrian basement and the Phanerozoic cover (Jones et al., 1980; Maughan and Perry, 1986; Greggs and Greggs, 1989). Their concentration along long-lived weakness zones is conformed by seismic, gravity and magnetic surveys (Penner and Mollard, 1991; Lyatsky et al., 1992; Edwards et al., 1996). But the relationships between topographic and structural tectonic lineaments are in many places loose, and it would be unrealistic to regard surface lineaments as direct indicators of faults at depth. Importantly, however, the Quaternary glacial deposits in Alberta and Saskatchewan do not mask the long and straight structural lineaments (also Misra et al., 1991).

The present-day elevation of the Alberta Plains, around 900 to 1,000 m above sea level, has been achieved party by post-glacial rebound, i.e. elastic and isostatic upwarping of the crust or lithosphere after the removal of the ice load. However, modern upwarping and downwarping expressed in today's topography does not coincide with the outline of Pleistocene glaciers, suggesting independent sources of tectonic movements. Analysis of river terraces confirms that differential regional uplift is still under way.

Figure 41. Surface lineaments interpreted from satellite images of the Western Canada Sedimentary Province (modified from Misra et al., 1991). Lineaments trending NE-SW/NW-SE are predominant across the region, though other orientations are also observed more locally.

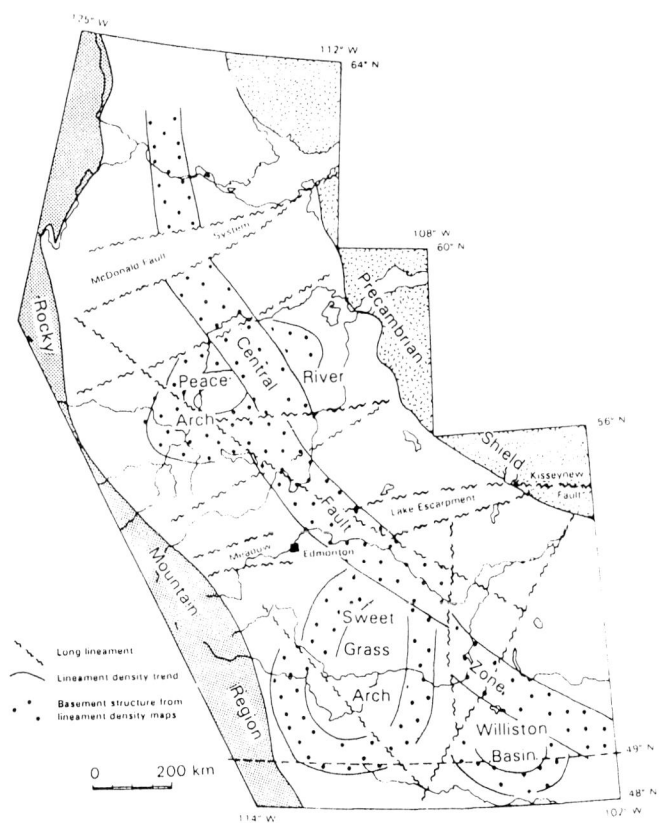

Figure 42. Main structural zones in the Western Canada Sedimentary Province expressed in remote-sensing images (modified from Misra et al., 1991).

4 METHODOLOGICAL REASONS FOR USING STRUCTURAL-FORMATIONAL ÉTAGES IN REGIONAL TECTONIC ANALYSIS

Methods of outlining tectonic regions in cratons

Regional tectonic analysis is made necessary by the practical need to understand the tectonic evolution of an area of interest. In the current practice, this is often reduced to the application of this or that existing generalized hypothesis or model, to which the regional facts are fitted by force. But the economic and social practice requires solutions derived from the observable facts and specific to the particular region. This is what practical tectonics aims to do.

Structural-formational étages are (partly eroded) records of the tectonic regime and structural configuration of basins and arches during their corresponding tectonic stages. To reconstruct these configurations and deep lithospheric processes that could have been responsible for them, each étage needs to be studied independently, in detail, primarily from outcrop and drillhole data. But even the original lateral extent of an étage, before it is reduced by subsequent erosion, may not provide a full outline of the tectonic region. Only based on a stack of étages can the contours of a tectonic region and tectonic units be delineated. From stacks of étages, it is possible to find out the extent and duration of various basins and arches, platforms and shields, and even an entire craton.

Numerous episodes of indigenous and internally-driven reworking occur in the continental lithosphere at different levels, under different pressure and temperature conditions. These episodes induce the structural reorganizations that account for the differences between the étages. To underestimate the external, extraregional and extracratonic, tectonic influences would be a mistake, but it would also be a blunder to overlook the internal, indigenous tectonic forces which are usually predominant in leaving their imprint on the structural-formational étages.

The North American continent, like other ancient continents, first appeared as a product of differentiation of a relatively felsic crust from the undifferentiated primordial mantle and lithosphere. Continued differentiated separated cratons from the continent's more-mobile parts. The North American craton was mostly high-standing in the Middle and Late Proterozoic. Its differentiation into shield and platform regions by the beginning of the Phanerozoic represented a fundamental restructuring of that craton. The appearance of a continuous sedimentary cover in the very late Neoproterozoic and earliest Paleozoic time, which occurred almost simultaneously in the North American, Russian, Siberian and other cratons, was a mark of worldwide restructuring. While broad cratonic interiors were still maintained as uplands despite erosion, since the beginning of the Phanerozoic large peripheral parts of the cratons subsided below sea level. The upland in the cratonic interiors shrank. At the cratons' peripheries, broad regions were beveled and mantled by flat-lying bedded sedimentary blankets. Long-lived pericratonic subsidence made room for vast epicratonic and epicontinental seas, where shallow marine sedimentation settings persisted for tens of millions of years. In regions of shallow seas and low-relief lands, even small tectonic and eustatic fluctuations were enough to produce vast transgressions and regressions (Eardley, 1962).

The sedimentary cover of the Interior Platform in western Canada is often divided into two main parts (Podruski et al., 1988, p. 14): Cambrian to Middle Jurassic, presumably related to a "stable passive continental margin *and* adjacent craton" (italics ours), and the overlying Late Jurassic to Tertiary, related to a "migrating foredeep created by the loading of the craton by the overriding tectonic mass of thrust sheets resulting from the collisional accretion of micro-continents to the western margin of ancient North America". A similar subdivision was also utilized in the recent *Geological Atlas of the Western Canada Sedimentary Basin* (Mossop and Shetsen, comps., 1994), where the entire history of this sedimentary province is divided into a "cratonic-platform stage" followed by a "foredeep stage".

Analysis of structural-formational étages (Chapter 3) shows that these regional tectonic stages are imaginary. The Paleozoic stratigraphic interval in this province is made up of carbonates, but this should not simply lead to its designation as a "passive continental margin". The Mesozoic and Tertiary stratigraphic interval is made up of clastics, but it does not mean this interval should be ascribed to just a foredeep or tectonically contrasted with its predecessor in a fundamental way.

In various influential summaries of the Western Canada Sedimentary Province, what some authors described as a foredeep basin (Podruski et al., 1988) others called a foreland basin (Ricketts, ed., 1989). Mixing together the settings of sedimentation and tectonics (which may be related but are not the same) Ricketts (1989, p. 6, 7) saw in this province a "continental terrace wedge" overlain by "a broad foreland basin superimposed on the cratonic succession". In fact, all these successions are cratonic and platformal. Podruski et al. (1988) divided the whole regional geologic record into two "major westward-thickening sedimentary wedges". But there is still no agreement on the stratigraphic content of the assumed "wedges": whereas Podruski et al. (1988) regarded the lower package as Late Proterozoic to Middle Jurassic, Mossop and Shetsen (comps., 1994) considered it Middle Cambrian to Middle Jurassic. In reality, not all the structural-formational étages in this tectonic region are wedge-shaped.

Confusion in the tectonic treatments of the Alberta Basin does not end there. Monger (1989, p. 9, 10) stated that since about 700 Ma, i.e. mid-Neoproterozoic, "the cratonic margin was probably an interplate boundary", whereas Price (1994) has argued that Windermere sedimentation occurred from 750 to 575 Ma in an intracratonic rift. Price (1994, p. 17) speculated that a subsequent continental separation occurred in "earliest Cambrian time", after which the geologic evolution in Alberta can be attributed to a "regional isostatic response of the lithosphere to loads imposed on it at the newly formed continental margin". He believed that further subsidence can be ascribed to "isostatic flexure of the North American lithosphere under the weight of the tectonically thickened supracrustal rocks of the foreland thrust and fold belt". In contrast, Struik (1987) supposed that continental separation in the eastern Cordillera could not have occurred till the late Paleozoic Carboniferous. As well, he recognized no continental margin along the old eastern Cordilleran miogeosyncline. These uncertainties in the applications of generalized models confirm that only the analysis of mappable tectonic, structural-formational étages provides real clues to the shapes of tectonic regions and permits to restore their tectonic history reliably.

Pericratons, forelands, foredeeps and their corresponding sedimentary basins

Many classifications of sedimentary basins were linked to momentarily fashionable hypothetical concepts. They have failed the test of time and are deservedly forgotten, but these abandoned classifications have produced some remaining detritus, including names and even notions. The useful notions, such as those of pericratonic, foreland and foredeep basins, still survive, although they are not well formalized. These basins are distinguished by their positions in marginal parts of cratons, defined variously in relation to shields or mobile megabelts. Sometimes, the terms *pericratonic*, *foreland*, *foredeep* are used interchangeably. This section clarifies these terms.

As was shown above, the corresponding basins appeared in the history of the western North American craton in a certain temporal succession and marked important, distinct tectonic stages in the evolution of the Alberta Platform. All these different basins are large rock-made bodies of first-order platformal scale. Each of them has its own structural-formational characteristics that record a particular tectonic regime.

Pericratonic sedimentary basins appeared early in the development of the cratonic platformal cover. The initial, extensive veneer-like sedimentary sheet which mantled the peneplained, beveled crystalline rocks that would become basement, marked the beginning of the pericratonic basins. This occurred in a similar fashion worldwide and at roughly the same time - in the late Neoproterozoic (late Riphean, Vendian, Ediacaran) to Early Cambrian. Reflecting differential crustal movements, thickness of the sedimentary cover became uneven, and individual broad basins were defined.

Recent hot discussions about the role of vertical tectonic movements vs. eustatic sea-level changes (Dott, ed., 1992) have overshadowed the significance of craton-wide and continent-wide epeirogenic movements which created the pericratonic basinal provinces. Pericratonic subsidence, pericratonic platforms and pericratonic basins have primarily been produced by negative, intermittent epeirogenic crustal movements. In the western Interior Platform of the North American craton, this early stage involved almost instantaneous inception of the Cambrian to Silurian structural-formational étage which is hundreds of meters thick; in the east, along the Appalachians, it began earlier and left behind a thicker tectonic étage. Both these pericratonic basins continued to developed thereafter.

Importantly, epeirogenic crustal movements are inherent not just to cratons but to continents as a whole. They continue to affect the continents at present, as they did in the past. Like block movements and broader upwarping and downwarping, continental-scale epeirogenic movements happen in cratons and mobile megabelts alike. They occurred before and during the deposition of the cratonic sedimentary cover.

Extensive pericratonic basins were at first relatively uniform, flat and little varied in their composition and structure. Quartzose sandstones at the base of platformal sedimentary strata indicate reworking of clastic material from exposed remnants of the Archean-Proterozoic crystalline terrains. During this tectonic stage, platformal settings of sedimentation often extended far beyond the initial platforms, into other parts of a craton (as in the Ordovician in North America) as well as into adjacent mobile megabelts (as in the early and middle Paleozoic in the eastern Cordilleran miogeosyncline).

The old North American craton was much wider than its successor today. The Grenville-Appalachian and Cordilleran mobile megabelts were formed over the old North American cratonic masses that extended much farther west and east than their modern successor. A single, shallow shelfal sea extended over a large part of that craton during much of the late Precambrian, as recorded by quartz clasts at the base of the Appalachian and Cordilleran volcano-sedimentary orogenic rock record (e.g., Reed et al., eds., 1993). Lithologic similarities of rock types in the basal horizons of these megabelts and the penecontemporaneous craton are notable. Analogous situations have been described for the initial stages of mobile megabelts on other continents as well (e.g., Labazin, 1963; Lyatsky, 1967).

Although mobile megabelts are distinguished from the stable cratons as zones of higher tectonic mobility and are separated from them by crustal- or lithospheric-scale boundaries related to major faults, in the paleogeography they were not always clearly separable. In western Canada, the Cordilleran mobile megabelt manifested itself clearly in the Neoproterozoic. The Windermere rift, having a typically Cordilleran NNW-SSE trend (Fig. 43), appeared at ~780 Ma. From its compositional and structural characteristics,

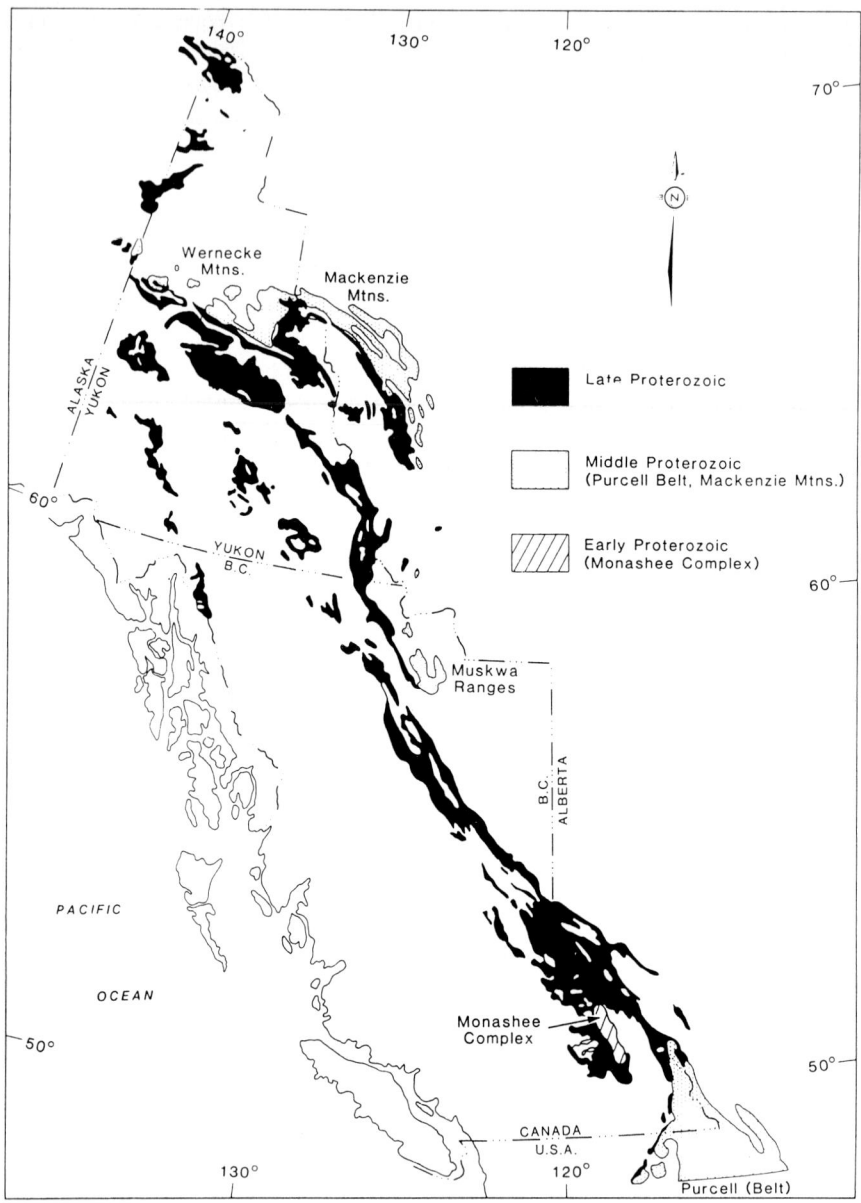

Figure 43. Distribution of Proterozoic rocks (Belt-Purcell, Windermere, their equivalents, and others) in the Canadian Cordillera (modified from Monger, 1989). Distribution of Windermere Supergroup rocks is shown in black.

the eastern zone of the evolving Cordilleran mobile megabelt has been recognized as a miogeosyncline. But during much of Paleozoic time, essentially similar, thick carbonate stratal successions spread from the inner parts of the Interior (Alberta) Platform into the miogeosynclinal zone without changes in lithofacies or thickness. Westward tilt of the Alberta Basin in the Cambrian was not prominent. No appreciable regional hinge line was apparent between the cratonic platform and the Cordilleran miogeosyncline throughout Paleozoic time. The platform and the miogeosyncline were crossed by old NE-trending fault systems inherited from the Early Proterozoic. In the Ordovician, the Northern Sweetgrass Arch, though still subtle, began to separate the platform in Alberta from the Midcontinent Platform, and the Alberta Basin from the Williston Basin. But no obvious differentiation in sedimentation settings occurred between the Alberta part of the Interior Platform and the Cordilleran miogeosyncline, which suggests a similar tectonic regime.

Stille (1941) distinguished miogeosynclines as cratonward zones of mobile megabelts lacking many tectonic manifestations observed in fully developed eugeosynclines: they are less magmatic, less metamorphosed. In North America, such a geosynclinal zonation was adopted by Kay (1951) and used in the Appalachians and Cordillera. With the rise of plate-tectonic concepts, miogeosynclines (renamed mioclines) were correlated with Atlantic-type (passive) continental margins (Dietz and Holden, 1967, 1974). Following this logic, the eastern Cordilleran miogeosyncline along the western margin of a Paleozoic craton was equated with an assumed continental margin of an ancestral North American continent (e.g., Gabrielse and Yorath, eds., 1991).

The tectonic-terrane concept, developed initially for the Cordillera (Jones et al., 1977; Monger et al., 1982), interprets as exotic (but "suspect") all fault-bounded blocks with composition and structure distinct from those in neighboring "tectono-stratigraphic terranes". This hypothetical concept assumed that an ancient continental margin of North America acted as a squeegee, to which newly-arrived far-traveled terranes, derived from unknown places of origin often thought to be Asia, were accreted in the Cordillera. Made up mostly of continental crust, they are nonetheless sometimes interpreted as former island arcs. In western Canada, the ancestral continental margin was assumed to be passive from the Late Proterozoic to the Jurassic, and active (with subduction) after that.

This conceptual tectonic template dominates the *Decade of North American Geology* volumes discussing the Cordilleran geology in the U.S. and Canada. However, it is poorly justified by real geologic facts obtained by field mapping from Alaska (Dover, 1994) through British Columbia (Woodsworth et al., 1991) to California (Ernst, 1988; Tagami and Dumitru, 1996). It is now thought that the idea of far-traveled and randomly accreted terranes is a mis-shaped offspring of the plate-tectonic concept, and many so-called terranes are regarded as native to North America. The Paleozoic and

Mesozoic Cordillera did contain local, marginal mediterranean seas, some of which had an oceanic-crust floor. But the most abundant were thick continental-crust blocks (cp., e.g., Clowes et al., 1995) which have been shown to have strong, intimate links with the rest of the continent. They developed *in situ*.

No ancestral continental margin is evident from the geology of the Canadian Cordilleran miogeosyncline and adjacent cratonic regions. Since the Windermere rifting at ~780 Ma, sedimentary basins along the miogeosyncline-craton transition zone were not one-sided, facing an open ocean to the west. On the contrary, they have been proven to be two-sided features. Clastic material for these basins was derived from continental-crust provenance areas which lay to the west as well as to the east. This has been reported for the Late Proterozoic (Struik, 1987; MacIntyre, 1991) and the Paleozoic Carboniferous and Permian (Henderson et al., 1993; Richards et al., 1993). The westerly provenance areas probably appeared due to uplift in the present-day Intermontane and Yukon-Tanana median massifs. Cecile et al. (1997, p. 56) noted that though a "cratonic hinge line" has to be a boundary between the miogeosyncline and the cratonic platform in western Canada, "in practice it is difficult to precisely locate hinge lines and define the outer edge of the ancestral shelf".

Thus, after much effort to find the western margin of the assumed ancestral Paleozoic North American continent (often incorrectly equated with the craton), detailed studies in the eastern Canadian Cordillera have negated this idea. In the two-sided basins which developed there in Late Proterozoic Windermere and Paleozoic time, tectonic and sedimentation settings were likely not just intracontinental but intracratonic (reminiscent of the much earlier, two-sided Middle Proterozoic Belt-Purcell Basin; Harrison, 1972).

In current practice, the term *eugeosyncline* is sometimes still applied to interiors of mobile megabelts where oceanic-crust slivers were emplaced by plate convergence, subduction and obduction. Bally (1989) argued that it is best to avoid this term, though he retained the term *miogeosyncline* (not *miocline*) as synonymous with passive continental margin. Trying to reconcile the evidence for two-sided Cordilleran miogeosynclinal basins with the terrane-tectonic template, Aitken (1993b) talked about "anomalous" properties of the eastern Cordilleran miogeosyncline: it is too narrow to fit into the crustal-stretching model of McKenzie (1978) and too long-lived for the thermal-subsidence model of Bond and Kominz (1984) and Bond et al. (1984); it lacks a "break-up unconformity" (Stott and Aitken, eds., 1993). Aitken (1993a-c) was inclined even to give up using the term *miogeosyncline* (or its reduction *miogeocline/miocline*) in the eastern Canadian Cordillera. It would, of course, be illogical to redesignate the tectonotype of miogeosynclines in North America only due to the subsequent interpretational misconceptions. In all aspects of their tectonic evolution - sedimentological, magmatic, metamorphic and deformational - miogeosynclines are intrinsic but distinct parts of mobile megabelts. They are also distinct from cratons, despite some shared characteristics. Miogeosynclines lie in transitional zones from mobile megabelts to cratons, and they have continental-scale tectonic significance.

Crystalline rocks similar in type and age to those in the cratonic crystalline basement have been recognized by mapping, drilling and seismic profiling in the Canadian Cordilleran miogeosyncline. Even in the Intermontane interior of the Cordillera, deep seismic surveys have revealed crustal characteristics similar to those in the craton to the east (Burianyk and Kanasewich, 1995; Cook, 1995b-c). Craton-like seismic properties were recorded deep in the crust across the Omineca miogeosynclinal belt and in the Intermontane median massif.

Mobile megabelts and their constituent belts, some related to subduction, others (such as median massifs or miogeosynclines) not, are separated by deep-seated and long-lived fundamental boundary faults. After their inception, mobile megabelts usually experience predominant differential subsidence. Variably subsided zones formed by rapid foundering were often influenced by subduction, as is typical in the Mesozoic-Tertiary European and some modern Far East mediterranean seas. Marginal sink-like seas are surrounded by steep bathymetric slopes (e.g., in the modern Black and southern Caspian seas). These local slopes should not be confused with the main continental slopes which separate continuous continental-crust masses from continuous masses of oceanic-crust affinity (Lyatsky, 1974, 1978). The faults separating mobile megabelts and cratons are often associated with rifts and rift-related marine basins and slopes. But these faults were not active always, and during periods of their inactivity the paleotectonic and paleogeographic conditions spread across them. The existence of a miogeosynclinal-platformal tectonic domain in the evolution of the Interior Platform reflects the fact that at deep crustal and lithospheric levels, tectonic processes do not stop even at the most fundamental crustal boundaries.

The distinction between former boundaries of marginal seas and oceanic continental margins in the geologic record may be elusive. Even at well-studied modern marginal slopes, the location of the continental-oceanic crustal transition is often known only very approximately, though commonly the continental-oceanic crustal boundary lies at the base of continental slopes or, where present, along deep-water trenches. No clear-cut correlation exists between bathymetry and crustal type. The inner boundary of a continental margin may be taken to be the shoreline, or the landward boundary of coastal plains underlain either by marine successions (e.g., a veneer of Quaternary, post-glacial marine sediments at continental margins) or by thick older paralic basins (e.g., along the northern Gulf of Mexico or the Beaufort-Mackenzie province). Crustal blocks on the periphery of modern continents are often influenced by tectonic movements of the nearby oceanic-crust plates (e.g., at the Vancouver Island continental margin; Lyatsky, 1996). The peripheral continental-crust blocks largely control the fluctuating position and extent of shelves, coastal lowlands and paralic sedimentary basins. Hinge zones with sharp changes in lithofacies and thickness of penecontemporaneous sediments commonly mark the inner boundaries of crustal blocks which defined the continentward extend of those sedimentary basins which formed due to block movements (rather than due to eustatic sea-level changes). Such pericontinental blocks are well recognized at the Atlantic margins of North America, where the distribution of corresponding lithofacies in the sedimentary cover is reliably mapped.

Though attempts to formalize the notion of *continental margin* have a long history, this term is still used in the literature freely. Practitioners of different disciplines - geographers, geomorphologists, hydrographers, geologists, geophysicists - fill it with different content. Without clear definitions, the statement that "the sedimentary cover of the North American craton in western Canada is in large part related to processes at continental margins" (Aitken, 1993c, p. 485) has little geologic meaning. Even more confusion arises from Aitken's (op. cit.) statement that sediments in the Paleozoic cratonic cover in the Interior Platform are "the proximal deposits of continental terrace wedges, laid down on subsiding, divergent margins (miogeosynclines)".

Importance of correct recognition of the tectonic nature of sedimentary basins for correct regional tectonic analysis

The two main mobile megabelts flanking the North American craton - Grenvillian/Appalachian in the east and south and Cordilleran in the west - were not exactly coeval: the initial subsidence did not occur everywhere at once. Also different was the timing of their magmatism, metamorphism, inversion. Even within the same mobile megabelt, vertical movements and shortening happened in different orogenic zones at different times. Interactions of miogeosynclinal or orogenic belts with the neighboring cratonic platforms also varied through the geologic time considerably.

Because cratons have their own energy sources for self-development, their tectonic stages reflect primarily their own, inherent tectonic history. Platform-miogeosynclinal tectonic domains were influenced by cratonic and later orogenic-sourced factors, whereas shield-platform tectonic domains are characterized by lack of orogenic influences. In much of the Paleozoic, the distribution of cratonic platforms was not the same as today. The huge Transcontinental basement arch was raised, and the platforms lay on its both sides: along the Appalachians on the east and along the Cordillera on the west. During most of the Paleozoic, the eastern (cis-Appalachian) and western (cis-Cordilleran, Interior) platforms spread into the craton interior. In the latest Mississippian, as the arch subsided, they merged. Since then, the Midcontinent Platform has been a single tectonic entity with its own timing of tectonic events. This joining of eastern and western platforms occurred in part due to an extensive orogeny in the area of the Marathon-Ouachita Mountains in the southwestern Grenville/Appalachian mobile megabelt. Its influence on the craton to the north is apparent, for instance, in the prominent structural trend that controlled the Anadarko Basin and some structural features (such as the Uncompahgre Uplift) in the Ancestral Rockies. Sloss (1988a) put all stratified rocks of Late Mississippian to Early Jurassic age in the Midcontinent Platform into a single Absaroka Sequence (in effect, étage). During that time, tectonic restructuring caused by vertical block movements affected large parts of the craton. The raised Canadian Shield provided a large volume of clastic detritus for the Midcontinent Platform. Carbonate and evaporite sedimentation at that time prevailed only in marginal basins of the post-Antler Cordillera (with the notable exception of the long-lived Oquirrh Basin, where very thick non-marine quartz sandstone and redbeds accumulated since the Permian). Sloss (1988a, p. 39) stressed that in the U.S., during the Late Mississippian to early Permian "the cratonic border of the Cordilleran trend was largely unaffected by extracratonic events".

By contrast, by the end of the Permian, the eastern part of the craton was strongly disrupted during the general inversion of the entire Appalachian mobile megabelt. The NE-SW Taconian and Acadian structural trends parallel to the Grenville-Appalachian orogenic grain were active in the Appalachian foreland, where a long, narrow foredeep basin developed along the foot of the rising mountains with abundant metamorphic clasts from the elevated Appalachian crustal blocks. These sediments were deposited commonly in deltaic environments (Sarwar and Friedman, 1995). Episodic marine incursions spread into the Appalachian Basin and smaller basins in the craton's interior (e.g., Illinois). Cyclic coal-bearing deposits were laid down in the Late Mississippian and Pennsylvanian in response to combined tectonic and glacio-eustatic fluctuations of the sea level (Veevers and Powell, 1987).

In the central Midcontinent Platform, the N-S structural trends that had been prominent in Early Proterozoic Hudsonian time were relatively dormant during most of the Paleozoic. However, they controlled the eastern and western extent of the transgressive sedimentary series in the Dakotas, Nebraska, Kansas and northern Oklahoma in the Cambrian. They left little evidence of their activity during the time when the region was dominated by the broad Transcontinental Arch. With the disappearance of this arch in the Late Mississippian, these ancient N-S structural trends revealed themselves again. In the late Paleozoic, they manifested themselves in the appearance of several N-S-oriented sedimentary basins (Levorsen, 1960; Sloss, 1988a).

Since the latest Paleozoic, the eastern part of the craton was relatively elevated: no continuous stratal rocks of Paleozoic age are found east of the rejuvenated N-S zone, where deposition was only local and intermittent.

Late Paleozoic sedimentation in the southern part of the craton declined steadily also: only within the Kansas-Dakota belt of active N-S structural trends did a sedimentary basin evolve during the Permian (e.g., Sloss, 1988a; Fig. 26). By the end of the Permian, expanding uplift caused a complete end of sedimentation in this entire region. Deposition went on locally till the Early Jurassic in the so-called Permian Basin in Texas and New Mexico, but that was related already to the subsequent Marathon-Ouachita foredeep subsidence.

The western part of the craton has since the middle to late Paleozoic obviously evolved differently from the cratonic regions to the east and south. The eastern Cordilleran miogeosyncline was orogenically immature even during the Antler orogeny, which happened without strong metamorphism or big magmatic events. King (1977, p. 95) noted that "if we define the eastern edge of the [Cordilleran] miogeosyncline as a line along which Paleozoic rocks thicken abruptly, with Lower Cambrian wedging in at their base, that line would not be at the eastern edge of the Rocky Mountains near Denver, but at the western edge of the Colorado Plateau in Utah". The foreland basin induced by the Antler Cordillera was neither large nor long-lived. By the beginning of the Mesozoic, a narrow foredeep developed in front of the orogenic Antler Highs over the westernmost foreland in Nevada and Idaho.

To the north, a late Paleozoic foreland basin was superimposed on the previous pericratonic, Cambrian-Silurian and Devonian-Early Mississippian étages only in the miogeosynclinal-platformal domain. The pericratonic basin of Cambrian to Silurian age in the western U.S. was called by Sloss (1988a) the "Cordilleran Shelf Basin", as distinct from the Late Devonian to Late Mississippian "Cordilleran Foreland Basin". In this scheme, the early Paleozoic shelf basin was related to a passive continental margin of ancestral North America, which after the Devonian was transformed into a foreland due to presumed plate convergence somewhere to the west.

The succession of Sloss' unconformity-bounded sequences records a history of the Midcontinent region which differed even from that of the Michigan Basin, and both are largely different from that of the Interior Platform. In the Alberta Platform, Podruski et al. (1988, p. 14) acknowledged that "not all his [Sloss'] boundaries correspond precisely to the major tectonic events".

The Antler orogeny, which marked the appearance of the first foldbelt in the Cordillera, is recognized compellingly only in the U.S. (Sando et al., 1990). The appearance of a foreland basin there was a response to this first general inversion in a large part of the U.S. Cordillera. In the northern part of the Cordilleran mobile megabelt, tectonic effects of the Antler orogeny were limited to a few discontinuous Middle(?)-Late Devonian grabens along a rift-like zone in western Canada (Struik, 1987; Gordey, 1991). Richards (1989) hypothesized that in the southeastern Canadian Cordillera during the Famennian to the earliest Mississippian Tournaisian, the North American continent extended westward at least to the western Omineca Belt and possibly beyond, including the Intermontane median massif. Conventionally, its western extent is marked by the parautochthonous "Cassiar terrane" (Klepacki and Wheeler, 1985; Smith and Gehrels, 1991). Aitken (1993b) included this "terrane" into the Paleozoic craton, so that the Carboniferous and Permian Prophet/Ishbel Trough developed over that extension.

During the Early Mississippian to Triassic tectonic stage, the Alberta Platform remained pericratonic, acquiring for the first time traits typical of forelands. Since that time, sedimentary basins there tended to have enlarged thicknesses on their west side, in a zone parallel to the Cordilleran mobile megabelt. The Carboniferous Prophet and Permian Ishbel basins (troughs) in the miogeosynclinal belt were just thicker parts of the platformal foreland basin. Some foredeeps in front of the discontinuous (in time and space) orogenic mountains could have also appeared at that time, although the general lack of molassic successions there suggests that the expanded foreland basin was not overprinted by fully developed foredeeps. A well-developed foredeep basin appeared in the Alberta Platform later, as a result of the Late Cretaceous-Tertiary Laramian orogeny.

In Late Cretaceous to Tertiary time, several thick sheets of rocks separated by low-angle thrust faults were displaced from the eastern miogeosynclinal belt towards the cratonic Interior Platform in the U.S. and Canada. Partly, they overlap the platformal cover, and partly involve it in the thrusting. The product is the broad Rocky Mountain Belt.

In the U.S., the western edge of the Colorado Plateau, underlain by cratonic continental crust (Beratan, ed., 1996), marks the position of the whole-crust miogeosynclinal roots. Miogeosynclinal rocks there were displaced to the east in several generations of thrust sheets: Paleozoic Antler, Mesozoic Sevier and especially Tertiary Laramian. In Montana, the Laramian deformation displaced to the east even a huge part of the Belt Purcell Basin, whose rocks now lie in the Rocky Mountains. King (1977) noted that in the western U.S., confusion commonly arises between the geographic extent of the Rocky Mountains, which include all areas of rugged relief west of the Plains, and the geologic Rocky Mountain fold-and-thrust belt. Many of the variously displaced successions in the Rocky Mountains are composed of cratonic platformal rocks of the previously wider Interior Platform (Peterson, ed., 1986; Sloss, ed., 1988; Stott and Aitken, eds., 1993).

In the western U.S. and northwestern Canada, the sinuous outline of Rockies forms deep protrusions towards the craton (Fig. 7, 44). The Rockies seem to form bulges just south of the U.S.-Canada border at 49°N and in the Mackenzie Mountains at ~60°N. These structural bulges have a very big radius of 400 km in the north and as much as 700 km in the south. Smaller (200-250 km radius) semi-domal structural shapes are also apparent in the topography in several areas: north of the Mackenzie Mountains and in southwestern Alberta. Yet, in general, the southern Canadian Rockies follow a fairly straight trend. The modern Rocky Mountains include complex mountain chains with variable orientations, configurations and shapes. The tectonic

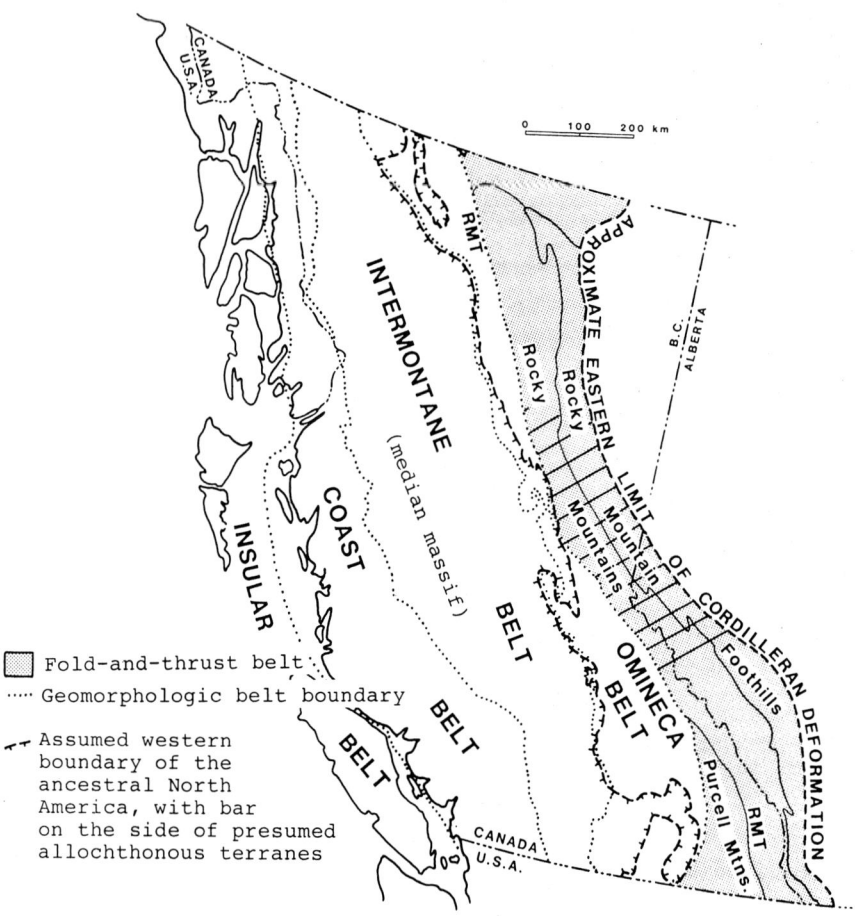

Figure 44. Main tectonic belts in the Canadian Cordillera (modified from McMechan and Thompson, 1989). The Rocky Mountain fold-and-thrust belt contains three main segments, from south to north: thrust dominated, transitional (ruled pattern) and fold-dominated. The amount of shortening decreases from south to north. RMT - Rocky Mountain Trench.

boundary between the Cordilleran mobile megabelt and the modern North American craton is not clear in this region.

In Canada, strikingly linear, fault-controlled, NNW-trending Tertiary trenches divide the entire eastern Cordillera into two physiographic domains: eastern, including the Front and Main Ranges, collectively called the Rocky Mountains or the Rocky Mountain fold-and-thrust belt; and western, containing the Cassiar, Omineca, Cariboo, Selkirk, Purcell and other mountains, jointly called the Omineca Belt (Fig. 45). In the Omineca Belt, miogeosynclinal rocks are preserved *in situ*.

In the 1960s, the concepts of Cordilleran evolution were based on variants of the geosyncline theory. In the 1970s, new concepts derived from plate-tectonic ideas were applied to the evolution of the North American Cordillera. Since then, it has become fashionable to interpret the entire structure of the North American continent as accidental collages of far-traveled accreted terranes. For the cratonic basement, a mobilistic terrane-tectonic hypothesis was employed explicitly and in full by Hoffman (1988, 1990).

A cautious revision of the terrane-tectonic concept in the Cordillera, however, began mainly in the late 1980s and early 1990s (e.g., Ernst, 1988; Woodsworth et al., 1991; Cowan and Bruhn, 1992). Although this concept still predominated as a rigid template in the volumes of the continental-scale synopsis published in the Decade of North American Geology series, growing difficulties with the application of these ideas have come from different lines of consideration of geological facts. Timing of presumed Mesozoic accretion of so-called "suspect terranes" in the Cordillera is found to be increasingly controversial (van der Heyden, 1992), and the petrological (Woodsworth et al., 1991) and geochemical (Ernst, 1988) evidence raises new doubts. No subduction activity along the Canadian Pacific continental margin has occurred in the late Cenozoic (Lyatsky, 1996). McMechan and Thompson (1993) have found no proof to connect the building of the Rocky Mountains with assumed terrane docking. No direct relationship has been found between the growth of Cordilleran mountains and subduction (Stock and Molnar, 1988; Lyatsky, 1993).

Platformal cratonic basins and arches, and their relationships through geologic history

A widespread fallacy holds that cratonized crust is incapable of internally sourced tectonic activity: As a result, cratons only carry the inheritance from previous, ancient orogenic activity or react passively to tectonic forces supplied from later orogens rimming the cratons. All tectonic activity is attributed to assumed zones (usually, to subduction zones) where plates interact with one another outside the continents. In the Decade of North American Geology volume about the platformal sedimentary cover in Canada, Aitken (1993a-c) supposed that Phanerozoic evolution of the North American craton

150

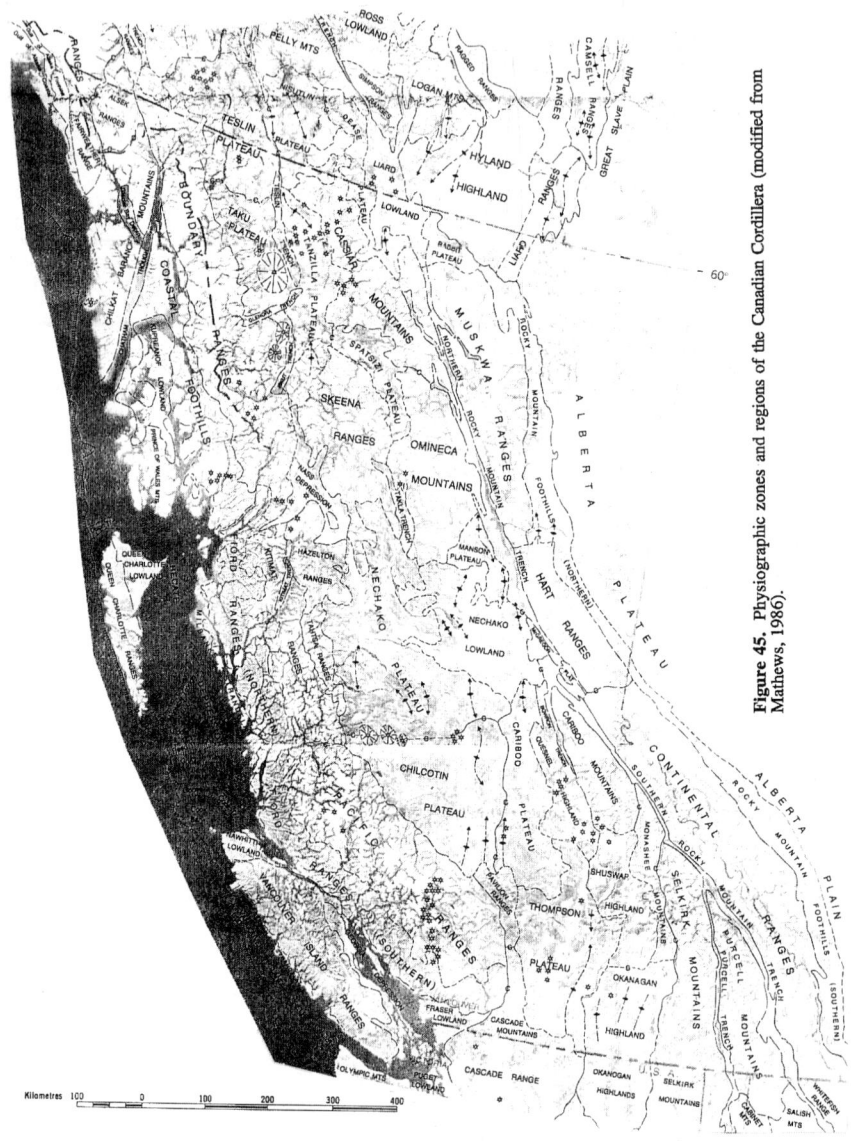

Figure 45. Physiographic zones and regions of the Canadian Cordillera (modified from Mathews, 1986).

was determined by extracratonic forces. He believed, in particular, that a "transcurrent" (this word was not clearly explained) tectonic regime was induced all over the North American craton from boundary zones of lithospheric plates in remote regions in the Atlantic and Pacific oceans. Without acknowledgment of indigenous cratonic tectonic activity, various phenomena - existence in cratons of round-shaped Phanerozoic basins like Williston, Michigan and others; activity of platformal arches and basins; intracratonic seismicity like that in the New Madrid and Ottawa zones; cratonic magmatism, including kimberlite, in the Mesozoic and Tertiary - are hard to explain. The proven difference of the cratonic structural patterns from the ancient, pre-cratonic ones was considered a "mystery" (Aitken, 1993c, p. 494).

Historically, pericratonic subsidence and basin development occurred in broad peripheral cratonic areas of long-lived crustal downwarping that have no relationship with subduction, collision or other extracratonic events. Later, foreland subsidence was roughly coeval with general inversion of the mobile megabelts. Even later, foredeeps formed due to subsidence in narrow crustal zones in front of growing mountain chains which overload the crust. Large marginal fold-and-thrust belts are also often coupled with narrow compensatory or flexural foredeep basins adjacent to overloaded areas. The Alberta Basin was originally purely pericratonic. Since the Late Mississippian, it has been affected by orogen-sourced influences. It shared some of its foredeeps with the miogeosynclinal zone nearby. But the evolving foreland basin, including the strongly developed orogen-induced Laramian foredeep, failed to overshadow the purely cratonic structural features such as the Peace River Arch/Embayment, North and South Sweetgrass arches and so on. Classification of sedimentary basins of cratonic tectonic platforms according to their real tectonic nature and timing permits to distinguish those platformal tectonic features which are intrinsically cratonic from the ones distorted by other influences and those which are entirely a response to external forces.

Sedimentary basins and arches in various étages of cratonic platforms are usually recognized from isopachs. The markers can be basement top, well-defined lithostratigraphic boundaries in the cover, even reliably calibrated seismic reflections. Contoured by these means, the arches may be elevated ridges of the basement or subtle rises in the cover étages. Basins are relatively thick and laterally extensive sedimentary lenses, surrounded by arches where the coeval rocks are of lesser thickness.

One can distinguish elongated arches and rarer round domes. Some of them are raised basement blocks traced through the entire sedimentary cover, others correspond to certain recognizable blocks in the basement, still others may be represented only in the upper étages of the cover and thus seem rootless. Syn-depositional uplifts are marked by zones of reduced thickness of coeval stratal rocks; post-depositional uplifts may or may not be eroded, depending on when the rise occurred. Syn-sedimentary and post-sedimentary differential tectonic movements can thus be reconstructed from geologic maps of marker surfaces. The more such maps are available, the more reliably can we reconstruct the structural and tectonic characteristics of tectonic étages and stages.

There are also variations in thickness of rocks within structural-formational étages caused by non-tectonic influences. Many of them are related to local passive draping over underlying features - structural highs, scarps, reefs - which occur even in tectonically quiescent conditions. In quiet marine sedimentation settings, such drape early in a new stage sedimentation may affect the distribution of topographic depressions. Entirely non-tectonic are post-depositional changes in the thickness of strata caused by variations in the compactibility of rocks, especially during early stages of their diagenesis. Compactibility varies greatly, for example, between clayey, sandy and carbonate lithofacies. Reef buildups are generally not compactible, but the clayey sediments that encase them usually are. Mud dewaters easily, losing much of its volume in the process. To estimate differential compaction is a hard challenge, but errors in these estimates may contaminate the reconstructions of tectonic movements.

Selective erosion, large-scale slumps on physiographic (chiefly bathymetric) slopes, karsts, salt dissolution, the associated synthetic and antithetic faults, all may affect the distribution of younger rock bodies, especially in fluvial and deltaic settings. There can also be non-tectonic structural deformation: collapse structures over reefs and zones of salt dissolution, syn-sedimentary listric growth faults, slides, and so on.

Tectonic influences can be direct (e.g., faults propagating upsection from the basement, or displacements of crustal blocks), but they can also be indirect and subtle (e.g., control on karsting or salt dissolution in the sedimentary cover by pre-existing basement faults). For tectonic analysis on the scale of a platform, at the basement top it is important to discriminate between paleotopographic features and those created by pre-, syn- and post-depositional cratonic tectonic movements. Often, pre-existing unevenness at the top of the basement, created by tectonic movements of various ages and selective erosion, is all lumped together as "basement relief". In well-drilled regions, discrimination between the types of unevenness can more easily be done from the study of the cover tectonic étages affected by block movements than from sparse basement-penetrating wells or basement-sourced potential-field anomalies. Results of geochemical, petrological and other studies of scarce basement cores offer only a few, non-systematic insights into the craton's post-orogenic development. Even studies in shield areas do not permit to decipher many characteristics of tectonic movements during the post-orogenic, cratonic mega-stage. The most reliable multi-aspect information about the cratonic stage of evolution may be derived from structural-formational characteristics of the platformal cover, not from the barren crystalline terrains of the shields.

In the evolution of a cratonic platform, tectonic deformation of the basement and cover is inseparable, except for marginal zones affected by thin-skinned fold-and-thrust disturbances. The structure of particular cratonic étages thus reveals the structural history of whole basement-cover lateral tectonic units. This is more easily done in platforms than in shields. Differentiation of the cratonized continental crust into shields and extensive platforms was a great break in the cratonic evolution. Shields and platforms were created by different tectonic processes deep in the crust or lithosphere. To some degree, they may be considered to represent the biggest of the crustal blocks that make up cratons. These great blocks are themselves composed of many smaller blocks.

After their differentiation, the shields and platforms have evolved as substantially different cratonic entities. At a lower hierarchical level, cratonic tectonic platforms developed differently from one another, and each of them has its own unique history. The Midcontinent Platform, for example, is different from the western Interior Platform, and the Alberta Platform is different from the Interior Platform's southern parts.

A striking similarity in the tectonic history of the modern cratonic platforms all over the globe is that they were initiated in the latest Precambrian and Early Cambrian. At that time, vast terrains of crystalline rocks, peneplained and topographically smooth, were suddenly lowered below sea level. Since the late Precambrian and Cambrian, they have been undergoing slow, fairly continuous tectonic subsidence, complicated by internal, low-amplitude differential platformal uplifts and depressions. On the North American continent, the first broad, low-amplitude arches (Transcontinental, Athabasca-Peace River-Skeena, Severn, Sweetgrass) of crystalline terrain were still broad early in the sedimentary-cover tectonic stage of the craton, but the general trend to subside eventually prevailed and the cover became more laterally uniform and widespread. The long-lived sedimentary basins such as Williston, Michigan, Hudson Bay appeared later, in the Ordovician. Further movements caused a new differentiation of the cratonic platforms, with the appearance of many intra-étage basins and arches. The sedimentary basins and arches subsided or rose differently, with different timing and amplitude.

The examples from the Alberta Platform (Chapter 3) show that the evolving Interior Platform was segmented by mid-Paleozoic time. The Alberta Platform retained its distinct character ever since. Some of its tectonic stages were inherently cratonic, others were influenced by the Cordilleran mobile megabelt and its main orogenies (Antler, Columbian/Nevadan and Laramian). A succession of pericratonic foreland and foredeep basins is distinguished in the corresponding étages. The first of the big sedimentary basins in the pericratonic Alberta Platform, the Elk Point Basin, developed in the Devonian and became broader with time. Carboniferous and Permian foreland basins appeared later in response to general inversion in the Cordilleran mobile megabelt. The younger tectonic foredeep in front of the Cordilleran orogenic mountains and cratonward-verging thrust belts appeared even later. Three principal orogen-related sedimentary wedges are recognized in the Mesozoic and Tertiary sedimentary succession in the cover of the Alberta Platform (Stott et al., 1993). As foredeep subsidence developed on the western platform margin, indigenous tectonic events that affected it were partly and locally masked. But the manifestations of inherently cratonic structural reorganizations in the Alberta foreland and foredeep were never eliminated.

According to Price (1973, 1994), all these basins in the Alberta Platform were "forelands", and all were caused by tectonic flexure of the crust along thrust-loaded zones. Bally and Snelson (1980) and Bally (1989) mentioned both "forelands" and "foredeeps", but used these words as interchangeable synonyms applied to "compressional" basins; they also invoked crustal flexure and associated these basins directly with "foreland folded belts". To ascribe the forelands and foredeeps completely to influences from orogenic foldbelts is misleading, as it overlooks the inherent cratonic tectonism. To do so is also confusing, because it lumps together sedimentary basins of different tectonic epochs recognized in the history of cratons. The evolution of mobile megabelts and cratons is continuous, but forelands and foredeeps are only local and temporary manifestations of the interactions of megabelts and cratons.

Tectonics in the form of vertical displacement of crustal blocks, known as block or Germanotype tectonics, is traditionally regarded as typical for cratons only. In fact, it occurs over entire continents, including mobile megabelts, where it is masked by the predominating orogenic tectonics. Fragmentation into blocks is a fundamental property of the entire continental crust, cratonized or not. In cratons, block movements are best recorded in the structural patterns of basins and arches, even though these are not the only possible kind of deformation. In mobile megabelts, motions of blocks are partly overshadowed by magnificent folding and thrusting, which may distract an observer's attention from the block structure which nonetheless exists.

The platformal block structure is expressed in the distribution of basins and arches in the tectonic étages of the cover-basement crustal entity (i.e. the tectonic platform as a whole). Blocks also dominate the structure of the cratonic shields, though the lack of a sedimentary cover makes them harder to detect. The basin-arch structure is a manifestation of differential block movements that take place over whole cratons, but such movements occur in mobile megabelts as well. The pervasiveness of block tectonics should not be overlooked anywhere on continents.

Other types of crustal deformation, besides block movements, occur on cratons as well. Very common are epeirogenic (i.e. continent-generated; Gilbert, 1890) movements and warping (creating crustal upwarps and downwarps). As a whole, shields can be considered persistent, long-lived upwarps, and platforms as extensive, long-lived downwarps. These continental-scale features are rooted deep in the lithosphere. Smaller upwarps and downwarps occur within these domains. Their lateral extent is more restricted, and they are limited to only the crustal horizons. Upwarping and downwarping with shorter wavelengths than big epeirogenic bulges are often less apparent in the geologic record. The basins and arches they create may be hard to distinguish from those caused by movements of fault-bounded blocks. But generally, block motions just disrupt the wavy background of crustal warping.

The term *sedimentary basin,* as used in this book, refers to a big, lens-shaped geologic body, as distinct from a depression or *basin of sedimentation* in which the sediments can, do or did accumulate. The large depressions and basins of sedimentation developed commonly in regions of prevailing subsidence. Many of them were caused by up/down movements and tilts of identifiable large crustal blocks at the time of deposition. Harder to decipher may be the response of basins and arches to post-sedimentary block movements. Even the round Williston Basin lies on a rectangular block well expressed in gravity and magnetic maps, but its rectangular isopach contours are apparent in only some of its structural-formational étages. The links of arches with blocks may be even harder to pinpoint. Practical difficulties in demarcating arches in the sedimentary cover were noted in the past (e.g., Aitken, 1993c). They arise in part because some of the presumed arches are just less-subsided parts of large sedimentary basins.

Gradual shifts of some depocenters might have been induced not by wavy or block movements but by the rise of arches in formerly downdropped areas. Downwarping and upwarping seen in the Alberta Basin may be rooted deeply in the platformal crust, as is suggested by Lithoprobe seismic data (seismic images of the mid-crustal Central Alberta Arch provide an example of an upwarp of platform-scale significance; see subsequent chapters). Indigenously cratonic was the warping in the Peace River Arch/Embayment, Northern and Sweetgrass arches, and other features of this scale. The timing and amplitude of vertical tectonic movements were specific for each of these areas. Importantly, all of them developed without noticeable extracratonic influences. All of them are restricted to the platform only, without involving the Canadian Shield.

From the analysis of structural-formational étages in the Interior Platform in general and in the Alberta Platform in particular, regional tectonic regimes were manifested variously in regional tectonic tilts to the west, north and south, and in the development of large sedimentary basins and arches of different shapes: the Devonian Elk Point Basin was oval; the Tertiary Deep Basin was asymmetric, west-facing, wedge-shaped. In the early Paleozoic stages, westward tilting in the Interior Platform was local, sporadic and short-lived (for instance, in the Cambrian and occasionally in the Devonian). Tectonic tilting to the north and south was more continuous, and at times it dominated the Alberta Platform.

Only later, since the Late Mississippian, did regional westward tilts produce the foreland and foredeep basins. But even that did not occur due to just the mechanical influences of the Antler and Columbian/Nevadan orogenies in the Cordilleran mobile megabelt as is often supposed, but due to deep lithospheric and crustal geodynamic processes which affectedthe entire miogeosyncline-platform tectonic domain. Mechanical loading, due to mountain-building and thrust-sheet progradation, depressed the platformal crust only in narrow marginal zones in the Mesozoic and Cenozoic foredeeps along the Cordillera. These foredeeps shifted gradually from the miogeosynclinal Omineca Belt eastward. But the influence of these orogenic processes did not extend far into the craton. The great stack of Laramian thrust sheets formed no appreciable bulge in the Alberta Platform. This has been explained by supposed lateral changes in the strength of the Alberta lithosphere (Wu, 1991). However, the Mesozoic foredeep did produce the Pembina bulge in Alberta, even though the crustal load at that time was smaller. This suggests deep-seated processes, not near-surface loading alone, controlled the formation and shape of foredeeps and bulges.

Since the first appearance of foreland and foredeep basins over the eastern Cordilleran miogeosyncline in the late Paleozoic, their axes have progressively shifted cratonward. This progression was particularly clear in Mesozoic through Cenozoic time due to intensified orogenic mountain-building in the Cordillera. Some modern hypotheses ignore this eastward progradation and overestimate the effects of presumed subduction-related orogenic enlargement of the North American continent. As the foreland and foredeep shifted eastward but the continent grew westward, the supposed links of the Alberta Basin evolutionary stages and plate interactions at the increasingly distant continental margin (e.g., Ricketts, ed., 1989; Macqueen and Leckie, eds., 1992) became more tenuous. These assumed links lack real geological justification, and the plate interactions themselves are often assumed from unreliable geophysical models that overlook contradictory geological facts (Lyatsky, 1996). It is hard to explain how eastward-migrating foredeeps in Alberta were induced by processes at a continental margin that was stepping in the opposite direction.

Consolidation of the cratonized continental crust of the North American continent into the post-Neoproterozoic craton, formation over this craton of the Grenville-Appalachian and Cordilleran mobile megabelts, and development of the Phanerozoic cratonic shield and platforms set the framework for the whole Neoproterozoic to Recent evolution of this continent. Comparison with other continents in the Northern Hemisphere suggests similar processes have operated there as well. Early Paleozoic regionalization of the North American craton into the shield and platform areas probably reflected a differentiation of the whole crust or lithosphere. Geophysical, chiefly seismic, observations, confirm that the modern crustal structure beneath the shield and platform is not the same. Surficially, the tectonic differenceswere expressed from the Cambrian to the Recent in the differential predominant tendencies of the shield to rise and of the platforms to subside.

Return to confirmed facts, away from speculative models and groundless hypotheses

Previously, Monger et al. (1982) discussed three lines of evidence which they believed support the hypothesis of terrane-collage composition of the Canadian Cordillera west of its miogeosynclinal Omineca Belt: (a) a sharp contrast in fossil assemblages of some assumed terranes in the Cordilleran interior (Monger and Ross, 1971) was thought to be decisive proof of exotic origin of some blocks; (b) paleomagnetic characteristics of some presumably far-traveled terranes were thought to indicate distant origin, possibly in South America or Asia (e.g., Monger and Irving, 1980); and (c) strike-slip displacements of thousands of kilometers were suggested across fault boundaries between the supposed terranes (Monger et al., 1972, 1982). Two decades later, new facts have negated these suppositions. Paleogeographic studies show that organisms like foraminifera and larvae of larger organisms can easily be dispersed over large distances in oceans (Imlay, 1984; Newton, 1988). Besides, in the Permian stratotypical Cache Creek section of the Canadian Cordilleran interior, the previously-assumed fundamental contrast in fusilinid assemblages was found to be gradational (Nelson and Nelson, 1985). Paleomagnetic results in many parts of the Cordillera are now recognized to be inconclusive. Fundamental uncertainties inherent to the current paleomagnetic techniques, related to inadequate definition of the paleohorizontal, errors in removing overprinting magnetizations, and so on (e.g., Butler et al., 1989; Beck et al., 1997). Geologic field observations rule out the grandiose tectonic displacements inferred from paleomagnetic results (see discussion by Monger and Price, 1996). Strike-slip offsets on major faults in the Cordillera have now been revised, and on the huge Rocky Mountain Trench fault they have been reduced sharply as a result of detailed field mapping (McDonough and Simony, 1988).

In the very mobilistic scheme promulgated by Monger (1989, 1993), in the Middle Jurassic the hypothetical western passive continental margin of an ancestral North America became active due to subduction of Pacific oceanic-lithosphere plate(s) under the continent. Some terranes arrived on these plates and became attached to North America, causing the continent's edge to jump stepwise to the west (also Coney et al., 1980; Monger et al., 1982). In this scenario, oceanic lithosphere arriving from the Pacific, ~13,000 km wide, has been consumed under the Canadian Cordillera, and subduction is thought to be continuing at present (Cook, 1995a-c; Hyndman, 1995). From models of plate reconstructions in the Pacific Ocean (Atwater, 1970, 1989; Engebretson et al., 1985), this is the amount of oceanic lithosphere that must have been subducted under the western North American continental margin since the Middle Jurassic, if the identification and age calibration of oceanic magnetic anomalies and hot-spot tracks is correct. However, the correctness and reliability of such reconstructions in the Pacific are still under discussion (e.g., Stock and Molnar, 1988; Babcock et al., 1992, 1994; Yañez and LaBrecque, 1997). Little mappable manifestation of subduction of such a huge volume of mafic material has been found in the Canadian Cordillera. Even at the Oregon-Washington continental margin, the modern subducting slab has been found with teleseismic data to underlie only a narrow zone not much farther east than the Cascade Ranges, and the dip of this downgoing slab is steep, in excess of 60° (Humphreys and Dueker, 1994). No ongoing subduction is occurring north of the Washington margin (Lyatsky, 1996).

Eocene basaltic magmatism in the Sweetgrass areas of the Alberta Platform is linked by some workers to low-angle subduction from the Pacific (e.g., Constenius, 1996), but this is just speculation: these cratonic igneous suites are distinct from the suites in the Cordillera (Collerson et al., 1990; Woodsworth et al., 1991). The Eocene mafic alkalic suites in the southern Alberta Platform include dikes, sills and plugs, with some volcanics. In the northern Interior Platform, these igneous rocks are found mostly in dikes. Cretaceous and Tertiary kimberlite pipes in the Northwest Territories, Saskatchewan and Alberta are in places diamond-bearing (Pell, 1997). This kind of magmatism is cratonic and deep-rooted. In the Siberian craton, a series of mafic igneous suites fall into several generations of trap magmatism from the Permian to the Early Tertiary, producing thick and laterally extensive lava flows. The famous kimberlite provinces in the African craton also contain Mesozoic and Early Tertiary magmatic diamond-bearing suites. No contemporary subduction is recorded in the vicinity of these provinces.

In studying a region, a geologist must free himself from pre-conceived models and make an effort to see the facts as they are. Regional analysis founded on examination of structural-formational étages provides the required, objective and reliable, factual information.

5   RECONSTRUCTION OF STRUCTURAL HISTORY OF   A   CRATON   BY   USING   STRUCTURAL-FORMATIONAL ÉTAGES IN ITS COVER AND PRE-COVER VOLCANO-SEDIMENTARY BASINS

Outlines of the Proterozoic Laurentian craton in North America

As a result of orogenic processes, the continental crust is cratonized. Heavily reworked at depth, it displays a radical restructuring at the surface. Regions of cratonized crust are typically characterized by considerable tectonic stabilization and subsequent differential vertical warp and block movements. In the cratonic platformal cover, we customarily recognize these movements from the history of sedimentary basins and arches.

The Archean continental crust on the site of the North American continent initially formed as low-density, silicic material floating over a little-differentiated primordial mantle. Since ~4,000 Ma, patches of early continental crust appeared in various regions in the Earth's perisphere. Enrichment with radioactive elements turned these patches into self-developing systems. The Archean continental crust was more mafic than the average continental crust today (Rudnick, 1995), and its evolution since the Early Proterozoic proceeded differently than at present (Hermes and Borradaile, 1985; Hamilton, 1993). In-situ processes caused further differentiation of lithospheric material under the combined influence from gravity and geothermal dynamics. Light, quartz-bearing rocks made up the buoyant continental masses, whereas olivine-bearing rocks became typical for low-standing areas of oceanic crust. Repeated melting of the lithosphere, with addition of light new material from the mantle and delamination of heavy material at the base of the crust, enabled light andesitic (granitic) rocks to concentrate in the growing continental-crust lithospheric entities.

Stockwell (1966, p. 34) argued that the early cratons in Canada, though relatively stable and rigid, had two other main distinctive properties: morphologically, they were "very broad regions quite unlike the mountain chains of later times", and tectonically they were less clearly differentiated than cratons today. Stockwell (op. cit.) noted also that "another difference is the very extensive overlapping of a younger orogen on an older one, and this occurs in regions not likely ever to have been the site of intervening geosynclinal deposition". Despite all this (see also Reed et al., eds., 1993; Stott and Aitken, eds., 1993), the oldest Archean-rock nuclei in the modern Canadian Shield are commonly metamorphosed only to the prehnite-pumpellyite and greenschist grades, whereas younger Late Archean and Early Proterozoic terrains contain rocks of much higher metamorphic grades, up to granulite. The preserved major Archean crustal blocks (denoted as Slave, Superior and Wyoming) are fault-bounded fragments of larger cratonic entities. If they were merged together is not clear, but no compelling evidence exists that these blocks were arranged relative to one another much differently than they are today.

Since the Late Archean-Early Proterozoic, it is possible to distinguish the tectonically relatively stable regions, denoted cratons, and the contrasting elongated mobile mega-belts. The Early and Middle Proterozoic craton(s) were already divided into raised shields and subsided platforms, although the latter developed rather locally and are often recognized as distinct volcano-sedimentary basins. Because the cratonic continental crust of that age was still thin and weak, these ancient platforms, including their cover basins, were easily subjected to profound reworking. Their basement and discontinuous cover were strongly metamorphosed and reworked by Early Proterozoic orogenies, which spread over large parts of the initial weak cratonic masses.

After cratonization but before the onset of continuous platformal sedimentation, several generations of basins and arches were formed. Also typical for that pre-cover time interval was the abundant extensional dike magmatism recorded in many distinct widespread suites, which represent episodes of cratonic tectonic restructuring.

Previously, these processes were thought to take place in a continental crust whose physical and geological properties are rather uniform all over the cratons. More-recent seismic studies suggest the crustal structure beneath shields and platforms is not the same, and even deeper levels of cratonic lithosphere under ancient shields may have their own specific properties.

The base of lithosphere beneath continents is still thought to lie at depths of 100 to 150 km, but it can also be considerably deeper (Lyatsky, 1994a). Jordan (1975) considered the "tectosphere" beneath continents to be very deep and stationary. Anderson and Dziewonski (1984) showed that high-velocity and probably low-temperature zones under continents may continue to great depths of some 400 km. Gossler and Kind (1996) reported that continental roots under North America reach depths of more than 400 km.

Rheological properties of the middle to lower crust and the lithospheric upper mantle are not known well, though it is apparent that at depth rocks should change their state and behavior from brittle to ductile. It has been estimated that, depending on the exact pressure and temperature conditions, creep can affect rock masses even in the middle and lower crust (Meissner, 1989). From petrological studies in orogenic belts, some rocks that were created as supracrustal are thought to have been placed by tectonic displacements at depths of up to 50-70 km. Illies (1981), like many others, emphasized that older strain generations in crystalline rocks may influence the subsequent tectonic deformation.

Because of these and many other complications, to interpret old structural relationships between tectonic units is often difficult, especially in terrains composed of ancient Archean and Early Proterozoic rocks. The indigenous cratonic, lithospheric and crustal energy sources from time to time cause reworking and restructuring, which is recorded in deep rock transformations, rifting, mountain building and other manifestations. Oroformal (potentially mountain-building) processes in cratonic regions are subdued, and instead platyformal, platform-making processes determine the tectonic evolution of platforms. In North America, the Laurentian craton was mostly elevated in Mesoproterozoic and Neoproterozoic time, but some of its large regions experienced long-term subsidence. In the west part of the Laurentian craton, since about 1,500 Ma (early Mesoproterozoic), the huge volcano-sedimentary Belt-Purcell Basin developed on the site of the modern Cordillera, straddling the U.S.-Canada border. After this basin had begun, large porphyritic adamellite bodies intruded its lower strata. Dated at 1,476-1,367 Ma, these plutons are coeval with anorogenic granites in the Midcontinent cratonic region. In what is now the western Canadian Shield, around that time the smaller but still very large Athabasca and Martin volcano-sedimentary basins developed (Campbell, ed., 1981; Armstrong and Ramaekers, 1985). Rifting and abundant rift-related magmatism (mostly extrusive), and coeval intrusion of anorogenic granites accompanied by widespread felsic volcanism, were manifestations of general cratonic reactivation. Such reactivations, without orogenesis, were also common on other cratons worldwide (e.g., Igolkina, ed., 1981).

The Belt-Purcell Basin (Figs. 43, 46) is heavily deformed and thrust-displaced by Laramian tectonism, but nonetheless it is well studied. Palinspastic restorations to correct for the Laramide shortening show that the Belt-Purcell Basin was roughly oval and surrounded by fault-bounded blocks (Finch and Baldwin, 1984; Winston, 1986). One such block ("Bell Island") lay west of the basin, somewhere in Oregon and Washington. A N-S-oriented rift containing over 12 km of deposits flanked this block on the east. This rift had eastward reentrants into the Uinta Mountains along the Wyoming-Utah state border, into the Grand Canyon area in northern Arizona, and into the Purcell Mountains area in Canada (Peterson and Smith, 1986; Reed et al., eds., 1993).

At the base of the Belt-Purcell Basin lie mature quartz arenites (thickness 250 m or more) derived from granitoid sources to the west and east (Harrison, 1972; Basu et al., 1975; Reesor, 1984). The total thickness of this basin is enormous, about 20 km. Thickness of the basal strata increases sharply to 1,800 m towards the center of the basin, and Schieber (1989) argued its origin was rift-related.

The Belt-Purcell Basin was intracontinental and intracratonic. It wedges out northward in the southeastern Canadian Cordillera (McMechan, 1991). The upper supergroup of this basin varies lithologically across the Belt Basin: fine-grained clastics and carbonates on the east, coarse siliciclastic deposits on the west. This indicates a source of

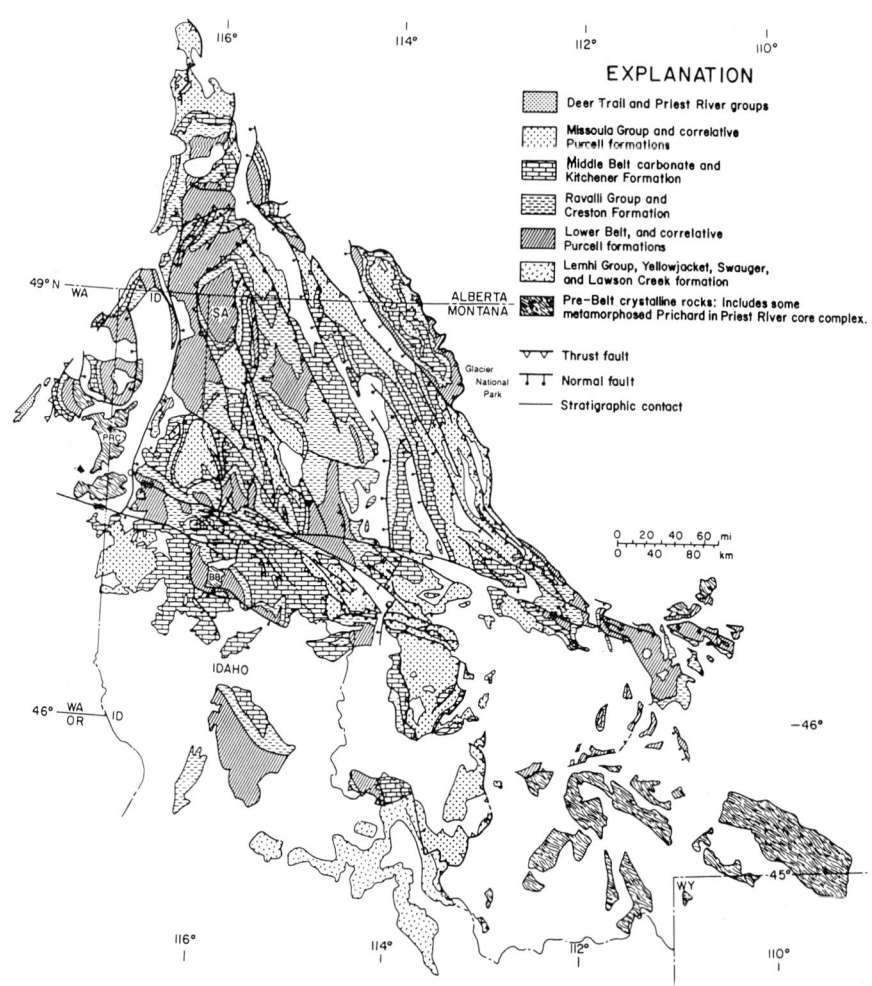

Figure 46. Generalized geologic map of the Belt-Purcell Basin, using the stratigraphic nomenclature accepted in the U.S. (modified from Link et al., 1993). Location of the Belt-Purcell Basin is given in Fig. 43. BB - Boehls Butte; PRC - Priest River Complex; SA - Sylvanite Anticline.

sediments lay in the west (Reesor, 1984), contradicting speculations that an ancestral North American passive margin was present there in the Middle Proterozoic.

Besides the Belt-Purcell Basin, other deep sedimentary basins developed on the western extension of the Laurentian craton, particularly in the Canadian Mackenzie Mountains. It is possible that a thick, mostly Neoproterozoic (post-1,000 Ma) sedimentary pile in that region (Douglas, ed., 1970; Aitken, 1993a-c) was another huge intracontinental and intracratonic basin (Young, 1981). Neoproterozoic platformal rocks have been mapped in a vast area east of the Tintina-Rocky Mountain Trench fault (including the Macken-zie, Selwyn and Rocky Mountains) as well as to the west of this fault (in the northern part of the Omineca Belt and Yukon-Tanana median massif in Alaska). The Protero-zoic craton thus seems to have extended far into the Cordilleran interior and into the Arctic (Plafker and Berg, eds., 1994).

McMechan and Price (1982) put the end of Purcell sedimentation at ~1,350 Ma and linked it with the onset of the East Kootenay orogeny (White, 1959). Obradovich et al. (1984) dated the uppermost Belt-Purcell Supergroup rocks at 900 Ma. Zartman and Stacey (1971) timed the end of Belt sedimentation, and the onset of orogeny, at about 820 Ma. It is also possible these events were not related (Link et al., 1993) at all. In either case, the ~800 Ma diabase sills and dikes seem to indicate a new stage of regional tectonic development in the western Laurentian craton (Harrison, 1972; Harrison et al., 1974).

The Late Neoproterozoic (after ~780 Ma) heralded rifting along the modern eastern Cor-dillera. The Windermere Supergroup (Fig. 43) accumulated in deep, narrow and dis-continuous crustal furrows, with unconformities at its base. In the Purcell and Macken-zie Mountains, these unconformities are related to the East Kootenay and Racklan oro-genic episodes (Gabrielse and Yorath, eds., 1991; Stott and Aitken, eds., 1993).

The rifting did not begin everywhere simultaneously. The most complete Windermere Supergroup developed in an elongated, narrow but 6-km-deep basin along the Rocky Mountain Trench and the Omineca Belt (Lis and Price, 1976; McMechan, 1987, 1991). The rocks are mainly thick-bedded and coarse-grained, terrigenous, petrologically im-mature: feldspathic grits, conglomerates, argillites, phyllites are predominant. Abun-dant feldspar grains, pebbles and boulders indicate proximal granitoid-rich sources. The age of basement rocks underlying the Windermere Supergroup and exposed at pres-ent in the Omineca Belt varies between 2,500 and 1,800 Ma (Parrish and Armstrong, 1983; Parrish, 1991, 1995). Continental-crust rocks are found in the exhumed base-ment best represented west of the Windermere rift, in the Malton, Monashee, Shuswap, Valhalla and other metamorphic core complexes along the southern Omineca Belt (Fig. 47).

164

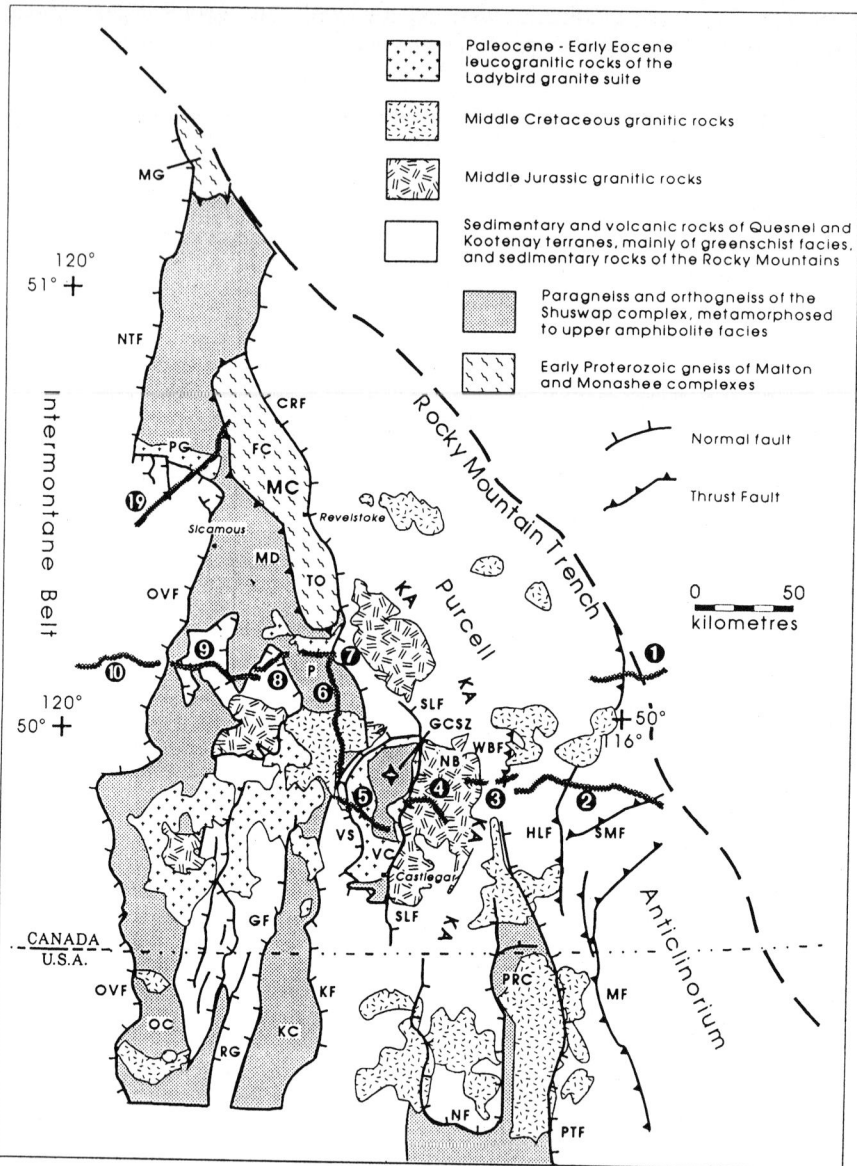

Figure 47. Locations of metamorphic core complexes in the southern miogeosynclinal Omineca Belt of the Canadian Cordillera (modified from Carr, 1995). Fault systems and shear zones: CRF - Columbia River; GCSZ - Gwillim Creek; GF - Granby; HLF - Hall Lake; KF - Kettle; MD - Monashee; MF - Moyie; NF - Newport; NTF - Adams-North Thompson; OVF - Okanagan-Eagle River; PTF - Purcell Trench; SLF - Slocan Lake; SMF - St. Mary; VS - Valkyr; WBF - West Bernard. Metamorphic core complexes and their culminations: FC - Frenchman Cap; KC - Kettle-Grand Forks; MC - Monashee; MG - Malton; OC - Okanagan; P - Pinnacles; PRC - Priest River; TO - Thor-Odin; VC - Valhalla. KA - Kootenay Arc; NB - Nelson batholith; PG - Pukeashun granite; RG - Republic graben. Numbers 1 to 10 and 19 mark Lithoprobe seismic lines.

According to Thompson (1989), Windermere deposition represents a "protracted period of crustal instability", from ~800 to 570 Ma, typified by "block faulting". Movements of crustal blocks separated by major NE-SW-oriented faults are well recognized in eastern Cordillera from sharp stratigraphic changes across these faults (Höy, 1982, 1984; Root, 1987; Cecile et al., 1997). Syn-sedimentary character of these block movements indicates that the old, Archean and Early Proterozoic major faults were active at that time in a fashion inherited from the cratonic tectonic setting.

The idea that the Windermere rift was caused by an episode of "continental separation" (Stewart, 1972; Sears and Price, 1978) and developed on a passive continental margin has not been proved. No eugeosynclinal-type rocks have been found in the Windermere Basin or in regions to the west (Struik, 1987). Presence of the cratonic basement and of the Neoproterozoic and Paleozoic platformal rocks on both sides of this rift (McMillan, 1991; Murphy et al., 1991) confirms its intracontinental and intracratonic character. In some of the recognized eastern Cordilleran blocks (cp. McMechan and Thompson, 1993; Cecile et al., 1997), an unconformity is present at the base of the Cambrian, in others Cambrian rocks overlie the Upper Proterozoic Windermere rocks conformably (Teitz and Mountjoy, 1989; McMechan, 1990).

Speculations exist that a certain type of zircon links the Windermere Supergroup to Australia (Ross, 1991), but zircons of the same type and age are readily available in the Canadian Shield inside North America. Another idea, that the eastern Cordillera and the supposed ancestral continental margin came into being only around 575 Ma (as modeled by Bond and Kominz, 1984), is also not justified by the geologic evidence (Henderson et al., 1993). Both the Belt-Purcell and Windermere basins were double-sided (see also Aitken, 1993b-c), intracontinental and intracratonic, and similarly two-sided and intracontinental were the early to middle Paleozoic basins that developed from time to time in the platform-miogeosyncline tectonic domain parallel to the eastern Canadian Cordillera (Chapter 3). A new feature in the Windermere Basin was its strong NNW-SSE structural grain, which is typical for the entire Cordilleran mobile megabelt.

On the site of the long-lived Grenville-Appalachian mobile megabelt on the eastern side of the old Laurentian craton (Figs. 2, 48), uneven subsidence began earlier, possibly as early as between 1,750 and 1,500 Ma and more clearly at ~1,300 Ma. As the initial rift-related faults cut the structural grain of the pre-existing craton (Rankin et al., 1993), elongated depressions provided room for accumulation of quartz-bearing sediments (Moore et al., eds., 1986). Thick volcano-sedimentary deposits formed as rifts grew along the incipient orogenic grain. These supracrustal rocks were then brought down to depths of 25-30 km, where they reached their high grades of metamorphism (Martignole, 1986). The main possible formation mechanisms of the Grenville-Appalachian orogen, as discussed by Woussen et al. (1986), could be (a) opening and closure, in a Wilson cycle, of a proto-Atlantic ocean; or (b) orogenized rifting followed by intracontinental compression (cp. Whitney, 1983). Moore (1986) attributed the Grenville orogenic compression to a collision of two continental-crust masses to the east and west of the initial rift. The western mass is preserved in the modern craton, while the eastern mass is now obscured by the Paleozoic Appalachian orogens.

Davidson (1986) noted that the only point of consensus amid all the hot discussions about the Grenville history seems to be that the Grenville belt represents the youngest orogen inside the Canadian Shield. But even that is debatable, because the Grenville belt runs parallel to and underlies the Paleozoic Appalachian-Ouachita mobile megabelt all across the North American continent as far as western Mexico. It was probably created by the first orogenic cycle in the two-stage Grenville-Appalachian mobile megabelt. Orogenic tectonism which affected the initial volcano-sedimentary grain peaked around 1,150-1,140 Ma, although the most common isotopic dates of the Grenville high-grade metamorphism are concentrated around 1,100-1,075 to 970 Ma (the Grenville orogeny; Moore et al., eds., 1986; Doig, 1991). Impressively, Grenvillian formations were thrusted on a vast scale westward and northward over the margin of the intact craton. The huge Grenville front is now traced from eastern provinces of Canada and northern New York state through the Midcontinent region into the Californian Cordillera. Unreworked blocks of Laurentian-craton affinity are well recognized along the outer Grenville fold-and-thrust belt (Fig. 48b). Detailed analysis of zircon grains from clastic formations of Pennsylvanian age in the central Appalachian Basin has revealed a broad scatter of their radiometric ages from 2,100 to just 400-340 Ma. Two distinct clusters (Gray and Zeitler, 1997) lie at 1,200-950 Ma (Grenvillian) and 450-400 Ma (Taconian); pre-Grenvillian dates, ca. 2,100 Ma, also exist.

Regional studies of paleocurrent directions show that provenance areas to the east have always been predominant, where two main episodes of uplift were Late Grenvillian and Late Taconian. Interestingly, there was no significant latest Proterozoic or early Paleozoic metamorphism in the eastern provenance areas. In some influential hypotheses (e.g., Rankin et al., 1993), some of the provenance areas are attributed to the West African craton that in Pennsylvanian time supposedly lay next to eastern North America. If, on the other hand, all the zircon families in this double-cycle megabelt had native sources, they could simply be correlated with the Appalachian and Grenvillian tectonic grain and the pre-Grenvillian (~2,100 Ma) basement.

The Ouachita-Appalachian orogenic belts developed parallel to the Grenvillian ones (Fig. 49), and they generally overlap the Grenvillian structures. Though the antecedent structures were reworked, in places they determined the younger structural patterns. Some synclinoria of Grenville affinity served as initial depocenters during the Taconic tectonic stage, and sediments for them were supplied from uplifted Grenvillian terrains.

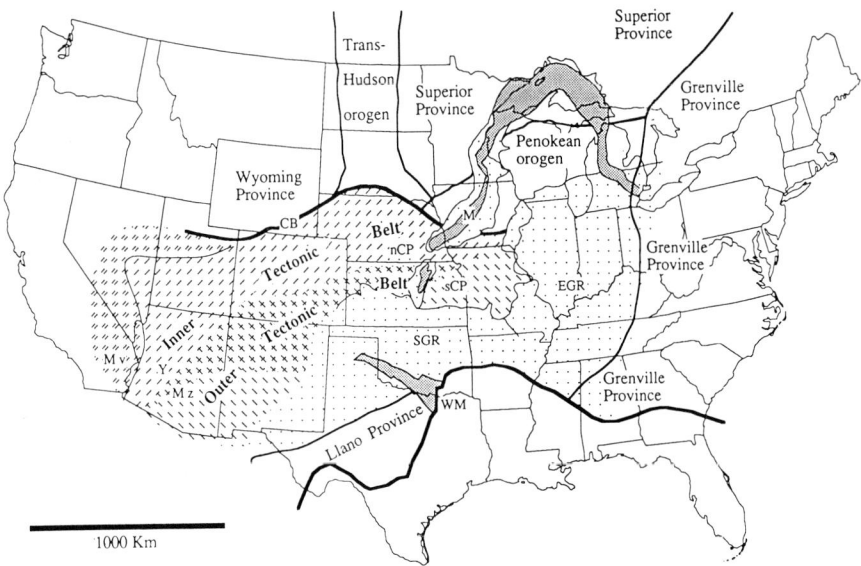

Figure 48a. Early and Middle Proterozoic crustal provinces in central North America (modified from Van Schmus et al., 1993). CB - Cheyenne belt; EGR - Eastern Granite-Rhyolite province; M - Midcontinent rift system; Mv - Mojave province; Mz - Mazatzal province; nCP - northern Central Plains orogen; sCP - southern Central Plains orogen; SGR - Southern Granite-Rhyolite province; WM - Wichita Mountains magmatic province; Y - Yavapai province. Llano Province belongs to the Grenville mobile megabelt. Overlapping patterns indicate the zone of tectonic reworking during the main 1,650 Ma tectonic pulse of the Outer belt.

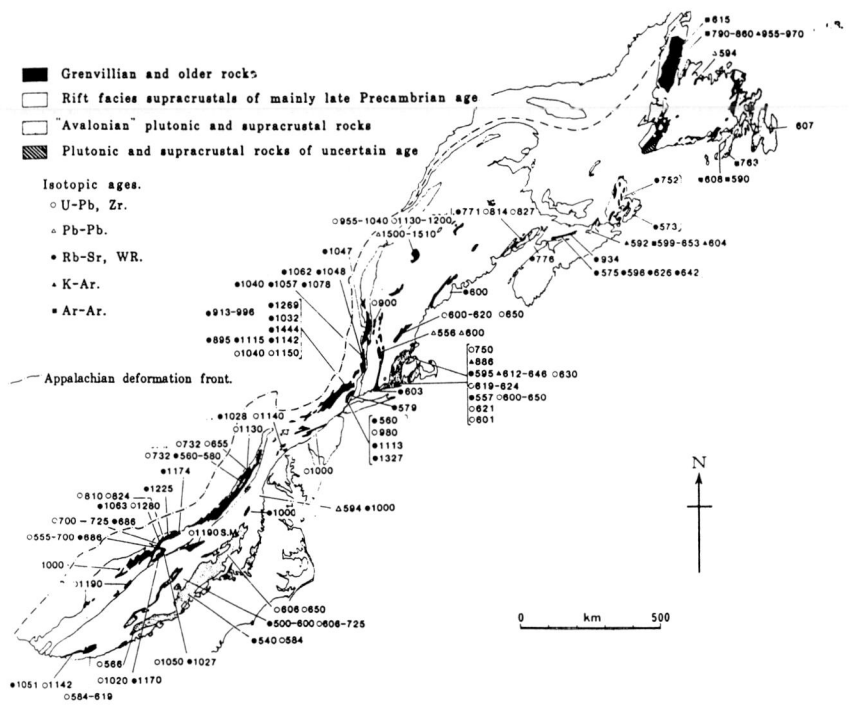

Figure 48b. Exposures of the crystalline basement rocks in the Appalachians, with their isotopic ages (simplified from Powell et al., 1988).

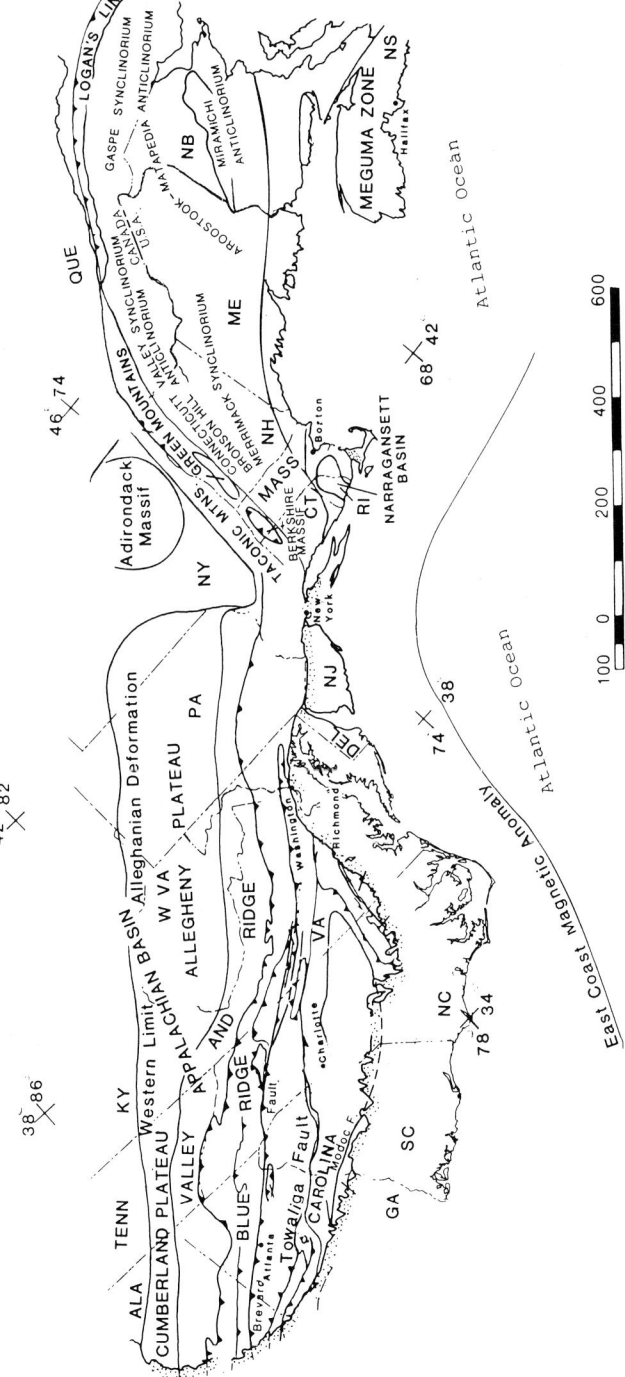

Figure 49. Main tectonic zoning of the Appalachians in the U.S. and Canada (modified from Hatcher, 1989).

Presence of Grenvillian rocks is typical in the Taconic Appalachian anticlinoria (Figs. 50, 51).

The Grenville belts run SSW from Canada into the eastern U.S., and swing to the west in the southern states. There, they are subparallel to the E-W Mazatzal structural trends of the Midcontinent Platform basement. Both the Grenville and Mazatzal trends continue westward into the Cordilleran mobile megabelt, where their structural fragments and geochemical signatures have been widely recognized (Reed et al., 1993).

In the cratonic interior, Late Proterozoic tectonism was expressed in the development of the Midcontinent rift system (van Schmus et al., 1993). A genetic connection of the Midcontinent rifting with the Grenville orogeny is sometimes proposed. But the timing of these events in the two distant regions is not the same. Their approximate age similarity indicates only that as the Grenville megabelt formed, other parts of North America experienced the last of their major Proterozoic tectonic episodes, with different manifestations in the evolving megabelt and the preserved craton. After that, the new craton delimited by the Grenville-Appalachian and Cordilleran megabelts appeared. The vast platformal sedimentary cover developed on it in much more stable conditions than ever before. This new, Phanerozoic sedimentary cover overlapped the older structures that had developed in the pre-cratonic and cratonic megastages, including structures of Laurentian and post-Laurentian affinities.

Tectonic zoning of the interior of the Laurentian craton

The first recognized extensive thermo-tectonic event in North America happened around 2,500 Ma. It was denoted as the Kenoran orogeny. The next continental-scale thermo-tectonic event was probably more restricted in space and time. Reaching its peak around 1,900-1,800 Ma, it was concentrated in some particular regions in Canada: the Churchill province SSW of Hudson Bay, the Bear province west of Great Bear Lake (Stockwell et al., 1970). During this Early Proterozoic, Hudsonian, thermo-tectonic event, the pre-existing proto-craton(s) experienced orogenic reworking whose initial surficial expression was rifting. Several Early Proterozoic orogenic belts are now distinguished (from east to west): Penokean, south of the Superior craton; Trans-Hudson, inside the Churchill province between the Superior and Slave cratons; and Wopmay, west of the Slave craton. They are nearly coeval.

In the 1980s, efforts were made to fit the Precambrian geology of the Canadian Shield and the entire modern North American craton to the plate-tectonic and terrane-tectonic conceptual templates (Hoffman, 1988, 1990). In the currently prevailing ultra-mobilistic interpretation, all Archean blocks are treated as far-traveled "microcontinents" or "suspect terranes" of uncertain origin, which were randomly amalgamated. Ill-recorded collision and subduction episodes are invoked casually, at

171

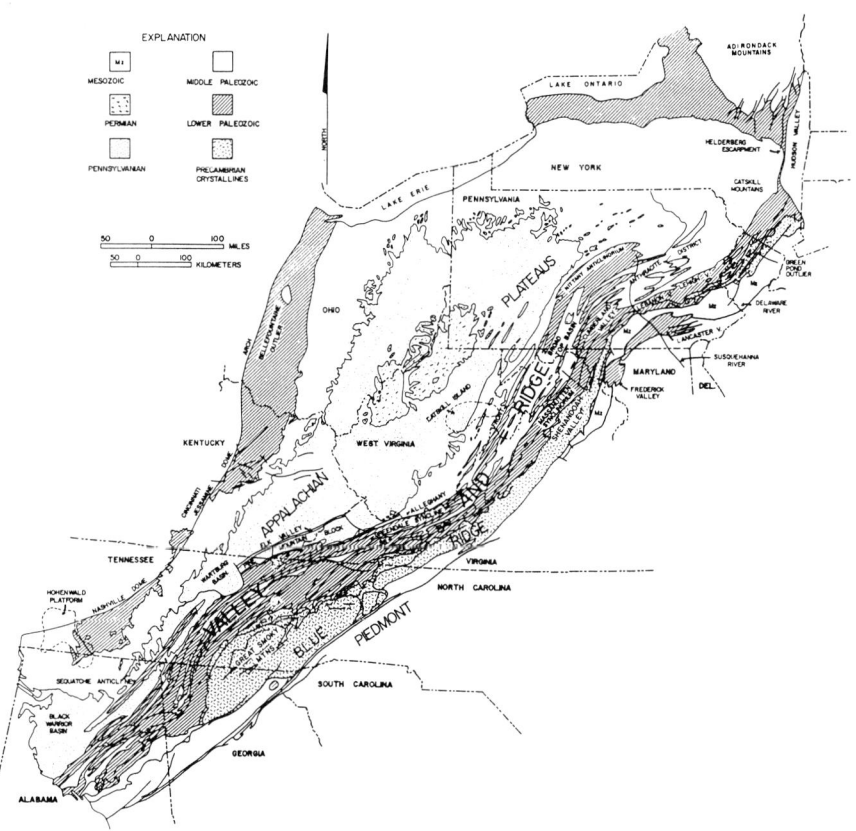

Figure 50. Generalized geologic map of the Appalachian basin region (modified from Milici and de Witt, 1988). NR - Newman Ridge; CM - Chilhowee Mountain; SB - Saltville Mountain.

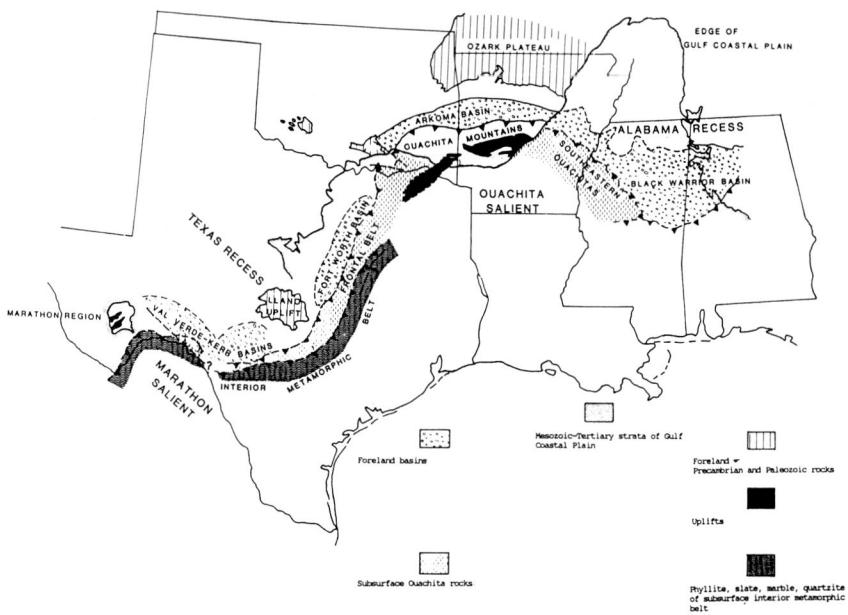

Figure 51. Generalized geologic map of the Ouachita orogenic belt in southern U.S. (modified from Viele, 1989).

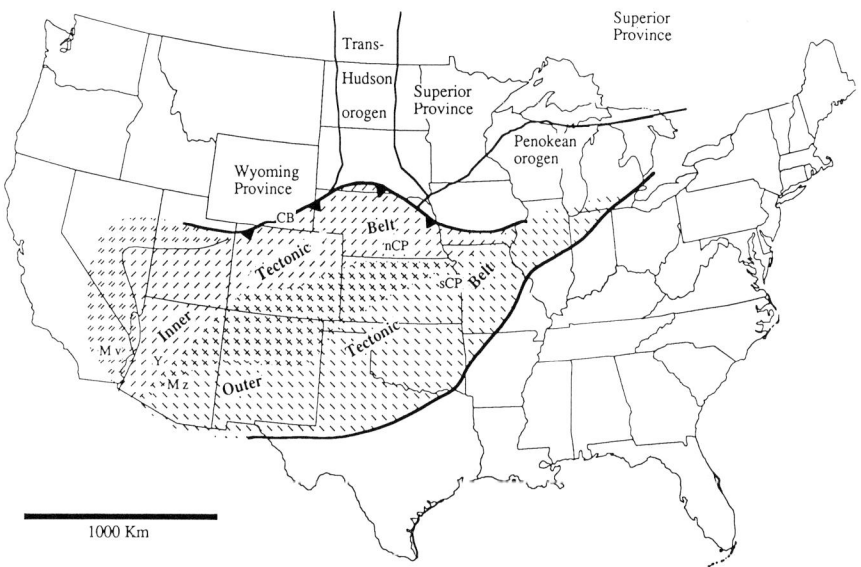

Figure 52. Presumed distribution of Early Proterozoic crustal provinces in central North America prior to the formation of Middle Proterozoic igneous and tectonic provinces (shown in Fig. 48a; modified from Van Schmus et al., 1993). CB - Cheyenne belt; Mv - Mojave province; Mz - Mazatzal province; nCP - northern Central Plains orogen; sCP - southern Central Plains orogen; Y - Yavapai province. Overlapping patterns indicate the zone of tectonic reworking of the Inner tectonic belt during the Outer belt tectonism.

different locations and with different polarities and timing. As the number of probable (and improbable) speculative scenarios grows, they do not become any more compelling. Recently, Zwanzig (1997, p. 90) acknowledged that in the so-called Trans-Hudson Orogen, assumed collisional belts "...fail to provide directly the polarity of subduction or configuration of early tectonic plates". In response, Ansdell et al. (1997, p. 91) agreed about "the dangers of inferring past plate configurations and subduction polarities", though they still believe a "need" exists "to compare [such paleotectonic events] to relevant modern analogues". Yet, ancient and modern tectonics are not simply analogous.

The currently designated "domains" or "terranes" in the Trans-Hudson Orogen (Cree Lake and Reindeer zones, "Western craton", Hearne and Rae platforms; Fig. 53) are mostly parts of the Archean craton(s) reworked to various degrees by ductile folding, intense faulting, metamorphism, extensive K-metasomatism, etc. (Fig. 54). Whether the Archean Slave, "Western", Sask (Saskatchewan), Wyoming and Superior cratons were previously separated or not remains a matter of speculation, there being no sufficient factual substantiation (Hoffman, 1988, 1990; Hajnal et al., 1996; Lewry et al., 1996). Petrological data are limited and inconclusive (e.g., Lewry and Collerson, 1990 vs. Ross et al., 1991), geochemical data permit multiple solutions (e.g., Collerson et al., 1990), paleomagnetic data are dubious (as discussed in previous chapters), seismic data interpretations are in dispute (Hajnal et al., 1996 and references therein). Archean cratonic blocks distinguished between the Superior and Slave, and Superior and Wyoming cratons (*sensu lato*) - Sask, Dakota - are big and continue into the basement of Early Proterozoic platforms and median massifs of the Trans-Hudson Orogen (e.g., Kisseynew block; see below).

The Lower Proterozoic sedimentary cover of paleo-platforms that have Archean basement and are attached to Archean shields is up to 8-10 km thick and consists of supracrustal metavolcanic and metasedimentary rocks, now metamorphosed. The most extensive are the Huron Supergroup and the Amer and Hurwitz groups. The Huron volcano-sedimentary assemblage is common in the east, in the Superior craton (*sensu lato*); the Amer and Hurwitz groups are best represented to the south of the Slave shield (which by itself is denoted commonly but incorrectly as a craton). These Early Paleoproterozoic (usually, 2,500 to 2,200 Ma) stratal rocks were deposited before and partly during the evolution of the Trans-Hudson Orogen (Figs. 55a-b). North of the area included in this orogen in Saskatchewan, protoliths comprising arkoses, quartzites, quartz-pebble conglomerates and so forth are combined into the Wollaston Basin (Delaney et al., 1996). Coarse-grained siliciclastic composition and great thickness of these rocks imply a high relief in the cratonic provenance areas and considerable subsidence of the platform. Paragneisses formed from the sedimentary protoliths, and subordinate mafic gneisses have volcanic protoliths intruded by huge granitic bodies produced by partial remelting of Archean basement rocks. This entire metamorphosed

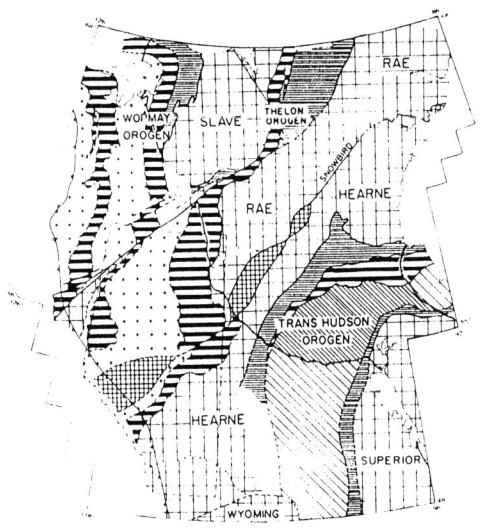

Figure 53. Distribution of Archean and Early Proterozoic tectonic provinces in central and western Canada and northern U.S., following the designation of Hoffman (1988). The reader is encouraged to review this and following diagrams in this chapter before reading further, to familiarize himself/herself with localities and place names that will be mentioned in the text.

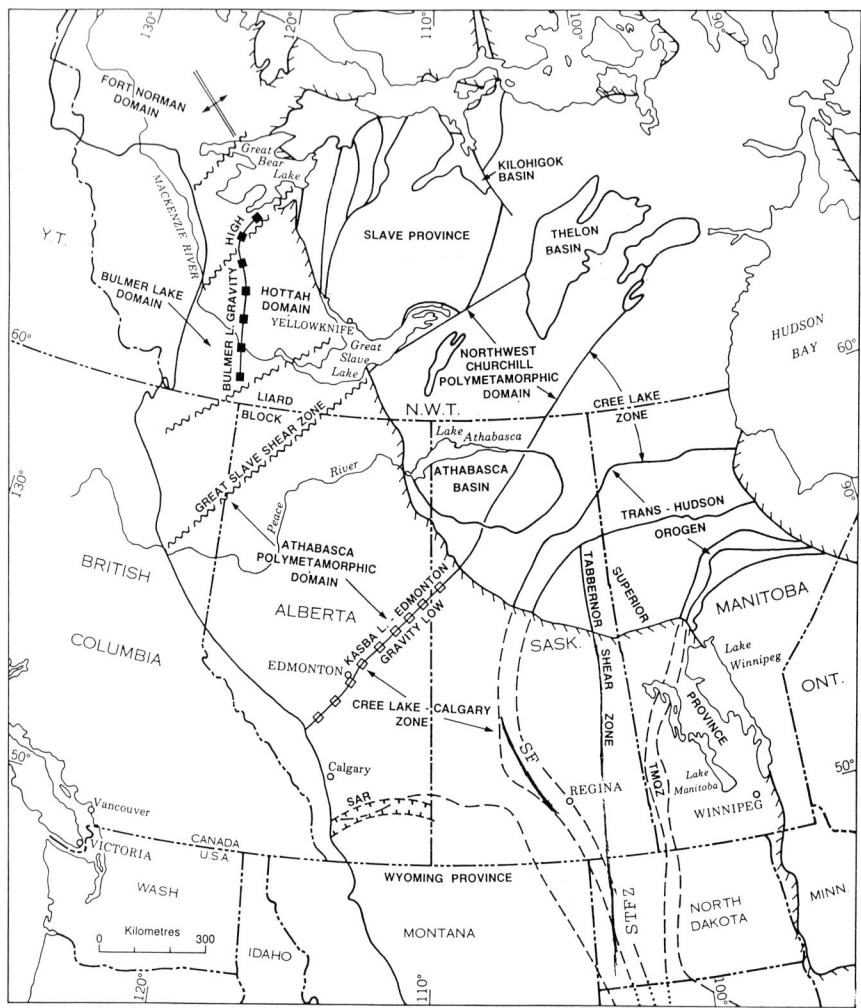

Figure 54. Tectonic provinces in the Precambrian cratonic crystalline basement of the Western Canada Sedimentary Province (simplified from Burwash et al., 1993). SAR - Southern Alberta zone of gravity lows; SF - Saskatoon fault; STFZ - Saskatoon-Tabbernor fault zone; TMQZ - Thompson Magnetic Quiet Zone.

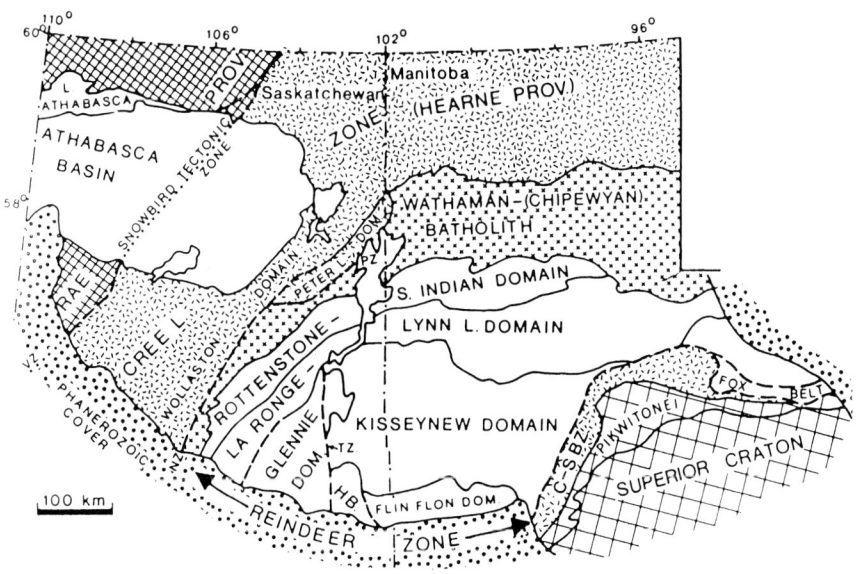

Figure 55a. Index map of the tectonic zonation in the Early Proterozoic Trans-Hudson Orogen (Reindeer zone) and surrounding Archean cratons in the Canadian Shield (modified from Bickford et al., 1990; Meyer et al., 1992). C-S BZ - Churchill-Superior boundary zone; NZ to PZ - Needle Falls-Parker Lake shear zone; TZ - Tabbernor fault zone; VZ - Virgin River fault zone (part of the trans-cratonic Snowbird-Virgin River fault system). The reader is encouraged to review this and following diagrams before reading further, as the names in them are used repeatedly in the text. Geographic index maps of western Canada are presented in Figs. 6a-b.

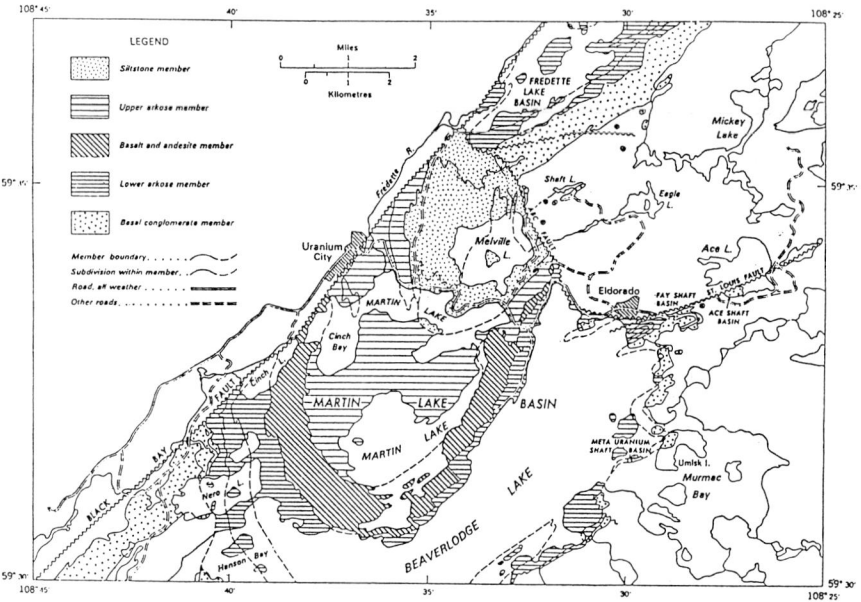

Figure 55b. Geologic map of the Proterozoic Martin Basin (modified from Tremblay, 1972). It is too small to appear in Fig. 55a, but is used here for comparison with the Athabasca Basin to the south.

A.

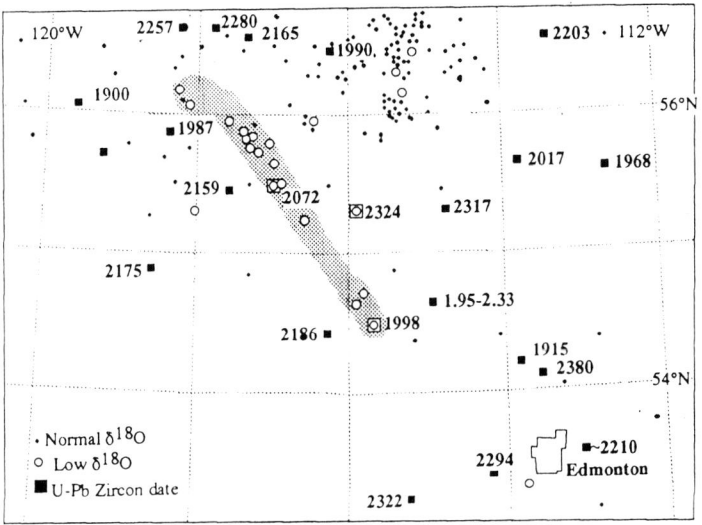

B.

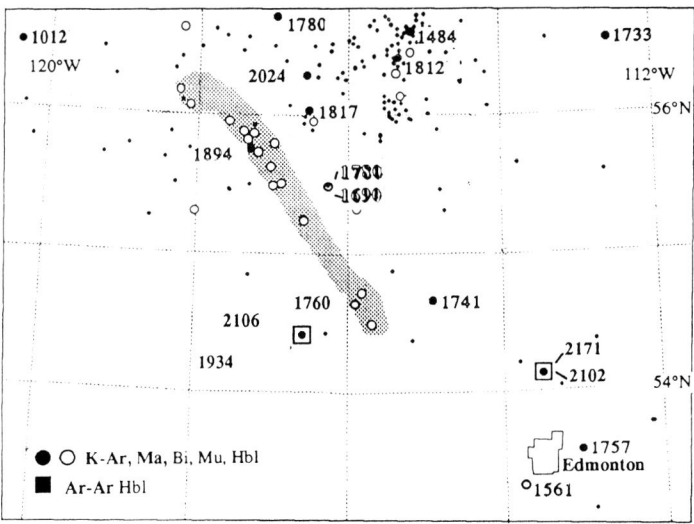

Figure 56. Position of the Kimiwan isotopic anomaly, age ~1,760 Ma, in the cratonic crystalline basement of the Alberta Platform (modified from Chacko et al., 1995). This anomaly seems to have formed by reheating of this zone in the cratonic crust. The NNW-SSE elongation of this zone of crustal reworking makes it the earliest known tectonic precursor of the Cordilleran mobile megabelt in western Canada.

complex of the Wollaston Basin, or domain, or belt (and the coeval Seal River Domain), exposed commonly southeast of the Late Proterozoic Athabasca Basin and east of the Phanerozoic Alberta Basin (Macdonald, 1987), has been assigned to the Hearne Platform (Hoffman, 1988). In various hypotheses, the Wollaston domain has also been interpreted as a miogeosyncline related to a cratonic rift, or as an Early Proterozoic continental-margin wedge.

South of the Wollaston domain lies the huge Wathaman-Chipewyan batholith, more than 900 km long and up to 150 km wide. It is made up of fairly monotonous granitoids, commonly megacrystic. Their age of 1,865 to 1,850 Ma slightly predates their deformation, which occurred at ~1,830 Ma (Meyer et al., 1992). In plate-tectonic terms, this batholith has been interpreted as an Early Proterozoic continental arc related to a subduction zone (cp. also Lewry and Sibbald, 1980; Gordon et al., 1990; Lewry and Collerson, 1990). Meyer et al. (1992) compared it with the modern Andean-type magmatic-arc batholiths, and stated (their p. 1083) that "the Wathaman batholith separates a miogeosynclinal domain from eugeosynclinal terranes..." and that immediately to the south an "...oceanic La Ronge arc formed outboard of the Churchill craton". However, the large Archean cratonic blocks there, Saskatchewan (or Sask), Kisseynew, Dakota, as well as the Wathaman-Chipewyan batholith itself, do not leave enough room for oceanic-crust areas, subduction zones, or accreted terranes (the La Ronge belt is in fact narrow, only ~50 km in width, though geochemically it is indeed juvenile). Yet, the rapakivi-textured granites in the Wathaman-Chipewyan batholith point to a quiescent rather than active-margin tectonic regime.

To the south, several exposed Archean blocks - Glennie, Peter Lake - have been mapped in the N-S-oriented branch of the Churchill or Trans-Hudson province. Their presence indicates that the Archean basement might continue under the Wollaston Basin and under the terrains of gneisses and schists of the Reindeer zone (or the Trans-Hudson Orogen). In any case, a tectonic setting in which the La Ronge belt formed seems more mediterranean than open-oceanic.

South of the Slave shield, the Rae and Hearne platforms contain an Early Proterozoic volcano-sedimentary cover of large thickness. All rocks in these platforms are metamorphosed and deformed. The structural, metamorphic and magmatic phenomena of the Hudsonian orogeny (imbricated folds, progressive and retrogressive metamorphism, formation of anatectic granites) are found all over the region between the Archean Slave and Superior shields. Meyer et al. (1992, p. 1084) acknowledged that the "internal structure of the [Wathaman-Chipewyan] batholith, depth of emplacement [estimated to be as much as 17 km], average rock composition, and lack of mafic facies are problems that must be considered in designing further studies".

The nature of Archean tectonism remains poorly understood (Kerr, 1991), but it was different from that in later time, even in Early and Middle Proterozoic (cp. Hamilton, 1993). A new, sober tendency in analyzing Precambrian tectonics is to take more notice of changes in the general conditions and processes in the evolving Earth. The usefulness of modern tectonics for studies of the Precambrian is less than full, and the farther back we go in time, the more tenuous such a uniformitarian approach becomes. Tectonic processes could have been repeated in a more or less similar fashion since the Proterozoic, but during their long evolution, the continental crust and lithosphere have changed progressively and in many ways irreversibly (Armstrong, 1991; Rudnick, 1995).

The Hudsonian orogeny led to further cratonization of the continental crust in North America. The product of this cratonization was the first post-Archean craton, called Laurentian. It incorporated the variously reworked and unreworked remnants of previous Archean cratonic masses and the cratonized parts of Early Proterozoic mobile megabelts. In Canada, this continental-scale event was followed by widespread differential crustal upwarping and downwarping. The first volcano-sedimentary basins were formed on the site of the Rae and Hearne platforms, where downwarping took place as early as in the Paleoproterozoic. The Wollaston Basin is an example of these early cratonic platformal basins. Many Early Proterozoic volcano-sedimentary basins are related to vertical displacement of blocks in the areas of the Trans-Hudson Orogen. During or soon after the orogenic Hudsonian time, with a transition to a cratonic regime, the new generation of basins developed over the old cratonic and newly cratonized crust.

The Archean Superior, Slave and Wyoming cratons, and the Early Proterozoic Penokean, Trans-Hudson and Wopmay orogens are recognized in the Canadian Shield and platformal basement of the North American craton as lateral tectonic units. The Trans-Hudson, Wopmay and Penokean orogens were likely roughly coeval (Hoffman, 1988, 1990), though the latter may be slightly younger (~1,860 Ma; Sims et al., 1993). They may be viewed as parts of a single mobile-megabelt system of the Paleoproterozoic North American continent. The Archean cratonic crust was tectonically relatively stable in comparison with these more-mobile crustal features. The appearance of the first shields (Superior, Slave, Wyoming) and associated platforms (Hearne, Rae and others) marked a differentiation of the cratonic crust of the North American continent. The Archean cratons, shields and platforms, and the newly cratonized crust in the extinct Early Proterozoic orogenic regions, together comprised the post-Hudsonian late Paleoproterozoic to Mesoproterozoic Laurentian craton.

A new tectonic differentiation began almost immediately after the Laurentian craton had come into being. In the east, around 1,453-1,445 Ma, strong orogenic contractional deformation affected the Early Proterozoic Huron Supergroup in the eastern part of the Canadian Shield (Fueten and Redmond, 1997). Subsidence on the site of the future

Grenville Orogen commenced at ~1,700 Ma and was especially dramatic at ~1,300 Ma. The ~1,100-1,000 Ma Grenvillian orogeny overprinted older tectonic structures; it culminated in the formation of the Grenville Orogen, which was in turn overprinted by the Phanerozoic orogenies in the Appalachian-Ouachita orogenic belts.

In the west, parallel to the future Cordillera, the Kimiwan isotopic anomaly of ~1,750 Ma age (Chacko et al., 1995) is now correlated with a large NNW-SSE-trending structural zone of crustal reactivation delineated in the basement of the western Alberta Platform (Burwash et al., 1996; Fig. 56). It overprinted different older domains recognized in the basement. This first, Kimiwan precursor of the Cordillera later shifted to the west, to the position of the Neoproterozoic Windermere rift initiated at ~780 Ma, which probably marked the beginning of the Cordilleran mobile megabelt.

South of 52°N, the main branch of the Trans-Hudson Orogen is marked by two parallel N-S zones of total-field magnetic anomalies collinear with the Saskatoon and Tabbernor faults. This band of anomalies and faults runs southward beneath the Phanerozoic sedimentary cover, but NNE and E-W anomalies correlatable with those in the Thompson fault zone and the Superior craton interfere with the Trans-Hudson anomaly pattern (Fig. 57). In South Dakota, the Trans-Hudson Orogen is terminated by the E-W trends of the Mazatzal orogen. Only an eastern belt of positive magnetic anomalies seems to continue farther, swinging SE around the western edge of the Superior craton (cp. Zietz et al., 1982; Bickford et al., 1986; Lyatsky et al., 1998).

A Mesoproterozoic episode of widespread magmatism (since ~1,540 Ma) has been recorded elsewhere in the western and central parts of the Laurentian craton under the Phanerozoic cover (Reed et al., eds., 1993; Fig. 58). In Wyoming and Montana, this episode was also marked by a radical rearrangement of isotopic systems in the crystalline basement, including whole-rock Sb-Sr re-equilibration (Redden et al., 1990). Another similar event, at ~1,370 Ma, affected a big part of the western Central Plains in North America, reaching as far south as Texas, New Mexico and Oklahoma. Modern data make it apparent that the region of this peculiar granite-rhyolite magmatism included the Midcontinent far beyond the preserved Early Proterozoic supracrustal cover. Commonly, the plutons and flows of these igneous suites are contaminated by Archean crustal material. The Midcontinent granite-rhyolite magmatism was very durable and diachronous: from around 1,750 Ma in Wisconsin to around 1,100 Ma in Texas (Bickford, 1988; Van Schmus and Bickford, 1993). The variety of Early and Middle Proterozoic tectonic episodes recorded from area to area resulted from the variability of local crustal properties and specifics of tectonic regimes. In general, the rocks attributed to this huge province are unmetamorphosed and undeformed, except for faults that accommodated the movements of basement blocks. These rocks and blocks are non-orogenic.

183

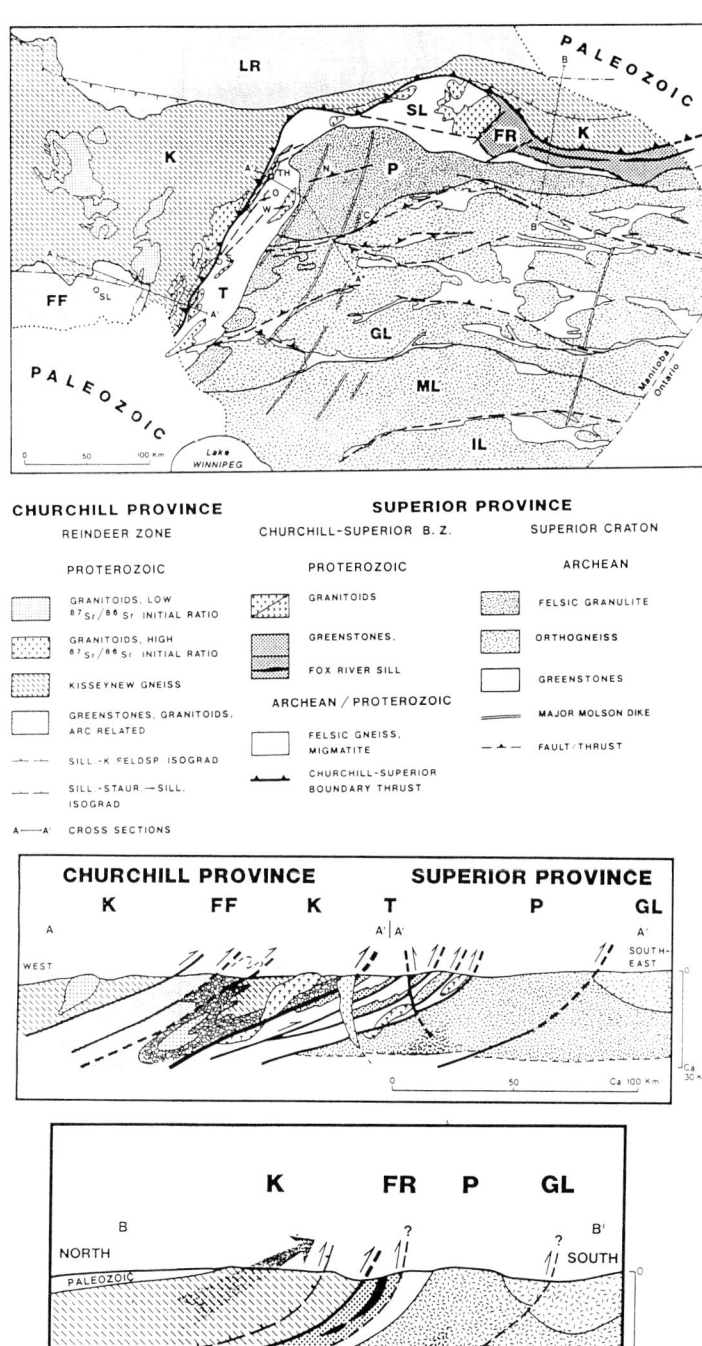

Figure 57. Structure of the Thompson tectonic belt at the boundary between the Churchill and Superior provinces (modified from Weber, 1990). Thrust faults are mapped at the surface, but their depth of detachment in the crust is not known. FF - Flin Flon domain; FR - Fox River domain; K - Kisseynew block; GL - Gods Lake domain; IL - Island Lake domain; LR - Leaf Rapids domain; ML - Molson Lake domain; P - Pikwitonei domain; SL - Split Lake domain; T - Thompson belt.

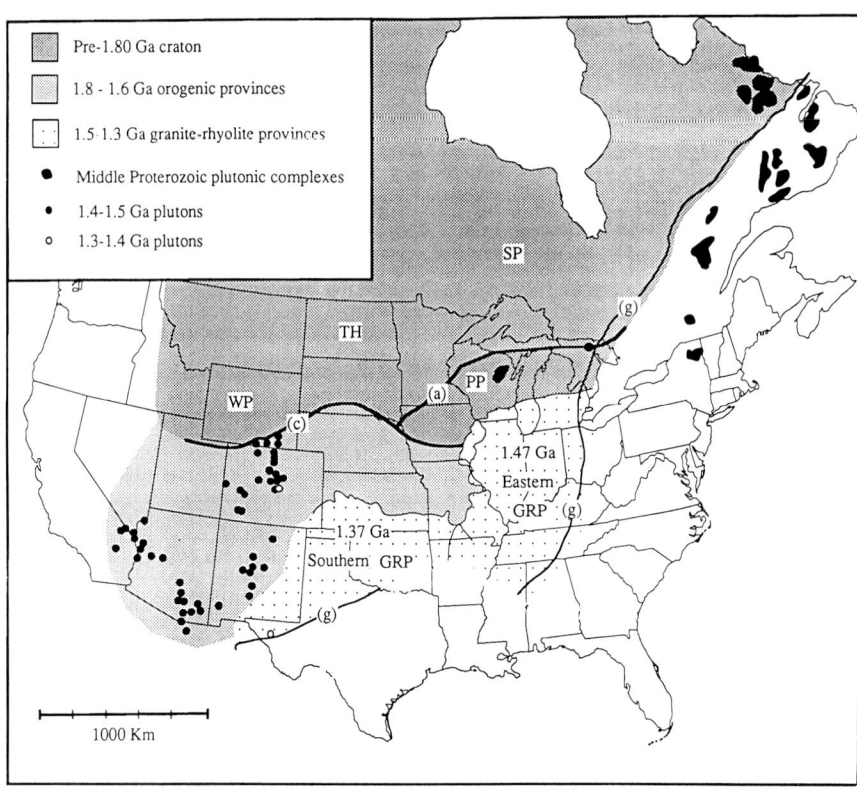

Figure 58. Distribution of Middle Proterozoic granite-rhyolite provinces in the central part of North America (modified from Van Schmus et al., 1993). GRP - granite-rhyolite province; PP - Penokean orogenic domain; SP - Superior craton; TH - Trans-Hudson Orogen; WP - Wyoming craton; (a) - presumed edge of the Superior craton in the Great Lakes area; (c) - Cheyenne orogenic belt; (g) - front of the Grenville megabelt.

Tectonic stages in the Proterozoic cratonic evolution of the North American continent were accompanied by warp and block movements of the crust which were of higher amplitude than today, as evidenced by the abundance of high-energy clastic sedimentary rocks, large thickness of stratified rocks, and extensive magmatism. The early volcano-sedimentary basins contained abundant volcanics and were cut by numerous dikes of several generations. Early and Middle Proterozoic time also saw voluminous magmatism in the form of flood basalts, extremely long and broad dike swarms (Fig. 59) and extensive layered intrusions (e.g., Friedman, 1955, 1957). One of the first such swarms of diabase dikes in the Canadian Shield is ~2,450 Ma in age (e.g., Halls and Palmer, 1990; Heaman, 1997). It recorded a widespread extensional cratonic tectonic regime, in which many deep fractures were opened and filled with abundant large dikes. There were several generations of such dike magmatism in the Early and Middle Proterozoic, suggesting that this tectonic regime recurred. Around 1,270 Ma, for example, it spread over much of the Laurentian craton.

Unlike the dispersed cratonic extension that had produced the dike swarms, localized extension along narrow whole-crust weakness zones that created rifts became common mainly later. In North America, the best known example of continental-scale rift system is the Late Proterozoic Midcontinent Rift, which cuts the basement across the Midcontinent Platform (Fig. 60).

Families of extension-related dikes appeared both before and after formation of the Kapuskasing Structural Zone in the Superior province around 1,930-1,920 Ma (Fig. 61). This structural zone contains a west-dipping reverse fault, called Ivanhoe Lake Cataclastic Zone, and Archean crustal blocks tilted and upthrown along it (Percival and Card, 1983). A ~20-km section of Archean crust is exposed in these blocks, bringing to the surface rocks that had been at mid-crustal depths. This indicates that localized compression alternated with dispersed or localized extension, induced by inherent deep-seated processes in the cratonic lithosphere and crust.

Warp and block tectonic movements were also common in the Proterozoic. Movements of many lateral tectonic units in the Proterozoic crust were large, up to tens of kilometers even in cratonic areas, as evidenced by the thickness of mapped basins and the grade of burial alteration and metamorphism of their deep horizons, as well as from fission-track studies. The oldest, Early Proterozoic cratonic basins in North America, such as Wollaston, are deeply metamorphosed, whereas later ones, such as Athabasca, are metamorphosed only slightly. Even coeval Proterozoic basins were disconnected at the time of their development, and crystalline-rock terrains in broad arches lay between them. Absence of a continuous cover is further evidence that vertical crustal movements in the Proterozoic craton were very intense. Only in the Late Proterozoic did the new tectonic regimes differentiate the barren shields from the vast new platforms, where a continuous and predominantly sedimentary cover was laid down.

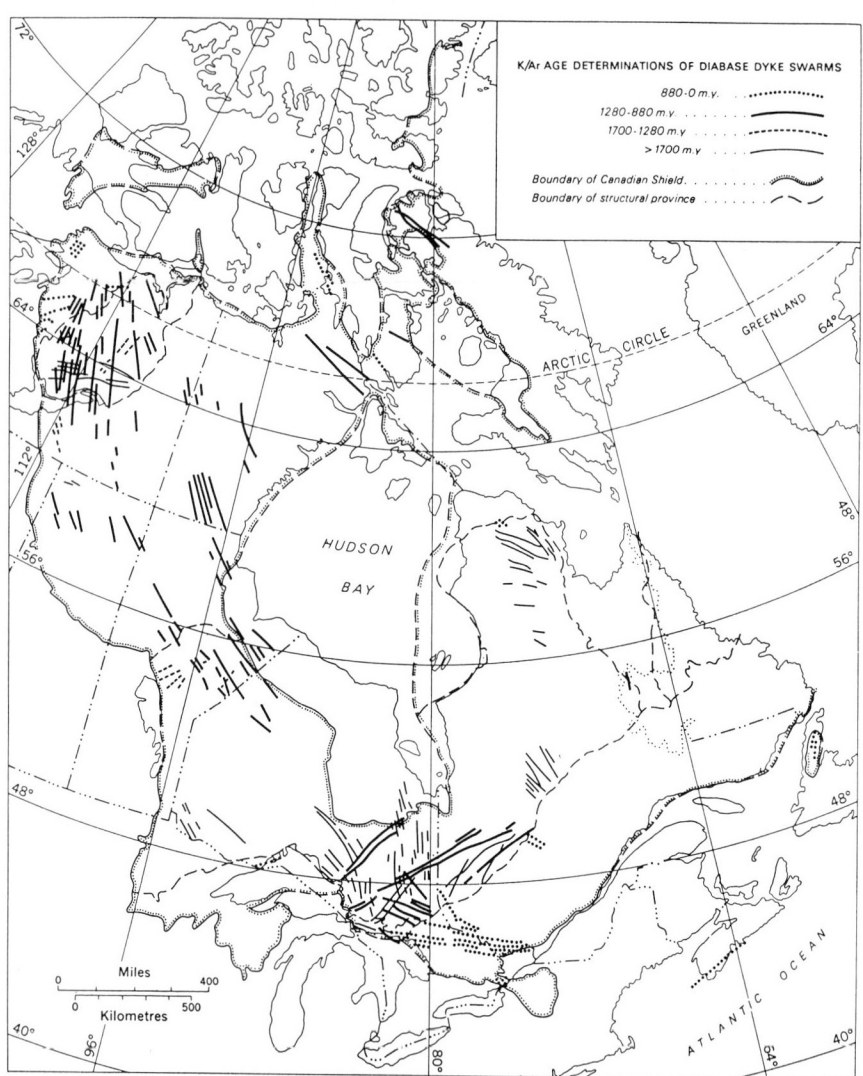

Figure 59. Precambrian continental-scale diabase dike swarms in the Canadian Shield (simplified from Stockwell et al., 1970).

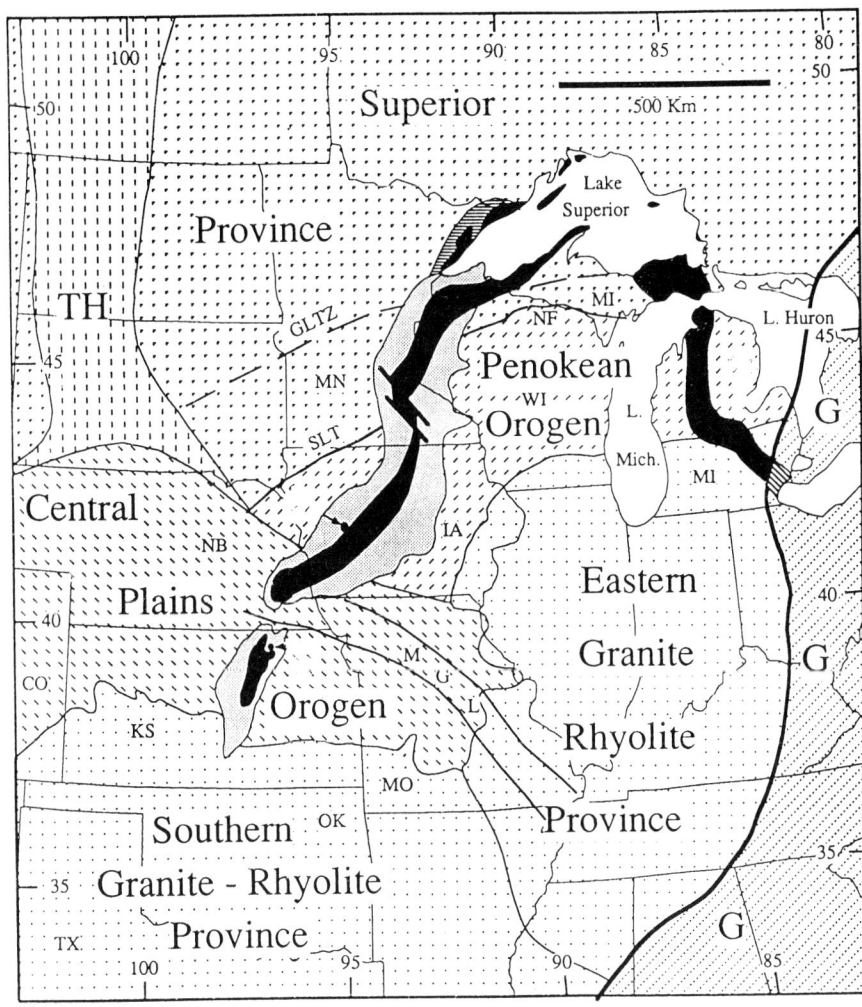

Figure 60. Generalized geologic map of the Midcontinent rift system and surrounding tectonic provinces (modified from Van Schmus et al., 1993; continental-scale location map in Fig. 48a). GLTZ, NF, SLTZ, MGL are major known or suspected faults and zones of crustal weakness in this region; G - Grenville megabelt. Black marks the part of the Midcontinent rift system where mafic igneous rocks are the most common; stippled areas are Proterozoic basins postdating the rift, filled mostly with clastic rocks; horizontal rule west of Lake Superior is the Duluth mafic-ultramafic body. Location of the gravity anomaly zone associated with the Midcontinent rift system is shown in Fig. 88.

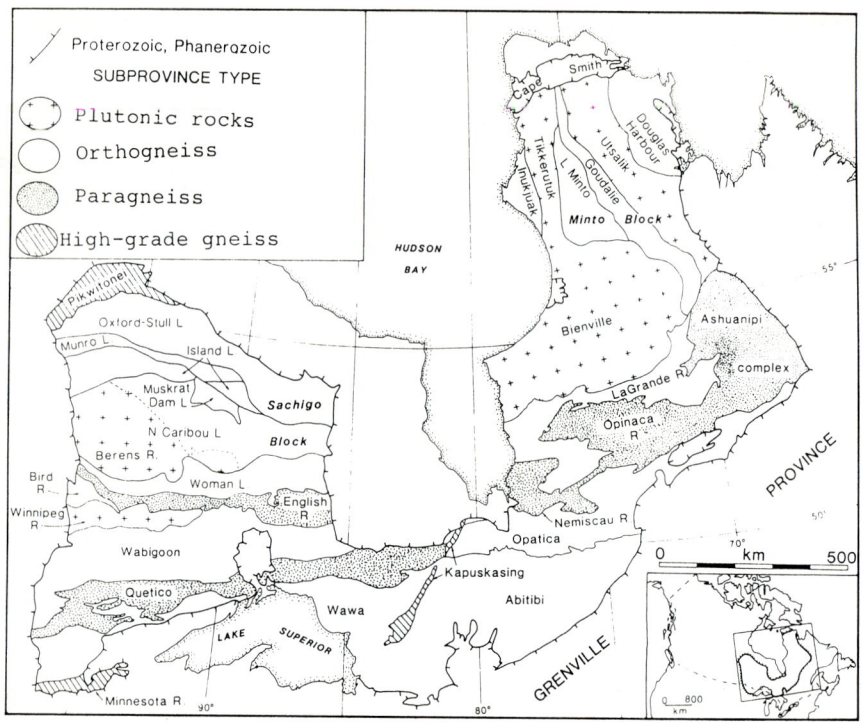

Figure 61. Generalized geologic map of the Superior craton, showing its principal subprovinces and tectonic belts (modified from Percival et al., 1992).

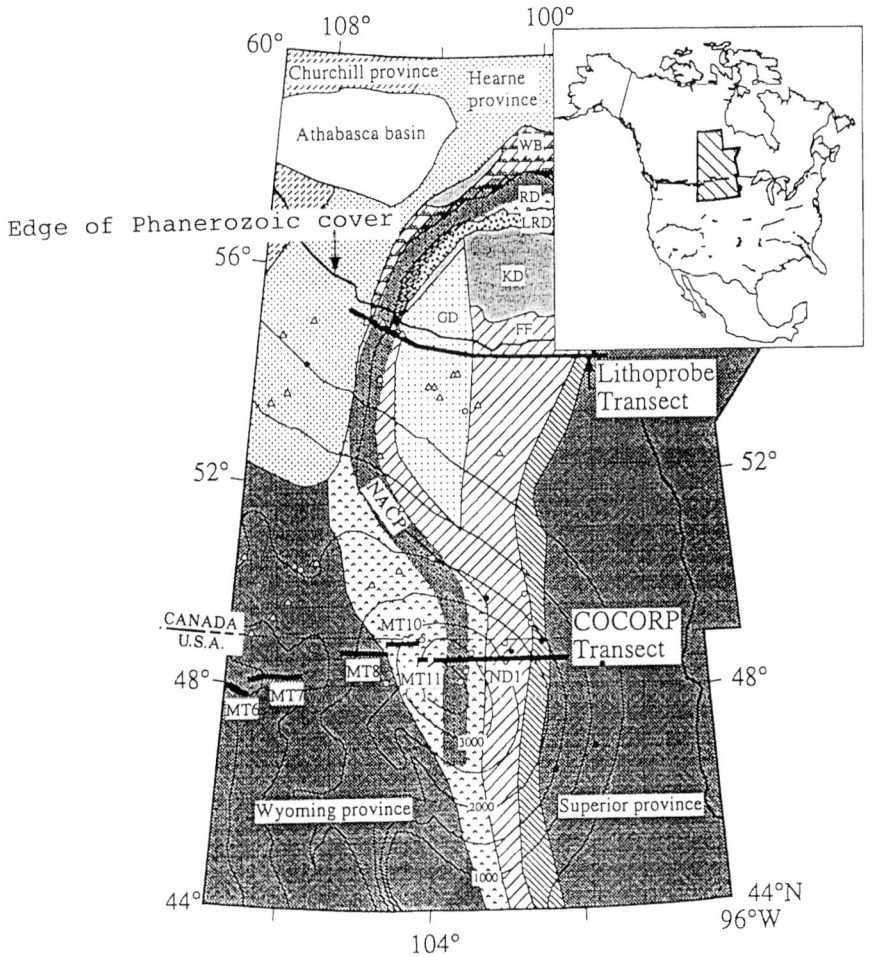

Figure 64. Location of the North American Central Plains (NACP) electrical conductivity anomaly, and of the Lithoprobe and COCORP deep crustal seismic transects across the Trans-Hudson Orogen in Canada and the U.S. (figure modified from Baird et al., 1996; see also Jones et al., 1993). GD - Glennie block; FF - Flin Flon domain; KD - Kisseynew Basin; LRD - La Ronge domain; RD - Rottenstone domain; WB - Wathaman-Chipewyan batholith. Contours, in meters, show thickness of the Phanerozoic sedimentary cover in the Williston Basin. Drillhole data: solid circles - Archean U/Pb ages; open circles - Proterozoic U/Pb ages; triangles - uncertain ages (see also Fig. 66b; Collerson et al., 1990).

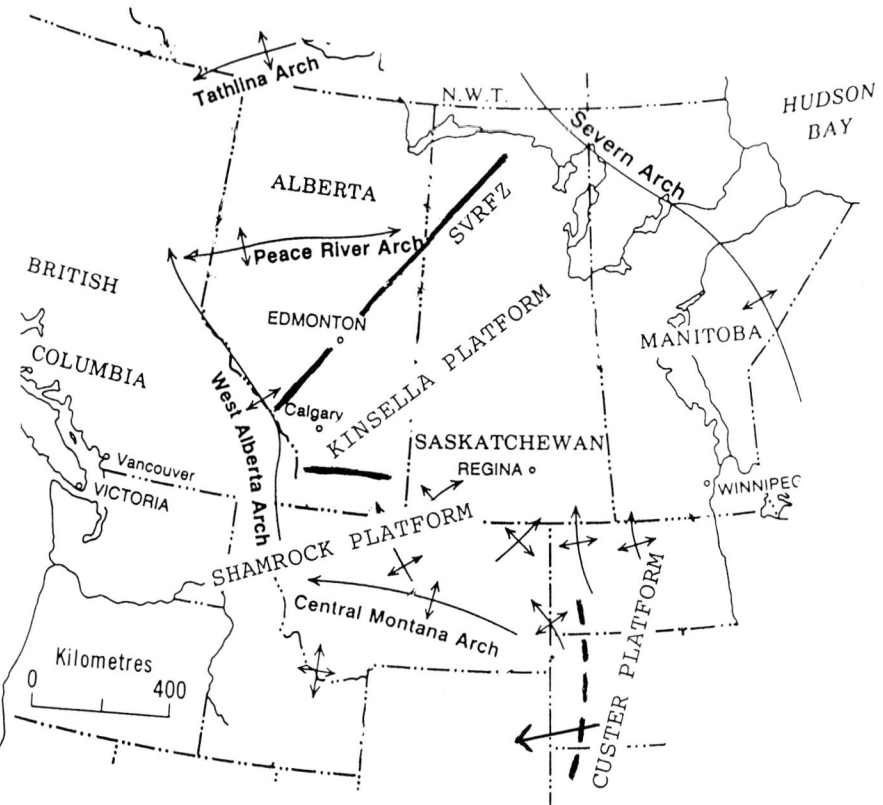

Figure 65. Distribution of Early Proterozoic platforms in the modern western North American craton, as designated in this book. Heavy lines mark crustal weakness zones that define platform boundaries. SVRFZ - Snowbird-Virgin River fault zone. These now-metamorphosed ancient platform rocks, and the Archean crystalline rocks in their basement, now form the basement of the Interior and Midcontinent cratonic platforms. Principal Phanerozoic basement arches (cp. Podruski et al., 1988, their Fig. 12) are shown for comparison.

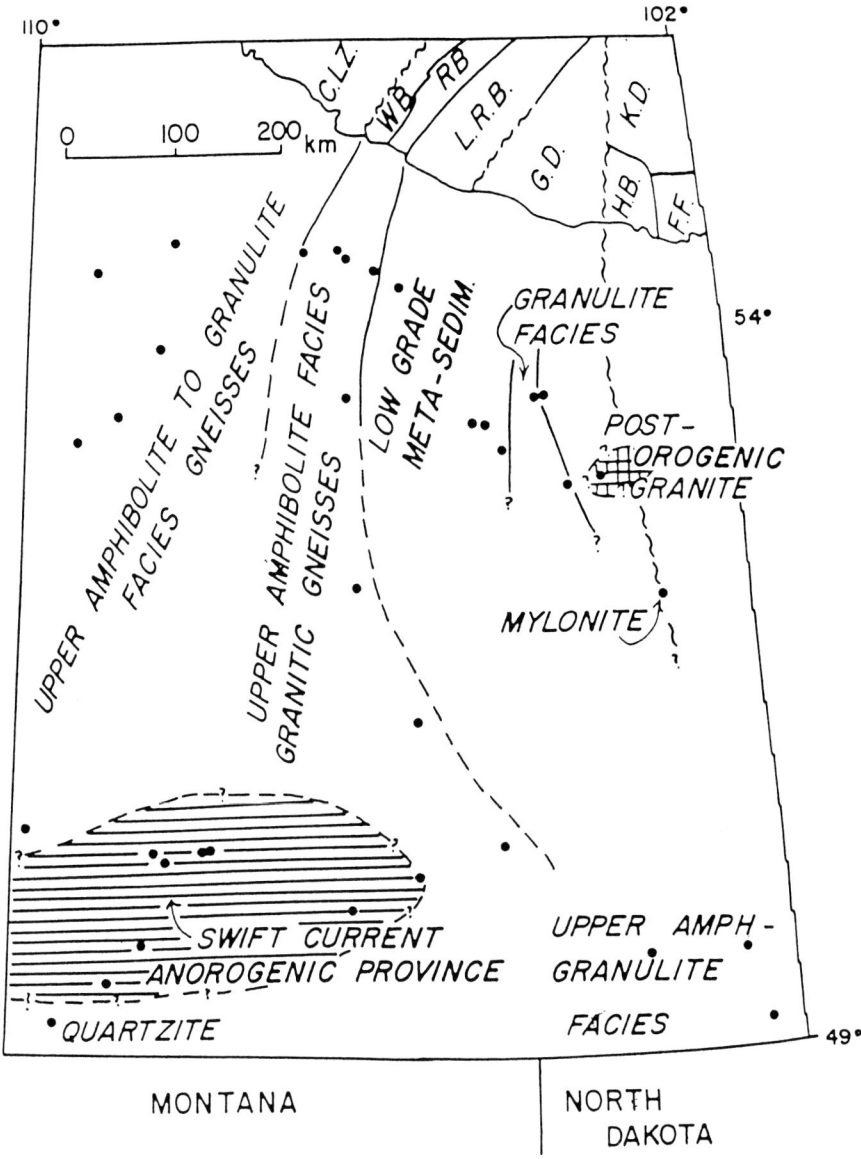

Figure 66a. Variations in metamorphic grade and lithology from crystalline-basement cores beneath the Williston Basin in Saskatchewan (modified from Collerson et al., 1990). CLZ - Cree Lake zone; FF - Flin Flon domain; HB - Hanson Lake block; LRB - La Ronge domain; KD - Kisseynew Basin; RB - Rottenstone domain; WB - Wathaman-Chipewyan batholith.

194

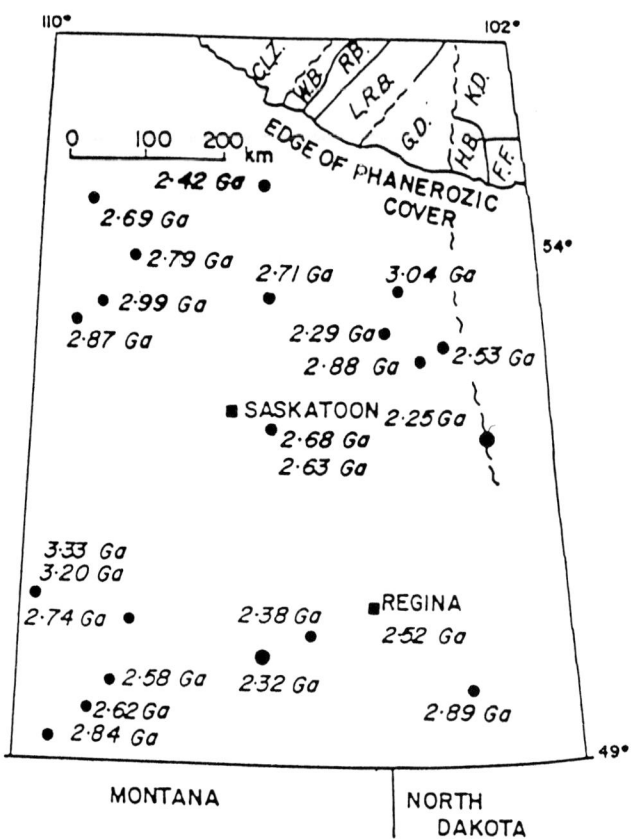

Figure 66b. Sm-Nd ages of crystalline rocks beneath the Williston· Basin in Saskatchewan (modified from Collerson et al., 1990). CLZ - Cree Lake zone; FF - Flin Flon domain; HB - Hanson Lake block; LRB - La Ronge domain; KD - Kisseynew Basin; RB - Rottenstone domain; WB - Wathaman-Chipewyan batholith.

Johnson et al. (1984) have modeled the edge of the Archean rocks 20 to 50 km south of the northernmost mapped Proterozoic outcrops, with a down-to-the-south step of 5 to 8 km. No Archean petrological or geochemical signatures are known in areas further south, where Middle Proterozoic orogens run E-W (Bickford, 1988; Van Schmus and Bickford, 1993).

In the Black Hills metamorphic terrains in western South Dakota, the Early Proterozoic rocks are more readily observed to overlap the Archean basement. Archean rocks, probably of the Wyoming craton, are recognized as sources for sediments of the Lower Proterozoic volcano-sedimentary cover (Redden et al., 1990). S-type granites were emplaced into the platformal cover after its metamorphism and folding, at ~1,715 Ma. A broad (more than 5 km) thermo-metamorphic aureole is recognized in a domal structure caused by a big stock, suggesting that the thermo-metamorphic episode at that time could have been unusually strong. From sedimentological, geochronological and isotopic studies, Redden et al. (1990) drew important conclusions that (a) the Archean-Proterozoic crustal structure in the Black Hills area is two-layered (similar to the Proterozoic platforms discussed above); and (b) the Proterozoic granites were probably produced by melting of the Archean basement as well as, to a lesser extent, of the Proterozoic supracrustal cover.

The least well defined is the northern boundary of the Wyoming craton (Bickford, 1988, his Fig. 12; Reed et al., eds., 1993), which is hidden under the Phanerozoic cover in the Montana and Alberta Plains and in the eastern Cordillera. Archean dates have been obtained in some samples as far west as the Idaho Batholith, from its most tectonized parts. Presence of a deep N-S crustal zone of weakness, active in the Proterozoic and rejuvenated later, is suggested there. In southwestern Montana, the oldest dated rocks are 3,300 to 2,600 Ma (Mueller et al., 1993), similar to the ages of some basement samples in Saskatchewan (Collerson et al., 1990).

The Hearne and Custer platforms differ in the age of their most abundant granitic magmatism. Lewry and Collerson (1990) proposed that the huge Wathaman-Chipewyan Batholith and the related smaller monzonite-granodiorite plutons were emplaced mainly in Hudsonian time, at about 1,850 Ma, but Halden et al. (1990, p. 201) noted that this huge batholith "cannot be unequivocally assigned to any one tectonic setting". Granitic magmatism in the Black Hills area, which resulted also from anatectic granitization of the deeply buried Archean rocks, is likely younger, ~1,715 Ma (also Duke et al., 1990). Plutonic bodies there have a different magmatic style, with irregular stockwork shapes, sometimes layered (Redden et al., 1990). They may represent a distinct post-Hudsonian period of tectono-magmatic activity.

Role of major Precambrian fault families in pre-cover tectonics of Canadian cratonic areas

Variable fault patterns are recognized in the Canadian Shield and the basement of tectonic platforms in North America. They include many fault families of different ages, nature and histories. In the western Canadian part of the modern North American craton, three generations of big fault families are most apparent (see reviews in Stauffer and Gendzwill, 1987; Edwards et al., 1996). Prominent N-S-trending faults, such as Tabbernor and Saskatoon, are typical in the Midcontinent. The major NE-SW-oriented faults, such as McDonald, Great Slave Lake and Snowbird-Virgin River, are well represented in western Canada - all across the Shield, Alberta Platform and Canadian Cordillera. The major NNW-SSE faults are often referred to as Cordilleran. The family of faults trending E-W is also persistent in the cratonic part of Canada, though it is less conspicuous and often overlooked. Many Precambrian faults attributed to fold-and-thrust belts were produced by ancient orogenic compressive forces. Most of the faults created in the purely cratonic tectonic regime are steep.

The straight, N-S-trending Tabbernor fault zone separates the Archean Glennie and Kisseynew blocks, whose dissimilar vertical movements are proven to be tens of kilometers in amplitude. The Tabbernor fault zone is 5-10-km-wide, with numerous fault strands. It has been active since before 1,735 Ma (Elliott, 1997) and is probably Late Archean, though its strands cut and offset many younger rocks and faults (e.g., the NE-trending Needle Falls, Stanley and other faults separating Early Proterozoic orogenic belts in the Trans-Hudson province). In the north, this zone meets the Snowbird-Virgin River fault system, with complex mutual interactions between them. To the south, in the U.S., the Tabbernor fault zone controlled the distribution of some sedimentary basins and arches in the Phanerozoic (Sloss, 1988a).

The NE-trending faults such as Needle Falls, Stanley, Virgin River in western Trans-Hudson structural province separate Early Proterozoic domains with different tectonic histories: La Ronge, Rottenstone, Wollaston, Mudjatik (Hearne), Rae. These domains have been distinguished in the exposed Canadian Shield, but their continuation into the subcropped parts of the craton are not clear (Fig. 67). The first three of them are narrow belts, only 50-60 km in width. These belts and their bounding faults have been traced mostly geophysically from the Canadian Shield in Manitoba and Saskatchewan under the Phanerozoic platformal sedimentary cover. The corresponding potential-field anomalies (Fig. 68) help to recognize lateral tectonic units in the basement beneath the cover. This is often difficult because the interpretation of geophysical data is not unique, especially where drillholes penetrating the basement are sparse. From its zero edge, the cover thickens to over 4,000 m in the center of the Williston Basin of the Midcontinent Platform in the south and in the Alberta Basin of the Interior Platform in

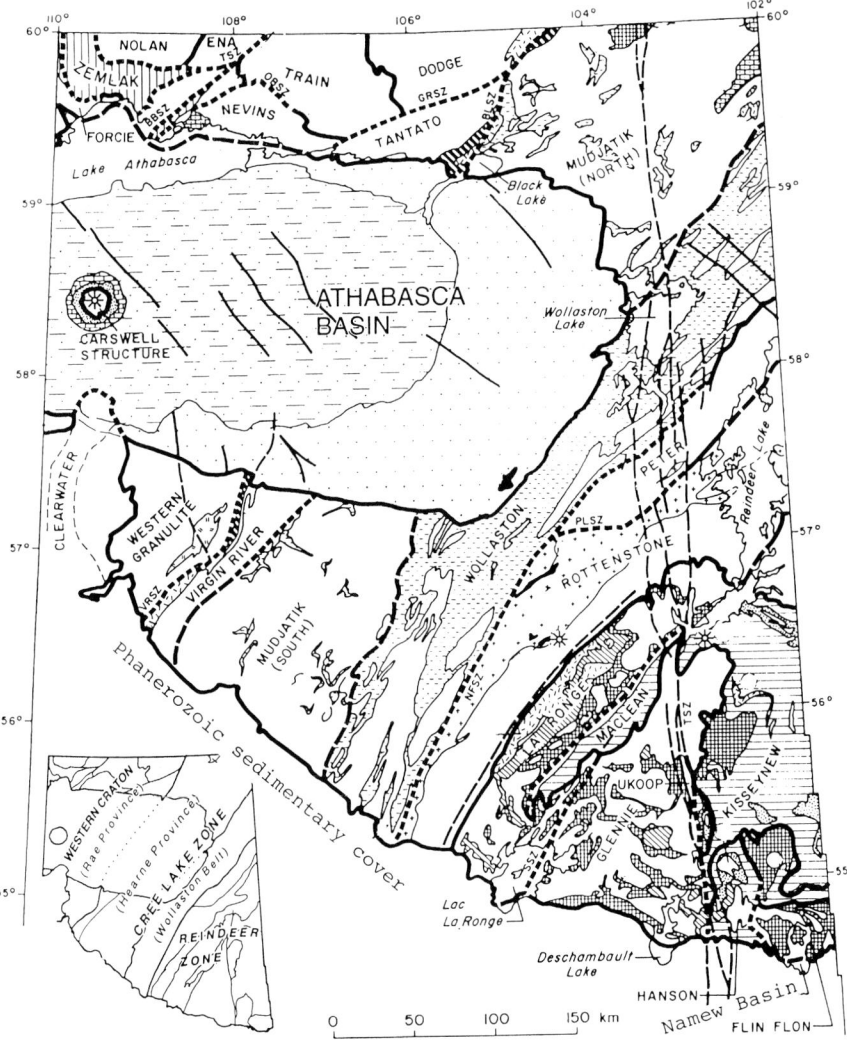

Figure 67. Geologic zonation of the Canadian Shield in Saskatchewan (modified from Saskatchewan Geological Survey, 1994). Fault and shear zones: BBSZ - Black Bay (part of the Snowbird-Virgin River fault system); GRSZ - Grease River; NFSZ - Needle Falls; OBSZ - Oldman-Bulyea; PLSZ - Parker Lake; SSZ - Stanley; TSZ - Tabbernor; VRSZ - Virgin River (part of the Snowbird-Virgin River fault system). The Carswell structure is conventionally attributed to a meteorite impact

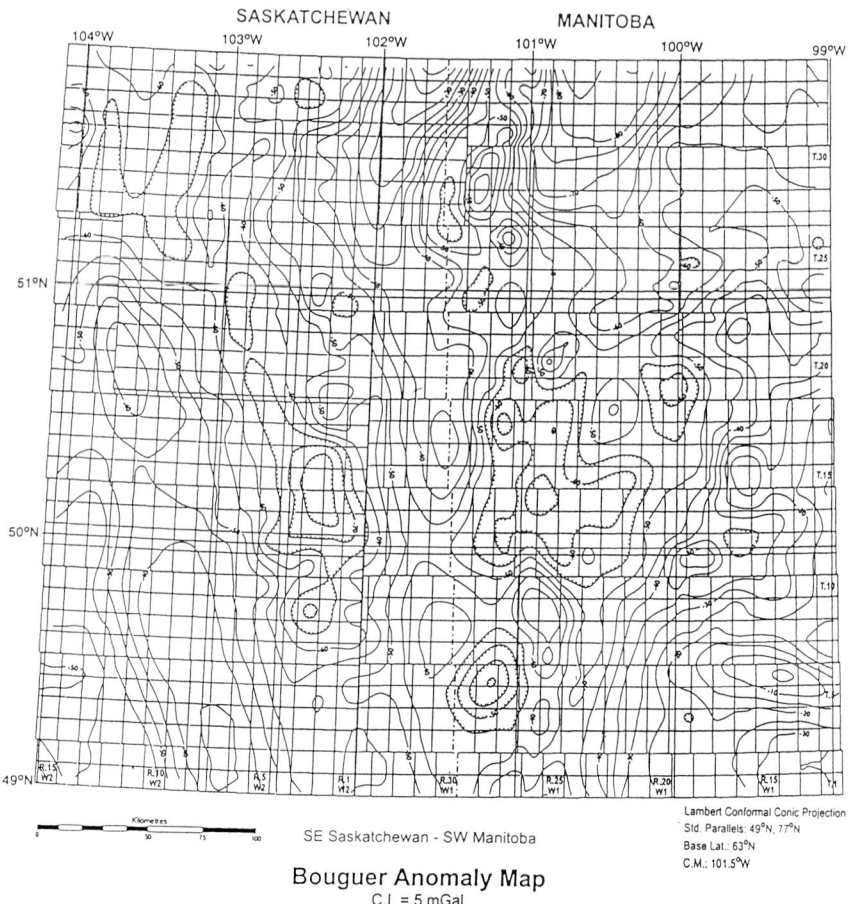

SASKATCHEWAN   MANITOBA

Bouguer Anomaly Map
C.I. = 5 mGal

Figure 68a. Bouguer gravity anomaly map of the buried Trans-Hudson Orogen in southern Manitoba and Saskatchewan (after Lyatsky et al., 1998). Plotting by the Geological Survey of Canada; contour interval 5 mGal. The system of townships (T) and ranges (R) is used conventionally in hydrocarbon exploration and agriculture in western Canada to identify localities; a township is a square six miles (about 10 km) on the side. C.M. - central meridian used in the map projection.

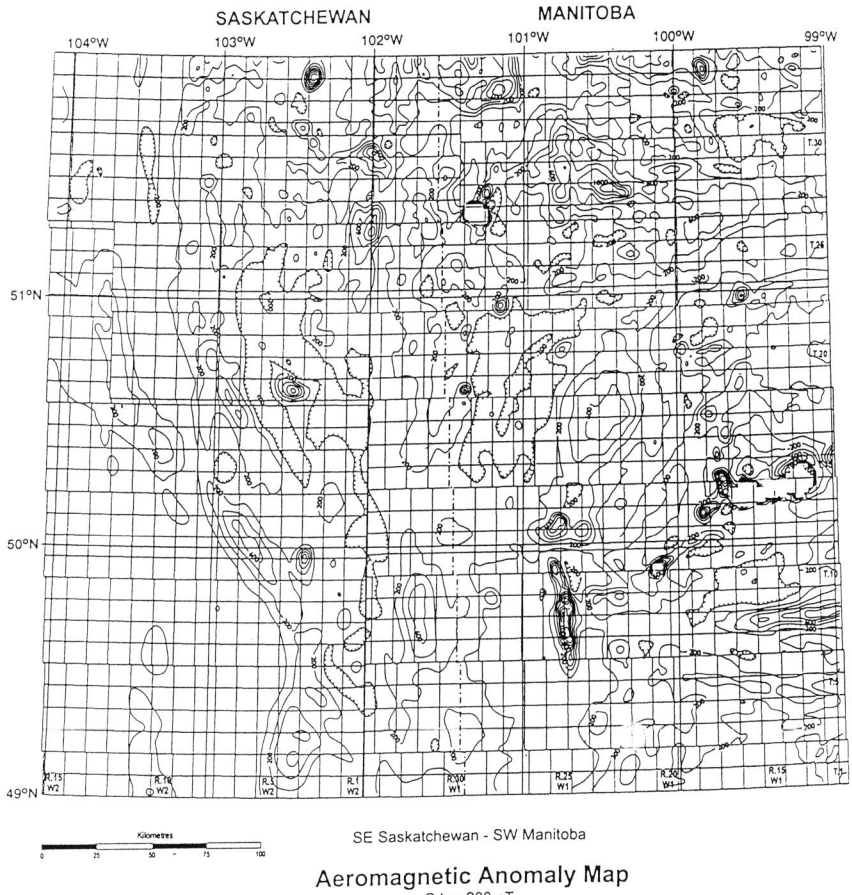

SASKATCHEWAN   MANITOBA

SE Saskatchewan - SW Manitoba

Aeromagnetic Anomaly Map
C.I. = 200 nT

Figure 68b. Total-field magnetic anomaly map of the buried Trans-Hudson Orogen in southern Manitoba and Saskatchewan (after Lyatsky et al., 1998). Plotting by the Geological Survey of Canada; contour interval 200 nT. Parts of three high-amplitude anomalies appear blank, as some data points were deleted to avoid contour overcrowding. C.M. - central meridian used in the map projection.

the west. South of the U.S.-Canada border, the Trans-Hudson Orogen is terminated by the E-W-oriented Mazatzal-Yavapai or Inner Midcontinental orogenic mobile megabelt.

In a recent interpretation of the basement in central Saskatchewan, Leclair et al. (1997) confirmed that between two major fault zones, N-S-trending Tabbernor and NE-SW-trending Thompson (referred to also as Churchill-Superior boundary) lie Early Protero-zoic Trans-Hudson orogenic rocks assemblages (Fig. 69). The most extensive of them, Flin Flon, is accessible for mapping only in a narrow zone north of the Phanerozoic cover edge and south of the Kisseynew block. The Flin Flon complex is generally regarded as a Mediterranean-type collage of various terranes bracketed by isotopic ages of 1,920 to 1,880 Ma (cp. Syme, 1995; Stern et al., 1995; Lewry et al., 1996).

Two structural domains, with contrasting fault patterns, lie to the east and west of a broad band of N-S faults including Tabbernor. The structural domain to the east is characterized by NE-SW fault trends and E-W trends typical for the Superior craton east of the Churchill-Superior boundary. The domain to the west has many NNW-SSE structures, corresponding to the Kimiwan-Cordilleran trend (cp. Leclair et al., 1997; Lyatsky et al., 1998). These trends are also common in many parts of Alberta. The broad zone between two contrasting structural provinces, informally denoted here as Superior-dominated and Kimiwan-dominated, may have a fundamental significance as a diffuse transition in the crust.

Deep Lithoprobe and COCORP seismic sections (White et al., 1994; Baird et al., 1996; Hajnal et al., 1996) display a broad whole-crust arch in southern Saskatchewan and North Dakota (the South Saskatchewan Arch, as we denote it herein; Fig. 70). The Archean Glennie block lies roughly along the N-S-trending apex and axis of this arch. The Glennie block (and its broken-off part known as Hanson Lake block) is an unre-worked piece of ancient (Superior, Wyoming?) cratonic basement later raised in the middle of the N-S branch of the Trans-Hudson Orogen. The Tabbernor and Thompson fault zones converge to the south, beneath the Phanerozoic platformal cover. There, the tectonic domains assigned to the Trans-Hudson Orogen wedge out. The Glennie block at first widens to the south, but then it wedges out too. The pre-existing basement was overprinted by Early Proterozoic orogenic processes, but the preservation of E-W, Supe-rior, structural trends in the Flin Flon domain suggests the Archean basement there was reworked only partly. The same is the case in the basement of the Williston Basin. In southwestern Manitoba, E-W potential-field anomalies are traceable as far west as the Saskatchewan border (Fig. 68; Lyatsky et al., 1998), and rock samples with Archean ages have been recovered by drilling basement in Saskatchewan (Collerson et al., 1990).

Of considerable importance for tracing crustal-scale faults may be the a zone of high elec-trical conductivity known as the North American Central Plains (NACP) anomaly (Fig.

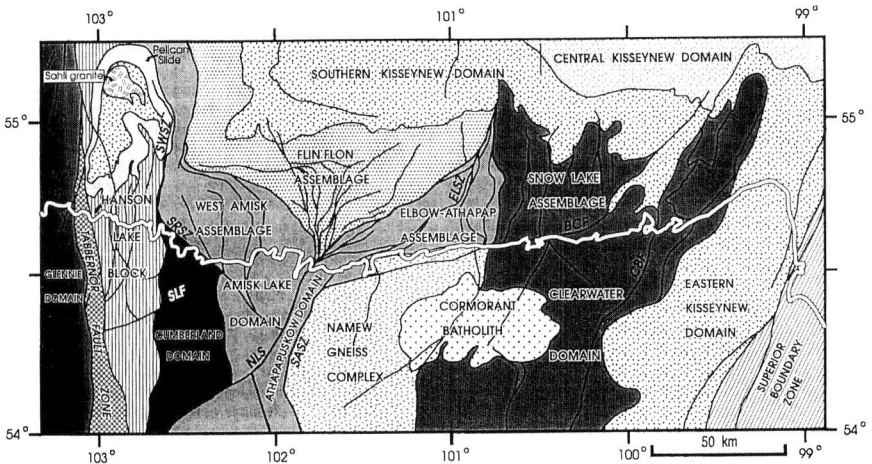

Figure 69. Major crustal faults and domains in the Trans-Hudson Orogen in central Saskatchewan and Manitoba, based on geologic field mapping in the Canadian Shield and interpretation of potential-field data under the Phanerozoic sedimentary cover (modified from Leclair et al., 1997). Faults and shear zones: BCF - Berry Creek; CBF - Crowduck Bay; ELSZ - Elbow Lake; NLS - Namew Lake; SASZ - South Athapapuskow; SLF - Siggi Lake; SRSZ - Spruce Rapids; SWSZ - Sturgeon-weir. White line marks the edge of Phanerozoic sedimentary cover.

202

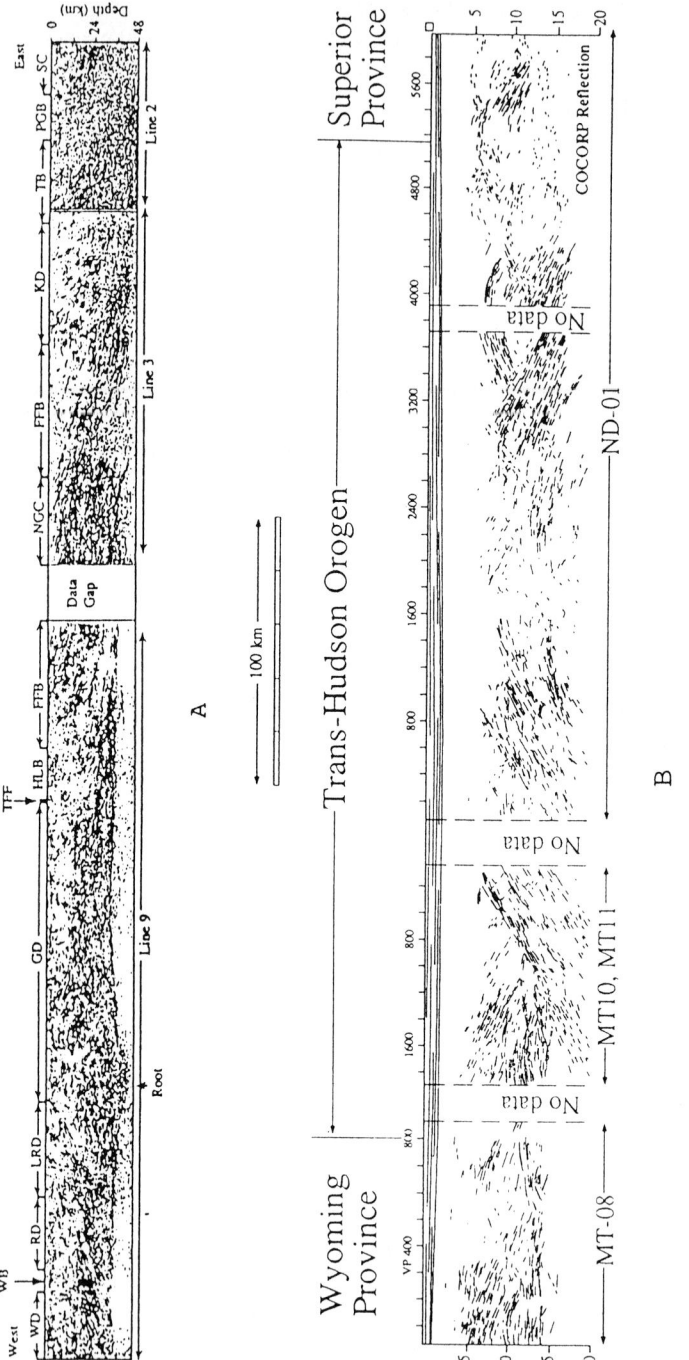

Figure 70. South Saskatchewan crustal arch in the core of the Trans-Hudson Orogen, as imaged in E-W Lithoprobe and COCORP deep crustal seismic reflection profiles (location in Figs. 64, 71). TWT - two-way traveltime (in seconds). GD - Glennie block; HLB - Hanson Lake block; FFB - Flin Flon domain; KD - Kisseynew Basin; LRD - La Ronge domain; RD - Rottenstone domain; WB - Wathaman-Chipewyan batholith.

64). It has been detected in several magnetotelluric surveys in Canada and the U.S. since the 1960s. It is known to continue for over 2,000 km, with a variable width. It is composed of several curved and straight segments. A big segment of it roughly follows the South Saskatchewan crustal arch. The NACP anomaly was at one time interpreted as resulting from a relic of oceanic crust containing preserved saline sea water, caught between collided Archean cratons (Handa and Camfield, 1984; Gupta et al., 1985; cp. Dewey and Burke, 1973). But such anomalies can also be caused by concentrations of graphite, which are common in fault zones in metamorphic schists (Lidiak, 1971), or by sulfides which are also precipitated often along big crustal faults.

Sources of the NACP anomaly and its curved segments have been modeled to lie at middle to upper crustal levels, mimicking the configuration of fault strands at the western edge of the Glennie block. Straight NACP anomaly segments are correlatable with Tabbernor fault zone on the east and with the Saskatoon fault on the west. Near the Canada-U.S. border, Jones and Savage (1986) moved the NACP anomaly some 75 km to the east from its originally inferred position: this increased its curvature around the Glennie block. At present, this anomaly is thought to be caused by a chain of several separate conductors rooted in the middle crust (Thomas et al., 1987; Jones, 1988). South of 51°30'N, the NACP is discordant with some magnetic anomalies, which are sourced near the top of the crystalline crust.

Many tectonic schemes of the crystalline basement in this region relied on plate-tectonic and terrane-tectonic ideas, and invoked "former open oceans", "suture zones", "Cordilleran-type arcs" and so on (e.g., Green et al., 1985a-b; Ross and Stephenson, 1989). The scheme of Green et al. (1985a-b) was based primarily on then-new aeromagnetic data, but some geologists cautioned against this "overly enthusiastic interpretation of geophysical data" and noted that making such models prematurely may be a "speculative and dangerous venture" (Lewry et al., 1986, p. 183, 187).

Drillhole geologic evidence in hand, Lewry et al. (1986) raised several important points of doubt. The magnetic signature of Archean terrains (e.g., Glennie block) may not be simple because of superimposed Proterozoic magmatism and tectonism. Though some of the basement rock samples in that area are Kenoran in age (also Collerson et al., 1990) some basement rocks west of the Tabbernor fault are low-grade metasediments with quartzite and iron formations. Post-Hudsonian rhyolitic lavas are widespread beneath the Phanerozoic cover in many parts of southern Saskatchewan and Alberta (Burwash et al., 1994).

Based on detailed mapping in the Archean Glennie block, Chiarenzelli et al. (1998, p. 247) have recently presented evidence for ancient links and common evolution of many parts of the Trans-Hudson Orogen and adjacent regions. They stated the following: "U-Pb geochronological results suggest calc-alkaline plutonism from 1889-1837 Ma, thrust stacking, peak metamorphism and associated anatexis between 1837 and 1809 Ma, isotopic closure of titanite at 1790-1772 Ma, and intrusion of late granitic rocks at 1770-1762 Ma. This is in agreement with ages from the Hanson Lake Block, and La Ronge, Kisseynew, and Flin-Flon domains in Saskatchewan and Manitoba, and from the Ungava-Baffinportion of the Trans-Hudson Orogen, suggesting broadly synchronous thermotectonic processes along a strike length of 2000 km". From this evidence, they suggested "that the Saskatchewan Craton, rather than representing an exotic continental fragment, rifted from the Superior and/or Hearne Provinces at ca. 2.1 Ga and that the Trans-Hudson Orogen is an internal orogen".

In the U.S., as was noted above, voluminous anorogenic felsic magmatism (granite-rhyolite, granite-anorthosite, rapakivi) occurred chiefly between ~1,500 and 1,300 Ma. Numerous dioritic samples from the basement in southern Saskatchewan, within and outside the Glennie block, have post-Hudsonian ages, and K/Ar studies have put the cooling ages of some of these rocks at about 1,200 Ma. It seems probable that though the exposed northern part of the Glennie block is tilted to the NE, its southern part is tilted to the south and SW: the slightly reworked Proterozoic cover is preserved there over the Archean basement. Such an interpretation is consistent also with the observation that magnetic anomalies generally decrease in amplitude towards the southern edge of this Archean block.

The Trans-Hudson Orogen wedges out towards the Penokean province, and it becomes narrow in the Dakotas. The E-W Mazatzal orogenic province cross-cuts all older tectonic units, including the Wyoming craton, the southern Trans-Hudson Orogen and the western Penokean orogen. The major Thompson, Tabbernor and Saskatoon fault zones are very old, probably Late Archean, but all of them were active later, under variable stress regimes, in the Laurentian and other cratons.

Guiding importance of Precambrian volcano-sedimentary basins for designation of regional tectonic stages during the cratonic megastage before the deposition of Phanerozoic platformal cover

Cratons are recognized as stabilized masses of continental crust, as opposed to mobile megabelts. Prior to the appearance of the first Archean cratons, the whole continental crust was mobile and, in the sense that it was being profoundly reworked, orogenic. The orogenic (reworking) processes in the Archean cratons continued, but they were subdued and restricted to deeper crustal levels. The first cratonic platforms, latest Archean and Early Proterozoic in age, were still quite mobile. They were reworked not just by orogenic processes in mobile megabelts, but also by the processes that continued in the ancient cratons. Only later, as the continental crust and lithosphere evolved further, were the orogenic processes clearly restricted to mobile megabelts. For the Proterozoic and Phanerozoic, such megabelts are often used with the adjective *orogenic*.

Only after 1,800-1,700 Ma, mainly in the Mesoproterozoic (Table IA), did the distinction between cratons and mobile megabelts become anything like as sharp as it is today. This change is illustrated in the development of the early cratonic volcano-sedimentary basins, whose rock record helps elucidate the regional tectonic history. Canada provides good examples of such basins, from early Paleoproterozoic to middle Neoproterozoic time, which had existed before the Phanerozoic sedimentary cover spread like a broad blanket over all the various basement formations in cratonic regions.

From the beginning, Proterozoic basins were discontinuous, surrounded by broad arches of crystalline rocks. These basins were still rather mobile, containing large amounts of high-energy clastic deposits interbedded with volcanic horizons. Yet, they contained distinct structural-formational étages, whose distinctiveness was later partly obliterated by strong metamorphism and deformation.

The early Paleoproterozoic Rae, Hearne and other platforms contain thick basins which are still poorly delineated. The Wollaston Basin in Saskatchewan is studied better than others, but even there the distinction of structural-formational étages in the pile of heavily metamorphosed and deformed rocks is a matter for future studies.

The huge Mesoproterozoic volcano-sedimentary succession conventionally assigned to the non-orogenic granite-rhyolite domains in the Midcontinent will someday probably also prove divisible into distinct basins with specific collections of étages. Thickness of these stratal rocks is in many areas very large, on the order of 10 km. Compositional variations in the rhyolite-dacite levels suggest variable tectonics, where more or less stable settings changed during several tens of millions of years (mostly, these events took place between 1,500 and 1,350 Ma).

Best-studied examples of Proterozoic volcano-sedimentary basins in North America have been described in the Canadian Shield in Saskatchewan and the Northwest Territories (Campbell, ed., 1981). Some of them developed over Archean basement, some over Hudsonian. Two of them, Athabasca and Martin (Fig. 55), are described in this section in some detail. Though they formed on a cratonic crust far more stable than in the preceding epochs, they still experienced some metamorphic and deformational tectonic reworking. But their lower extent of thermal alteration sets them apart from the strongly metamorphosed and folded Wollaston Basin.

The Athabasca Basin lies mostly in northern Saskatchewan and partly in northeastern Alberta. It overlies high-grade, deformed rocks, and is partly overlain by the unmetamorphosed, flat-lying sedimentary cover of the Alberta Platform. In the basement, Archean gneisses, Hudsonian granitoids (~1,900 Ma) and Early Proterozoic metamorphosed supracrustal rocks have been found, in part by drilling for radioactive ore deposits (Saskatchewan Geological Survey, 1994). The Athabasca Basin was originally not much wider than at present (Ramaekers, 1981; Wilson, 1986). It comprises up to 2,000 m of fluvial, lacustrine and (in the west) marine clastic rocks, as well as thick flows of basalt and andesite. Many faults cut the Athabasca Basin, in the middle of it as well as on its flanks. This basin's polygonal to oval shape is defined by these faults.

Indicating tectonic quiescence and a subtropical climate, a thick layer of weathering developed on the basement crystalline rocks (Wanless and Eade, 1974). The layer of saprolite and laterite was later partly stripped off, but in places its preserved thickness is up to 50 m (Wilson, 1986). The saprolite has K/Ar ages of 1,632±32 Ma to 1,411±36 Ma, which represent resetting episodes as overlying basal rocks of the Athabasca Basin yield older ages of 1,700 to 1,600 Ma (see, e.g., Fahrig and Loveridge, 1981; Saskatchewan Geological Survey, 1994). Cumming et al. (1987) and Carl et al. (1992) concuded that deposition of the Athabasca Group had ended by 1,700-1,650 Ma.

The western part of the basin was tectonically more unstable, as indicated by thick layers of conglomerate found at the bottom as well as at higher stratigraphic levels in the basin. Depocenters in the basin shifted with time, mainly in the E-W direction. The most durable and greatest total subsidence took place in the east. Sporadic syn- and post-sedimentary block tilting occurred several times, with inversions, and is well recorded in lateral lithological variations. Many mafic dikes that cut the Athabasca Basin have a NE-SW trend, and they were intruded between 1,350 and 1,000 Ma (Carl et al., 1991). One suite of these dikes has Rb/Sr ages of 1,430±30 Ma (Armstrong and Ramaekers, 1985). Another dike suite, dated at 1,310±70 Ma, is perpendicular to the older pattern. Yet, both these sets of dikes were offset by E-W faults. This indicates that the regional stress regime from 1,700-1,600 to after 1,300 Ma changed at least three times, reflecting distinct tectonic stages in the evolution of the western part of the craton in Saskatchewan and Alberta.

The Athabasca Basin is elongated in the E-W direction which is highly discordant with the regional network of NE-SW and N-S faults. A similar E-W trend is typical for the Athabasca-Peace River-Skeena zone of crustal weakness, where this structural trend persisted into the Phanerozoic. In the Athabasca Basin, however, very prominent were the NE-SW faults, which influenced the basin's sedimentation patterns. Reactivated several times, the major Virgin River fault is apparent also from a sharp change in the fabric of magnetic anomalies across it (Pilkington, 1989).

The most akin to the Athabasca Basin is the small Martin Basin to the north. It formed between ~1,700 and 1,450 Ma (Tremblay, 1972; Ramaekers, 1981), and is as much as 5,000 m thick. Basalt and andesite flows up to 1,500-2,000 m thick intercalated with clastic sedimentary rocks. Due to gradual, syn- and post-depositional westward tilting, the basin is the thickest on its western side. On the east, thicknesses of all stratigraphic units are reduced and the upper part of the basin is eroded. No saprolite beds are found at the base of the Martin Basin, but paleosols did develop after its cessation. Large gabbroic dikes, probably comagmatic with some of the volcanic rocks, intruded the Martin Basin no later than 1,490 Ma (based on K/Ar dating of whole-rock dike samples). The entire volcano-sedimentary assemblage was only gently folded and slightly altered. It is overlain by redbeds.

Both the Athabasca and Martin basins are intracratonic in origin. Thick volcanic layers and abundant dikes point to a still-heightened tectonic activity during their formation. The outlines of their flanks are variously polygonal and oval: oval basin outlines indicate local sagging, but angular flanks reflect fault movements. Networks of steep faults largely account for basin shape and internal deformation. Another basin in the western part of the Shield is the Thelon Basin, which was distorted to a much greater extent by later cratonic tectonism.

All these basins are intracratonic, but differences in their tectonic history indicate a great variety of crustal regimes even in neighboring regions. Local sagging and differential block downdrops are inferred from these basins' geologic record. Less well known are examples of Proterozoic cratonic crustal upwarping and horst movements.

Peterman and Hildreth (1978) have described one such upwarp in a 120-km-wide zone in the southern Wyoming shield, adjacent to the Cheyenne foldbelt. This E-W-trending zone contains Archean rocks whose K and Ar analysis suggests their elevation and cooling took place sometime between 1,500 and 1,300 Ma. Assuming normal cratonic geothermal gradients, the magnitude of uplift and erosion has been estimated at 10-12 km. Because the Mazatzal (or Central Plains) Orogen was much older, 1,760-1,680 Ma (Sims and Peterman, 1986; Bickford, 1988), there is no reason to connect this crustal upwarp with any orogenic activity nearby. Rather, this long-wavelength upwarp was created by an intracratonic displacement of the crust.

Advanced methods of rock analysis, such as fission-track, offer new prospects for studies of extensive Precambrian arches. These methods are still imperfect and costly, but in the future, refinements will surely make them more accurate and cheaper. For now, though, elucidation of regional tectonics of cratonic areas is the most reliable when it is based on detailed studies of volcano-sedimentary basins. The geologic history of Precambrian basins is instructive not just for areas of advanced cratonization but also for areas assigned to Proterozoic orogens.

The Kisseynew Basin developed over a big crustal block that may be at least in part Archean and which acted as a buttress or median massif in the Trans-Hudson Orogen (Figs. 55, 69). It was buried considerably, first to make room for the accumulation of a volcano-sedimentary succession and then to permit these rocks to reach the amphibolite and even granulite grade of regional metamorphism. As early as around 1,850 Ma, this basin had already reached great depth, and its peak of metamorphic reworking was achieved at ~1,815 Ma (Gordon et al., 1990). The granulite facies is conventionally thought to require pressures of 7-9 kbar (700-900 MPa) and temperatures of 700°-800°C, suggesting crustal depths of 25-30 km or more. This burial and heating was associated with the ductile deformation recorded in the Missi Group rocks, and with the appearance of palingenic granites which intruded this basin (Lewry et al., 1990; Zwanzig, 1990; Zwanzig et al., 1996). Later, as the tectonic movements were inverted, the Kisseynew block rose, and the high-grade gneisses formed from supracrustal sedimentary rocks were brought to the surface. Gordon et al. (1990) estimated that, following the peak of metamorphism (around 1,815 Ma), this uplift was rather continuous. Leucogranites, common for extensional tectonic regimes associated with uplifts (cp. Colpron et al., 1996), yield ages between 1,820 and 1,790 Ma (Zwanzig et al., 1996). As the Missi Group was elevated, it cooled past the temperature of 280°C at 1,705 Ma, and the deeply metamorphosed rocks were juxtaposed against neighboring blocks whose rocks were metamorphosed to much lower grades. As the Kisseynew block rose, numerous brittle faults with estimated offsets of 2 to 4 km developed on its perimeter (Bleeker, 1990; Stauffer, 1990); some of them followed old zones of mylonitization (Bleeker, 1990).

The rise of the Kisseynew block ended only by late Mesoproterozoic time. Then, at ~1,270 Ma, a new, extensional stress regime was marked by emplacement of huge NNW-trending dike swarms over most of the Canadian Precambrian terrains (Heaman and Le Cheminant, 1988). These dikes mostly ignored the older structural fabrics, and in Saskatchewan they cross-cut major Hudsonian-age faults. No tectonic offsets of these dikes are reported.

The Kisseynew Basin developed after the basins of the Wollaston generation. It was roughly coeval with other volcano-sedimentary basins recognized in the Trans-Hudson Orogen, such as the Paleoproterozoic Namew (Leclair et al., 1997) Basin to the south (Figs. 69, 71). The Namew Basin marks the termination of orogenic activity in the Flin Flon domain. Mafic to ultramafic igneous protoliths, dated at ~1,890 Ma, are found in this basin's basement. The basement was probably covered by basal sedimentary rocks of the Namew Basin around 1,850 Ma. These late Paleoproterozoic sedimentary rocks have been described as similar to those in the Missi Group of the Kisseynew Basin (Ashton et al., 1996): arkoses, intercalated with volcanics, transitional to the Kisseynew wackes.

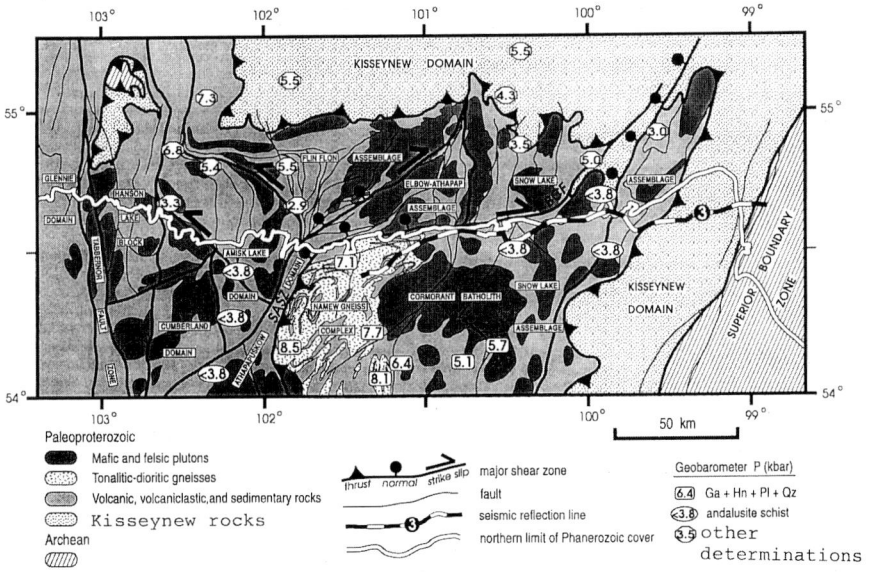

Figure 71. Distribution of rock lithologies and metamorphic grades in the Namew Basin area, central Manitoba and Saskatchewan (modified from Leclair et al., 1997). The seismic line is the Lithoprobe reflection profile in Fig. 70. BCF - Berry Creek fault; SASZ - South Athapapuskow shear zone. White line marks the edge of Phanerozoic sedimentary cover.

Largely, the Namew Basin is composed of gneisses formed from quartz-bearing rocks, igneous and sedimentary, including granites and psammitic and pelitic clastic rocks. Penetrative deformation makes it difficult to construct an accurate stratigraphic column. The age of some basin rocks (quartz diorite samples) is probably 1,880 to 1,850 Ma. Monzogranites of a big batholith and its derivatives, which intruded the basin, have been dated at 1,831 to 1,828 Ma. On the north and west, the extent of the Namew Basin is defined by big faults (shear zones; cp. Syme, 1995) which juxtapose the basin against a greenschist-grade terrain. In contrast to the greenschist-grade rocks in the basin's northeastern part, its southwestern part contains rocks metamorphosed to high amphibolite grade.

Metamorphism of the Namew Basin requires large vertical displacements, first to make room for the deposition of the supracrustal protoliths and then to rework them at depths exceeding 10-15 km. Later, tectonic inversion brought these rocks back to the surface. Leclair et al. (1997) summarized the existing geobarometric data for the Flin Flon tectonic wedge between the Thompson and Tabbernor fault zones (Fig. 71; see also Gordon et al., 1990, 1994; Digel and Gordon, 1993). The large variations in estimated values (3.0 to <3.8 kbar or 300 to <380 MPa; >5.5 to 7.3 kbar or >550 to 730 MPa; and even 8.5 kbar or 850 MPa) across several mapped faults indicates that different crustal blocks were buried to very different depths. Whereas the Kisseynew Basin has fairly consistent geobarometric values of ~5.0-5.5 kbar or 500-550 MPa, rocks on the southern rim of the Namew block have variable values of 5.0, 3.5 and 2.9 kbar (or 500, 350 and 290 MPa). More differences are seen in the distribution of geobarometric values between the Thompson belt and the South Saskatchewan Arch (the Glennie-Hanson Lake horst system). Low values of 3.0 to <3.8 kbar (300 to <380 MPa) are found in the east, but just ~100 km to the west, the values reach 7.7 and even 8.5 kbar (770 and 850 MPa).

The Namew Basin was buried to at least the middle crust in Early Proterozoic time, and was uplifted back to the present-day topographic surface in post-Hudsonian time. From the estimated P-T-t paths, the vertical block movements had magnitudes of 10 to 20 km (Digel et al., 1991). Cooling of the elevated rocks occurred from 1,820 to 1,730 Ma (also Cumming and Krstic, 1991). Leclair et al. (1997, p. 631) related it to a "differential uplift during postcollisional deformation".

The geology of this big part of the so-called Trans-Hudson Orogen is still understood poorly. In the terrane-tectonic template, it is fashionable to define zones of ancient subduction and collision and to assume that tectonic movements, metamorphism and deformation in the Kisseynew and Namew basins were induced by crustal-scale thrusting. Mafic and ultramafic rock complexes in the Flin Flon domain are attributed variously to ophiolitic or Alaska-type formations, though these are very different in terms of their tectonic setting (cp. Patton et al., 1994). On the other hand, felsic to ultramafic rocks in the Kisseynew area has been attributed variously to subduction magmatism or asthenosphere-induced magmatism in a tectonic setting of Mediterranean type (see, e.g., Zwanzig, 1997).

Even regarding the relatively well-studied Kisseynew Basin, Ansdell et al. (1995, 1997) often used world like "enigmatic" and "problematic". Among many other unsolved problems are the structural relations between this area's different domains. Kisseynew gneisses proper have been found in south-verging thrust sheets overlying rocks of the Namew Basin in the Flin Flon domain (Norman et al., 1995). Like in the Thompson zone on the eastern boundary of the Kisseynew block, it is hard to determine whether the main process was underthrusting (from the south) or overthrusting (from the north), but in any case, the end of this thrusting came at ~1,800 Ma (Gordon et al., 1990). Later, some tectonic deformation there occurred around 1,690 Ma, near the Paleoproterozoic-Mesoproterozoic boundary (Fedorowich et al., 1995).

Numerous steep faults, in this region and beyond, bound big and small crustal blocks on which the Proterozoic basins rest. Though mapped on the ground, these faults are commonly missed in the interpretations of seismic sections, including the deep Lithoprobe transects (White et al., 1994; Clowes, 1996). To identify steep faults in seismic profiles is often difficult, and they are often overlooked (a recent exception is the work of Hajnal et al., 1996).

In Saskatchewan, steep crustal faults are well expressed in gravity and magnetic maps, where many of them are associated with pronounced, straight gradient zones (Leclair et al., 1997; Lyatsky et al., 1998). The cratonic structural pattern in this region has been superimposed on the pre-cratonic one probably since ~1,840-1,830 Ma (Gordon et al., 1990). The fault-bounded Namew Basin is one of the late Hudsonian to early post-Hudsonian block-related features, formed on a SW-tilted crustal block just south of the Archean Kisseynew block. In the E-W direction, in a row with the Namew block lie the Archean Hanson Lake and Glennie blocks. All this evidence suggests the Namew Basin itself may be underlain by another Archean block. Such an interpretation is consistent with the abundance of quartz-rich crustal material in the basin's orthogneisses and younger batholithic granitoid intrusions.

High-angle faults were active in the Proterozoic and later, and they are traced in the basement under the Phanerozoic platformal cover. Fault and block tectonics determined the sedimentation and structural settings in different times and regions, including during several tectonic stages in the Williston Basin. Major N-S faults bounding the Archean Glennie and Hanson Lake blocks are expressed prominently in potential-field maps, in places because volcanic belts are aligned along them. Some of the N-S faults (such as Sturgeon-Weir and Tabbernor) might be as old as Late Archean; faults within the NNE-SSW-trending Thompson boundary zone are younger, Early Proterozoic.

Some of them penetrate through much of the lithosphere, others only the crust; some offset the Moho, others do not (Hajnal et al., 1996).

The original Lithoprobe interpretations of deep seismic data in terms primarily of low-angle crustal thrust stacking is now being revised all over the platform and shield areas in Canada (e.g., Hajnal et al., 1996). In the Archean terrains in Ontario, many subvertical faults, even those intruded by dikes, were originally "overlooked or misinterpreted" (Zaleski et al., 1997). But so strong is the bias towards low-angle detachments that even a proven post-orogenic strike-slip fault system in the Trans-Hudson Orogen has been interpreted by Leclair et al. (1997, p. 631) as "linked to a low-angle detachment at depth (cf. Hajnal et al., 1996), parallel to the pervasive reflectivity in [Lithoprobe] line 3, in order to accommodate its postmetamorphic vertical throw (ca. 0.45 GPa) without offsetting deep crustal reflectors or the Moho". This supposition mistakenly subordinates the results of direct geological observations to a preferred but non-unique interpretation of indirect seismic images.

These matters are important for developing a methodology of tectonic analysis in cratons, and we will return to them in other chapters. Suffice it to say for now that Proterozoic basins in the pre-cover interval are actually discontinuous structural-formational étages which developed during various tectonic stages of the crust. They developed at different times in different regions, on top of both unreworked Archean and reworked Hudsonian basement. Like with the better-defined structural-formational étages in the Phanerozoic sedimentary cover, these Precambrian basins offer the most reliable geological information about local upwarping and downwarping and about more-regional tectonic reorganizations recorded in many generations of volcano-sedimentary basins and their internal étages. Both oroformal and platyformal compartments of the Proterozoic continental crust experienced variable epeirogenic warping (e.g., sagging or doming) and block movements, whose great amplitudes exceeded the Phanerozoic amplitudes suggested by the preserved record of younger basins and arches.

Designation of lateral tectonic units in the Alberta Platform basement: a discussion

A well-founded geological framework makes it easier to interpret geophysical data in terms of the lateral tectonic units that make up the basement of the Alberta Platform. This should help constrain the ongoing controversies surrounding the designation of these units. In the 1950s and 1960s, the early geophysical gravity and magnetic surveys in the Western Canada Sedimentary Province were used for this purpose in large part by projecting the faults and geologic domains mapped in the shield areas (e.g., references in Burwash et al., 1993). In the 1970s and especially 1980s, more common became genetic constructions which took the place of just defining the crustal zonation. This new approach was more speculative, inspired by ideas about convergence and collision of continent-bearing lithospheric plates and subduction of oceanic plates beneath them. Mobilistic hypotheses about the formation of the North American craton (Dewey and Burke, 1973; Hoffman, 1988) often distracted geologists and geophysicists from the more-cautious and more-reliable studies of the craton's platforms.

The main contributions during that time came from new mapping and drilling and from increasingly detailed petrological analysis of Precambrian metamorphic rocks in the Canadian Shield and the Western Canada Sedimentary Province basement. Joining together the petrological, geochronological and structural considerations has allowed to establish the protoliths and age of some basement metamorphic rocks with more confidence than ever before. New structural reconstructions were produced, though some of them were rather generic and subjective. Even for unmetamorphosed rocks, paleogeographic and paleotectonic situations can often be restored only with difficulty and only approximately (cp., e.g., the discussions about the Pennsylvanian flysch rocks in Oklahoma and the so-called turbiditic sequences in the North Sea; Slatt et al., 1997 vs. Shanmugam and Moiola, 1995, 1997; Hiscott et al., 1997 vs. Shanmugam et al., 1995, 1997; Shanmugam, 1997). Much more difficult to derive from rock types are the original and subsequent tectonic settings (even the modern tectonic nature of the Vancouver Island margin has been reinterpreted several times in recent years; cp. Clowes et al., 1987, 1997; Hyndman et al., 1990; Dehler and Clowes, 1992). From the greatly reworked, altered and contorted Precambrian crystalline rocks, such reconstructions are even harder to make unambiguously.

Long, narrow gravity anomaly belts, positive and negative were at one time interpreted in the Canadian Shield as zones of former subduction (Gibb and Thomas, 1988). The assumed ancient sutures were placed along the La Ronge and Wollaston foldbelts, Fond du Lac gravity anomaly, Virgin River fault, NACP conductivity anomaly, and so on (Fig. 54). Shortly after, the NACP anomaly was relocated 70-80 km east from its originally inferred position, and was found to be segmented (Thomas et al., 1987; Jones, 1988). The Virgin River fault was included into a bigger Snowbird-Virgin River fault system, which is traced across most of the Canadian Shield - but is was found not to be associated with the Trans-Hudson Orogen or a collisional suture (Hoffman, 1990; Lewry and Collerson, 1990). The Fond du Lac gravity anomaly in Alberta marks the southern boundary of the Athabasca polymetamorphic province in the basement (Burwash and Power, 1990), not a subduction-related suture. Other paired gravity anomaly zones in many regions mark nothing more than boundaries between crustal blocks with different thicknesses and internal density distributions (Lyatsky, 1996). The ideas of grandiose "frozen" subduction zones or "cryptic" sutures are now going out of fashion in the interpretation of gravity and magnetic data, but they still crop up in some speculations about the history of the Alberta Platform basement. In practical interpretation, the potential-field data remain underused.

Coupled with other geological and geophysical information, these data have nonetheless proved exceptionally valuable for studies of the structure of cratonic crust in platformal areas (Hinze and Braile, 1988; Mariano and Hinze, 1994). Examples of successful implementation of such a combined approach come from the Interior Platform on both sides of the Canada-U.S. border (Wold and Hinze, eds., 1982; Zietz et al., 1982; Geological Survey of Canada, 1990). Green et al. (1985a-b, 1986) and Thomas et al. (1987) provided an early interpretation of a then-new set of regional magnetic and gravity data over the northern Interior and Midcontinent platforms, and offeredcomprehensive schemes of structural partition for their basement. Among the important achievements were, in particular, interpretational insights from the distinctive magnetic and gravity anomaly fabrics of the Wyoming craton. Their continuity northward into Alberta and Saskatchewan was described. Unlike in neighboring regions, this craton has a distinctive ENE/NNW magnetic anomaly pairset (see also Zietz et al., 1982). In the peripheral zones of the Wyoming shield, this anomaly fabric is harder to distinguish, because it is overshadowed by other anomalies related mainly to a network of younger big faults. Magnetic anomalies of the ancient Wyoming craton are commonly due to the huge superimposed Laramian thrusts. Their load also complicates the interpretation of regional gravity anomalies.

Major faults bound the Wyoming shield along the Cheyenne belt in the south, Black Hills in the west, and the Great Falls Zone in the north (Fig. 72; O'Neill and Lopez, 1985; Collerson et al., 1990; Klasner and King, 1990). In the states of Wyoming and South Dakota, these bounding faults sometimes give an impression of semi-concentric curves, although in detail, these shapes are polygonal. To the north, in Montana, the straight NE trend prevails. The elevated Wyoming shield block is generally tilted to the NNW, towards the Cordillera. In the south, the Wyoming craton (*sensu lato*) is downdropped stepwise in the basement of the southern Interior Platform. Similar stepwise drops of crustal blocks are now apparent to the north of the Wyoming shield, in the Archean basement of the southern Alberta Platform.

Like the Slave shield, which is surrounded by Early Proterozoic platforms (Rae, Hearne), the Wyoming shield is also surrounded by ancient platforms (Fig. 65). Its Custer platform lies to the east, continuing to the Saskatoon fault. Another such platform, named herein Kinsella, is inferred to lie to the north. Partial tectonic reworking of the Wyoming craton's platforms in the Early Proterozoic produced metamorphic and structural signatures strongly expressed in the patterns of potential-field anomalies there. This holds true for the platforms of the Slave craton, as well as for the newly-designated Proterozoic platforms of the Wyoming craton. Thus, the standard practice of delineating structural provinces based only on their potential-field anomaly patterns, useful in younger platformal tectonic provinces, may be misleading in the case of old cratonic platforms which were reworked considerably by the still-strong Proterozoic tectonism.

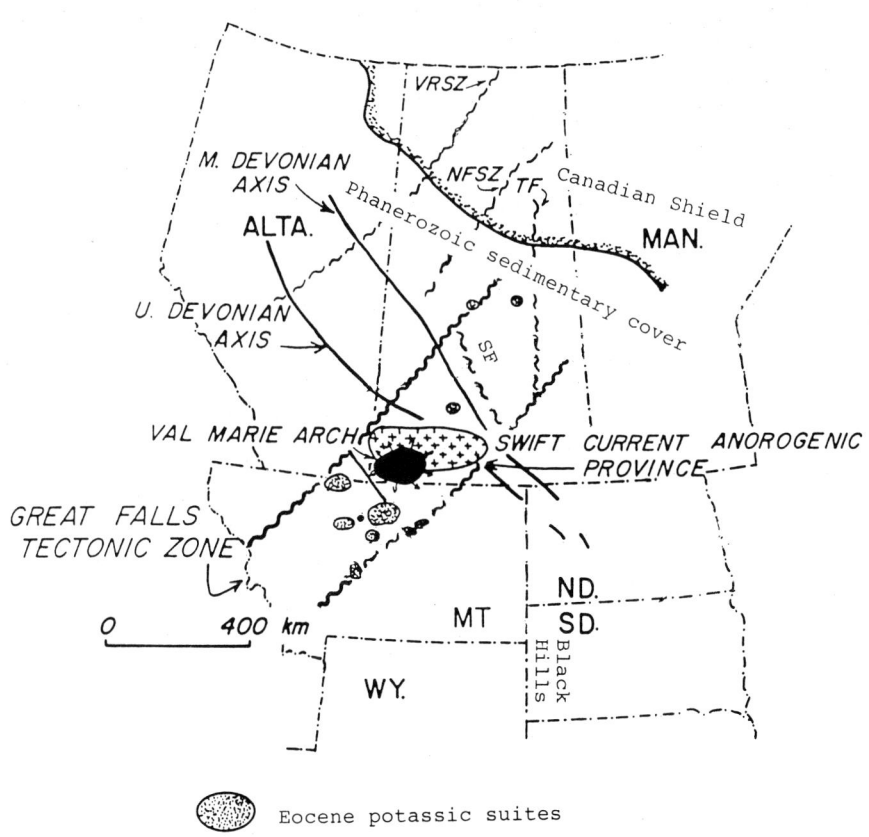

⬭ Eocene potassic suites

Figure 72. Principal structural features in the cratonic crystalline basement of the Williston Basin (modified from Collerson et al., 1990). Faults and shear zones: NFSZ - Needle Falls; SF - Saskatoon fault; TF - Tabbernor; VRSZ - Snowbird-Virgin River.

Green et al. (1985a-b) extended the "Wyoming basement" into southwestern Saskatchewan and southern Alberta, almost as far as Calgary, from their analysis of magnetic anomalies. Based on interpretation of gravity anomalies, Thomas et al. (1987) also confirmed the continuity of the "Wyoming province" into southern Canada. Later, however, Stephenson et al. (1989) reduced the Wyoming potential-field domain to the area south of a prominent E-W-trending gravity low in southern Alberta, which had been interpreted as the so-called Southern Alberta (or Brooks) rift or aulacogen in the basement (Kanasewich et al., 1969). Though Frost and Burwash (1986) showed that this anomaly may merely reflect a fault with an associated band of granitic plutons, the rift/aulacogen interpretation continued to be used by some workers (Ross and Stephenson, 1989). But even from sparse drillholes, presence of Lower Proterozoic platformal supracrustal rocks in southern Alberta is well proven (Burwash, 1993), and the Kinsella platform of the Wyoming craton may continue as far north as the Snowbird fault. In accordance with the available data, the northern boundary of the Wyoming shield is probably the Great Falls fault zone in north-central Montana (cp. Hoffman, 1990). A down-to-the-north step seems to separate the Wyoming shield from the Proterozoic Kinsella platform to the north.

All over the Custer (to the east) and Kinsella (to the north) platformal regions of the Wyoming craton, the gravity and magnetic signatures are fairly consistent between several fundamental potential-field lineaments. One of these lineaments is associated with the NNW-trending Saskatoon fault, which separates the Custer platform from its Hearne counterpart of the Slave craton in Alberta and Saskatchewan. Another such NNW trend, interpreted as the eastern boundary of the Wyoming shield, lies in eastern Montana and southern Saskatchewan is, parallel to and some 50 km west of the Lloydminster fault. Aeromagnetic data over this part of the Western Canada Sedimentary Province were acquired only in the last several years, and current ideas about the basement structure of this region are considerably different from earlier schemes (e.g., that of Stephenson et al., 1989).

From gravity and then-incomplete magnetic maps in southern and central Alberta, Stephenson et al. (1989; also Ross and Stephenson, 1989) "postulated" (sic) several basement domains, from south to north: Medicine Hat block, Southern Alberta rift (renamed as Vulcan low), Matzhiwin high, Loverna block, Lacombe domain, Rimbey high, Thorsby low, Wabamun high. Criteria for defining these features were inconsistent: mainly gravity, but sometimes magnetic; sometimes purely geophysical, but sometimes geological (rift, block); some regional (Loverna block, Lacombe domain), but others local (Thorsby low, Vulcan low). The mixed nomenclature did not follow the principle of priority in assignment of names. Surprisingly, for the subsequent 10 years this scheme has been used as an unchanged template for Lithoprobe interpretations.

A big shortcoming in the interpretation of Stephenson et al. (1989) was a non-discriminant approach to regional potential-field anomalies, whose causes may in fact be very different:(a) pre-cratonic, orogenic; or (b) post-orogenic, truly cratonic. Lack of consistency in the definition of potential-field domains has caused them to include areas with different regional anomaly patterns, or their boundaries to run through areas with internally consistent patterns (Fig. 73). As a result, the application of that scheme has failed to meet the needs in the petroleum industry and academia alike.

Despite such fundamental shortcomings, the initial scheme (which at first had the benefit of just 57 age determinations of recovered basement rock samples) was transformed into a tectonic framework of terranes surrounded by numerous postulated zones of collision, subduction and suturing (Ross et al., 1991, 1997). Geological, including tectonic, interpretation of the Alberta Platform basement is indeed difficult, because drill-hole penetrations are sparse and irregularly distributed (most of them are in the Peace River Arch area). Some of the postulated "domains" have only one or two studied basement samples (Villeneuve et al., 1993), not enough for strong conclusions. Still, Ross et al. (1997, p. 493-494) stated that this tectonic framework for the Alberta basement had been "...established through synthesis of geochronologic and isotopic analyses... combined with interpretation of regional aeromagnetic and gravity data" and that "these models expand on the theme of tectonic assembly for the Canadian Shield..." These authors also stated that "much of the present Canadian Shield took shape" between 2,000 and 1,800 Ma, that the 500-Ma cover "...offers a unique laboratory for determining how the antecedent structure affected the evolution of the sedimentary basin" and that "the region still bears the scars of the ancient collision and resulting plateau-like uplift...", but they noted that "perhaps surprisingly, these fossilized, deep-seated structures have exerted only minor influence on the evolution of the [Western Canada Sedimentary Province] over the last 500 Ma".

This scheme has proved hard to apply productively to practical exploration, environmental and research activity. The idea that North America was assembled randomly from far-traveled terranes (Hoffman, 1988, 1990) finds little factual confirmation (for a healthy skepticism about such interpretations, see Zwanzig, 1997). Besides, orogenic-age structures were ductile, and most of them were not reactivated later; their influence on the cover is restricted mostly to the lowest horizons which were affected by the erosional topography of the basement. Soon after deposition began, by the end of the Ordovician, most of these erosional highs and lows were no longer apparent. In contrast, the influence on the structural-formational characteristics of the cover from cratonic tectonics remained strong throughout the Phanerozoic (Chapter 3).

Though the unevenness of the basement top affected the lower Paleozoic carbonate and evaporite sedimentation, the top of the Cambrian to Silurian tectonic étage in many parts of the Western Canada Sedimentary Province is much smoother. Unfortunately,

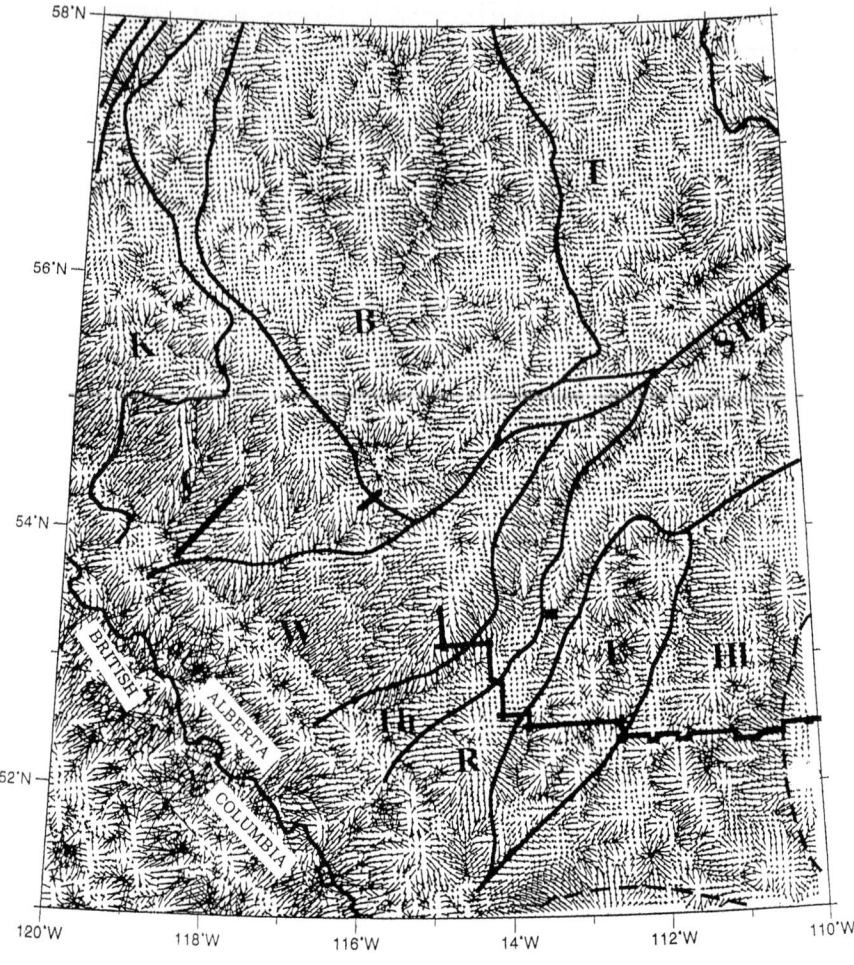

0   50   100 km

Figure 73. Mismatch in the central Alberta Platform of assumed tectonic domain (terrane) boundaries of Ross et al. (1991) with real potential-field anomalies, illustrated with a Bouguer gravity horizontal-gradient vector (Lyatsky et al., 1990, 1992, 1998; Edwards et al., 1998) map (for comparison, a conventional horizontal-gradient map of these data is presented by Goodacre et al., 1987b). Data from the Geological Survey of Canada, gridded at 5 km; Bouguer gravity anomaly data in Fig. 63a. Vector arrows are plotted away from local maxima or "downhill". STZ - Snowbird-Virgin River fault system ("tectonic zone"). Postulated tectonic domains: B - Buffalo Head; C - Chinchaga; Hl - Loverna; K - Ksituan; L - Lacombe; R - Rimbey; T - Taltson; Th - Thorsby; W - Wabamun. The Alberta-British Columbia border is marked by the sinuous line in the southwestern corner of the map; zig-zag line in the southeastern part of the map shows the location of the Lithoprobe seismic reflection profile (Fig. 79).

these rocks have acoustic impedances similar to that of the metamorphic crystalline rocks in the basement, and the high reflectivity within the lower levels of the Alberta and Williston basins scatters the seismic signal. As a result, the basement top is usually ill-defined with seismic surveys, even ones designed specifically to image the sedimentary cover (the top of the crystalline basement has proved even harder to depict with deep crustal seismic surveys carried out by Lithoprobe). Usually, only those fault structures are apparent in seismic sections which coincide with detectable vertical offsets of stratified rocks in the sedimentary cover. These faults were active during the platformal cratonic stage of basement-cover evolution. The pre-cratonic structures, faults and folds, usually have no direct relationship to the cover. To distinguish these two different crustal structural patterns is a fundamental task of regional tectonic analysis in a platformal cratonic region. This cannot be done with geophysical and geochemical methods alone, and it requires geological control from drillholes in the cover.

In outlining lateral tectonic units, it is particularly wrong to rely on the zero contour (or any other contour) in gravity and magnetic anomaly maps (contrary to Ross et al., 1991). Contours help to highlight the anomalies, but anomaly amplitude contours tend to be smoother than these anomalies' geologic sources. These sources, especially those of tectonic origin, are commonly polygonal. Not curved contours but linear gradient zones, straight contour segments, sharp deflections of contours are of the most value for detection of faults. Horizontal-gradient maps, especially those which treat the gradient as a vector (Figs. 73, 74, 75; Lyatsky et al., 1990, 1992, 1998; Edwards et al., 1998), help to highlight the lineaments bounding areas of distinct anomaly patterns, which may be interpreted as structural boundaries between major tectonic blocks. For this reason, potential-field lineaments, not just zero contours, need to be identified in gravity and magnetic maps, and horizontal-gradient vector maps are a useful aid in this work.

Of crucial importance are drillhole samples from the crystalline basement and the cover étages. On this criterion, Dietrich and Palmer (1996) found no correlation between the hypothetical sutures in the basement of the Alberta Platform and its Cambro-Ordovician tectonic étage. In the sedimentary cover, Eaton et al. (1995) also noted little evidence of activity on the postulated terrane boundaries. Misra et al. (1991) noted that they had been unable to a correlation of the originally postulated potential-field domains with remote-sensing images.

Despite these questions regarding their scheme, Ross et al. (1997, p. 493, 494) claimed that the Lithoprobe Alberta Basement Transects "revealed ancient, now eroded mountain belts", even though seismic data do not display mountains, eroded or standing; that these results changed "our understanding of continental dynamics of western Canada", even though geophysical data may reveal information only about deep structural characteristics of a region, not about its dynamics. These authors (op. cit., p. 493)

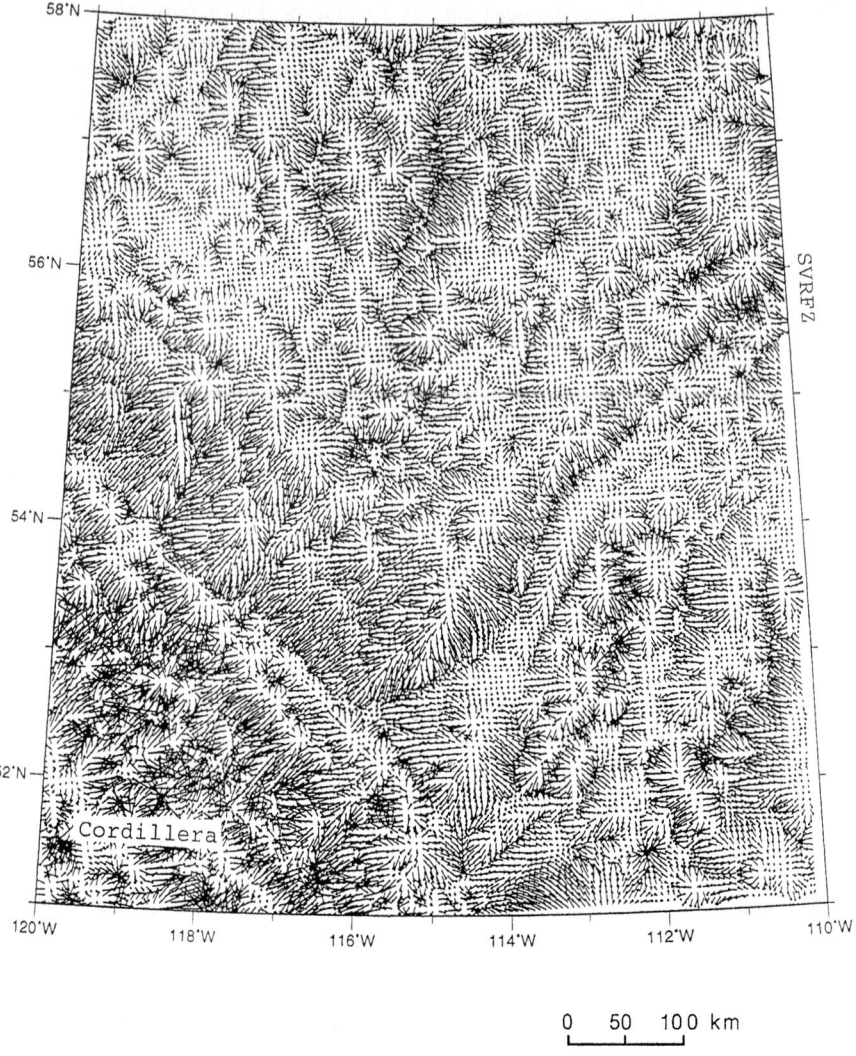

Figure 74. Bouguer gravity horizontal-gradient vector map of the central Alberta Platform, as in Fig. 73. Vector arrows are plotted away from local maxima or "downhill". Data from the Geological Survey of Canada, gridded at 5 km; Bouguer gravity anomaly data in Fig. 63a. SVRFZ - Snowbird-Virgin River fault zone. Note the predominance of NE-SW and NW-SE anomaly trends, and the continuity of cratonic NE-SW trends into the Cordilleran interior.

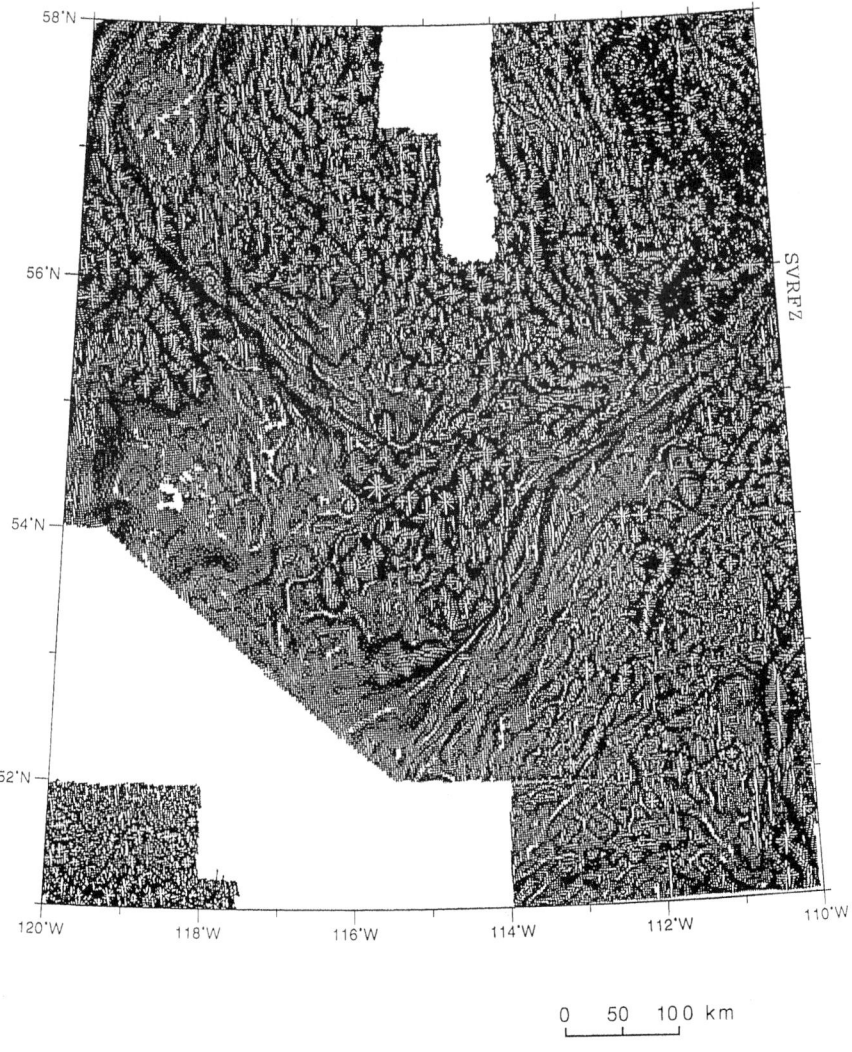

Figure 75. Total-field aeromagnetic horizontal-gradient vector map of the central Alberta Platform (same area as in Figs. 73, 74). Vector arrows are plotted away from local maxima or "downhill". Data from the Geological Survey of Canada, gridded at 2 km; total-field anomaly data in Fig. 62. SVRFZ - Snowbird-Virgin River fault zone. Note the predominance of NE-SW and NW-SE trends, and the colinearity of major NW-SE anomalies with the Proterozoic Kimiwan isotopic/structural anomaly in the Alberta Platform's cratonic crystalline basement (Fig. 56).

stated that through the Lithoprobe program, "armed with a variety of techniques, the scientists are resolving features within the sedimentary basin and unravelling the genesis of the underlying Precambrian crust and upper mantle". But it is oil-industry seismic surveys, not deep Lithoprobe ones, that are specially designed to image the sedimentary cover, and anyway, geophysical data are unable to reveal rock genesis. These authors (their p. 493) assumed that areas of Paleozoic sedimentation in the Western Canada Sedimentary Province "...lay well inboard of active basin-forming mechanisms", which makes the existence of the basin hard to explain, and the origins of the Williston Basin in their scheme are unexplained, "enigmatic".

Contrary to the meaning contained in the notion of *craton*, these authors (op. cit., p. 493) stated that they had probed a "buried craton in western Canada", even though the Interior and Midcontinent platforms are integral parts of the North American craton itself (the truly buried cratonic remnants are found in the mobile megabelts like Cordilleran and Appalachian). The statement that they were able to "examine the ca. 500 m.y. sedimentation history [of this province] as a paleoseismologic [?] record of reactivation of the craton during the Phanerozoic" is confusing.

Lessons from the failure of initial attempts to unravel tectonic history of the structural framework in the basement of the Alberta Platform

A practical need exists in resource exploration and tectonic research to distinguish cratonic-stage structures from pre-cratonic ones, structures coeval with the evolution of the cover from pre-cover cratonic structures, and typically platformal structures from those common in shields. This need, however, remains underappreciated in the North American geological literature.

During early stages of exploration, before extensive drilling, the leading role in the tectonic interpretation of the structural framework of platforms normally falls to gravity and magnetic methods, which provide regional coverage. The inferred geologic structure of the crystalline basement and the estimated relief of its top were used to roughly predict the structure of the sedimentary cover. In well-explored provinces, a reverse approach is taken: well-studied peculiarities of pre-cover volcano-sedimentary basins and of the cover itself are used to unravel the tectonic events that affected the basement. The structural scheme of the Alberta Platform basement constructed from only gravity and incomplete magnetic anomaly maps (Stephenson et al., 1989) was purely geophysical. Supplemented with a limited number of radiometric age determinations of basement samples, it acquired a more-geological sense. The path astray began when the still poorly delineated potential-field domains of uncertain nature were proclaimed to be "tectono-stratigraphic terranes" (similar to the presumed terranes in the Cordillera).

Such an adventurous interpretation stands in contrast to the traditional caution used in combining geophysical (mainly potential-field) data with the geological information about the sedimentary cover and the basement gathered in the Western Canada Sedimentary Province since the 1950s (see, e.g., reviews in Burwash et al., 1993, 1994). Errors in the initial scheme of Stephenson et al. (1989) and its successors (Ross and Stephenson, 1989) were not corrected. By now this scheme is obsolete, as it does not rely on the standard, time-tested methods of delineating potential-field domains and correlating them with geologic data. As an example, in the scheme of Stephenson et al. (1989; also Ross and Stephenson, 1989), the Snowbird-Virgin River fault zone was associated with a broad belt of variable potential-field anomalies in central Alberta (Figs. 62, 63, 74, 75). The width of this belt (and fault zone) was shown to rapidly increase westward and reach ~300 km near the Cordillera. This contradicts other interpretations (among others, Burwash et al., 1994) which recognize a narrow, transcurrent Virgin River anomaly zone from the Canadian Shield, along the NE-SW segment of the North Saskatchewan River near Edmonton, towards about 52°N/116°W (Fig. 54).

Field mapping in the Shield and drillhole data in the Alberta Platform show the Virgin River fault may indeed continue all across Alberta and into the Cordillera. It is one of the chain of faults in the transcontinental Snowbird fault system mapped which runs as far east as Hudson Bay (see Hoffman, 1988, 1990). In western Canada, to the north of it lies the similarly NE-SW-trending Great Slave Lake Zone (Hanmer, 1987, 1988; Hanmer et al., 1994). Both these fault systems cross the modern Alberta Platform as well as the eastern Cordilleran miogeosyncline (cp. Cecile et al., 1997). Both of them appear in gravity maps of the Alberta Platform (Fig. 74) as narrow strings of lows (Burwash and Culbert, 1976; Edwards et al., 1998). In horizontal-gradient and other derivative gravity and magnetic anomaly maps (Geological Survey of Canada, 1990; Lyatsky et al., 1992; Edwards et al., 1998; Figs. 74, 75), they are well expressed as long lineaments of regional significance, some 20-40 km wide. They separate large distinct potential-field domains with sharply dissimilar anomaly signatures. Numerous deviations of potential-field anomaly amplitude contours, and sharp changes of contour curvature or linearity, indicate the position of these fundamental NE-SW structural zones, which coincide also with known geologic or topographic lineaments (e.g., Misra et al., 1991). Cecile et al. (1997) again confirmed the significance of the major ancient NE-SW-trending faults, Snowbird-Virgin River, Great Slave Lake and some others, for the entire depositional and structural history of the Interior Platform in Canada. Major faults with this orientation are also well known in western U.S., as far south as the Colorado mineral belt which includes the Colorado lineament of Warner (1978). In the Alberta Platform, during its tectonic stages, big crustal blocks defined by these huge faults predominantly subsided and were variously tilted.

Practical geologists need solid facts and fact-based assumptions to meet the demands of their work, which include risk reduction of the costly exploration and environmental work. Exploration and environmental companies and agencies require realistic tectonic predictions. Regional analysis of tectonic étages and stages, as outlined above, provides the most reliable constraints, which are derived from observed properties and characteristics of rocks and rock bodies. Each (volcano-)sedimentary étage contains its own, specific structural fabric expressed in the distribution of basins and arches during the corresponding tectonic stage. This puts factual limitations on the subjective imagination and *a priori* prejudices. With objective clues obtained from proper identification and delineation of tectonic étages and accurate definition of particular structural features in them, it is possible to better understand the tectonic history of the entire crust in a particular cratonic area during the particular tectonic stages without excessive reference to "far-acting basin-forming mechanisms", "far-field stresses", mysteries and enigmas.

In a recent summary of the geology of the sedimentary cover of the North American craton in Canada, Aitken (1993a-c) rightly pointed out that the common but loose term *basement control* is vague and therefore hard to apply properly. Indeed, the notion of basement control appeared from early applications of gravity and magnetic methods intended to predict the structure of the sedimentary cover ahead of drilling. These geophysical techniques, especially the magnetic ones, mostly provide information about the basement. Some of it, concerning basement relief and fault offsets, may be extrapolated upward and used to predict the structural and reservoir characteristics in the cover. *Basement control*, in this sense, was and remains important, but linguistically it is no more than colloquial jargon. Aitken, apparently taking it too literally, supposed that basement control requires a "near 1:1 correspondence throughout the length of basement belts".

To expect such correspondence is unrealistic, and it is not apparent from correlations of the structural characteristics of platformal basement with basins and arches in the cover étages. Ancient orogenic belts of the pre-cratonic stage are fossilized in the shield and platformal crystalline basement areas. After cratonization, the continental upper crust lost its ability for folding. Its purely cratonic tectonism was manifested instead mainly in brittle deformation commonly expressed in block movements. Buckling and warping which occurred in the crust and lithosphere contributed to the mechanical breakage of the crust into blocks. The block division of the cratonized crust is caused by major steep, deep-rooted faults in the crust and even subcrustal lithosphere. Big brittle fractures propagated variously from bottom to top and from top to bottom.

Radiometric and fission-track studies have shown that cratonic vertical movements were substantial, commonly many kilometers in amplitude, and they began long before the development of a continuous sedimentary cover. Proterozoic cratons contained the pre-cratonic network of Early Proterozoic mobile megabelts and the preserved Archean proto-cratons between them. However, the new stresses reactivated only a few of the ancient crustal zones of weakness. Most of the Proterozoic cratonic brittle faults, however, were new. Unfortunately, although they are of primary importance for reconstructions of post-orogenic tectonism, cratonic fault systems are less obvious in gravity and magnetic anomaly maps because they are not always associated with intrusive magmatic bodies.

Even the crustal structure of modern cratons may be partly and locally inherited from recognizable remnants of ancient structural features of the pre-cratonic stage. But mainly, the cratonic reorganizations created a new fabric, which formed due to movements of blocks bounded by brittle, steep, cratonic faults. The products of crustal warping are less explicit, except for some volcano-sedimentary basins, but warp and block movements have continued episodically during the entire cratonic evolution. Vertical cratonic crustal deformation, not pre-cratonic foldbelts, created the pre-cover and syn-cover cratonic topography and cratonic basins and arches. For this reason, correspondence between the orogenic basement structural fabric and the structure of the platformal sedimentary cover is encountered only rarely (e.g., Eaton et al., 1995; Ross et al., 1997).

Even the correlation of cratonic-stage basement structural irregularities with the platformal cover is never 1:1, because many cratonic faults were active before but not during the deposition of the cover. A close correspondence with the cover is to be expected only with syn-depositional and post-depositional faults, which affected their contemporaneous and older étages. Tectonic structures in the basement and sedimentary cover may coincide because the cover has no internal sources for a tectonic activity of its own. The deeply rooted tectonism in the cratonized continental crust affects both the basement and the cover, giving them common features.

Variations in the arrangement of crustal warps and blocks from one tectonic stage to another were caused by these deep-seated processes of episodic crustal reworking and restructuring, with resulting tectonically-induced changes in paleogeographic settings and deposition patterns. This permits to use the distribution of basins and arches in different structural-formational étages to reconstruct the history of tectonic movements in the shields and platforms. There is nothing mysterious about vertical tectonic movements and the resulting development of platformal arches and basins, as long as the observable geology is analyzed without pre-conceived biases.

Construction of depth-to-top maps of regional marker surfaces, and their tectonic utility

To aid the restoration of structural history of the Alberta Platform, a series lithofacies, thickness and depth-to-top maps was used for structural-formational étages separated by or contain within them reliable, distinct and regionally persistent markers (Chapter 3). In constructing depth-to-top maps, an interpreter has to decide whether the top of the underlying older rocks or the base of the overlying younger rocks in an unconformity is less diachronous and thus more appropriate for mapping. Flattening on a selected marker surface restores the older surfaces to their position at the time of the marker. Such flattening (cp. Fig. 76) helps reveal where the warps were located, which blocks were up and down, which faults were activated to accommodate the crustal movements during that time, and so on. Unfounded prejudices about a persistent westward tilt of the Alberta Platform, or about westward-thickening stratigraphic wedges all through the stratigraphic column of the Alberta Basin, are dispelled by these maps.

In the sedimentary cover of the Alberta Platform, there are several such pervasive stratigraphic markers with little diachronism, which permit to reconstruct the structural fabric during the demarcated tectonic stages. One such marker is the unconformity capping the distinct étage of Cambrian, Ordovician and Silurian age. It corresponds to a stratigraphic break spanning some 50 m.y. prior to the Devonian étage. An abrupt break change separates the Middle and Upper Devonian strata. This boundary is well recorded in a lithologic and biostratigraphic change, and is easily traceable across the region. The third reliable marker is the unconformity at the base of the Lower Cretaceous Mannville Group. It spans ~100 m.y. Depending on the depth of erosion, it marks the erosional top of the Mississippian or Upper Devonian.

Reliable regional markers are lacking for post-Neocomian time in Alberta. Extreme variability of paleogeographic environments and lithofacies at that time allows to compile only conceptual paleogeographic and sequence-stratigraphic maps. Lacking reliable markers, these schemes are rather subjective.

Accuracy of depth-to-top maps is reduced where well penetrations are sparse, as is often the case with the basement top in the Alberta Platform. Automatic contouring may either smooth out the undersampled real features or create artificial angularity where none exists. Where well data are sparse, computer-generated artifacts may contaminate the contour maps considerably, or leave real fault offsets unresolved. When viewing contour maps, one needs to mind if the apparent breaks and gradient zones are justified by the data. Despite these shortcomings, correlation of such maps with gravity and magnetic data, and with detailed seismic data, is very helpful. By using these data sets together, it is possible to locate even those faults whose offsets are small, which are often overlooked in the seismic profiles. For practical tectonic analysis of platforms, one needs to know the position and structural role of prominent and subtle faults. Vertical offsets can be restored by flattening the stratigraphic horizons on marker surfaces representing the time of interest. In exploration, these maps help to understand not only the pattern of active faults but also the former distribution of provenance areas, direction of sediment transport, location of areas of deposition, and the possible timing and routes of migration of hydrocarbons.

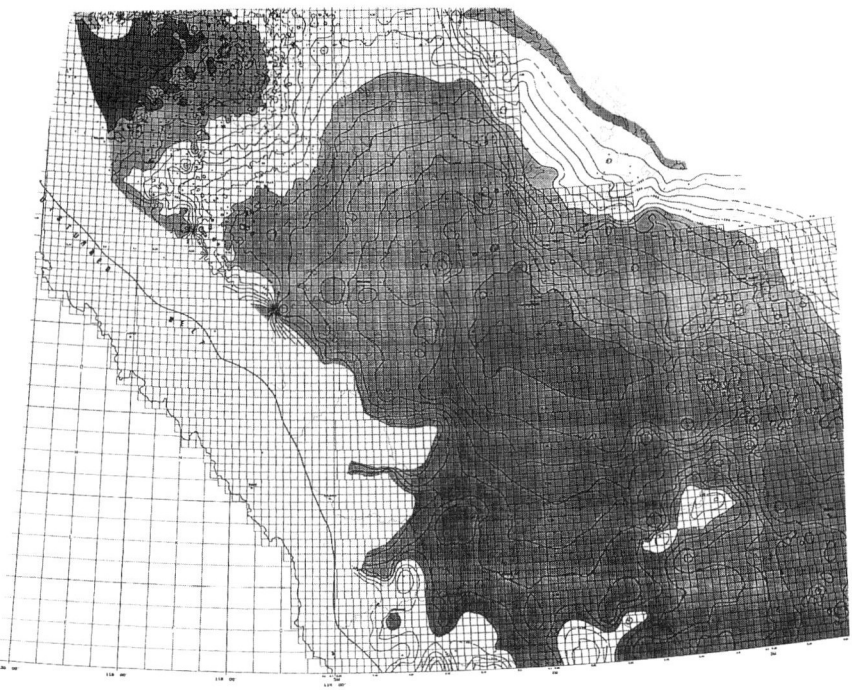

Figure 76a. Map of the relief of the top of crystalline basement in the Western Canada Sedimentary Province, flattened on the top of Slave Point. This map reveals basement depressions and rises that existed at that time.

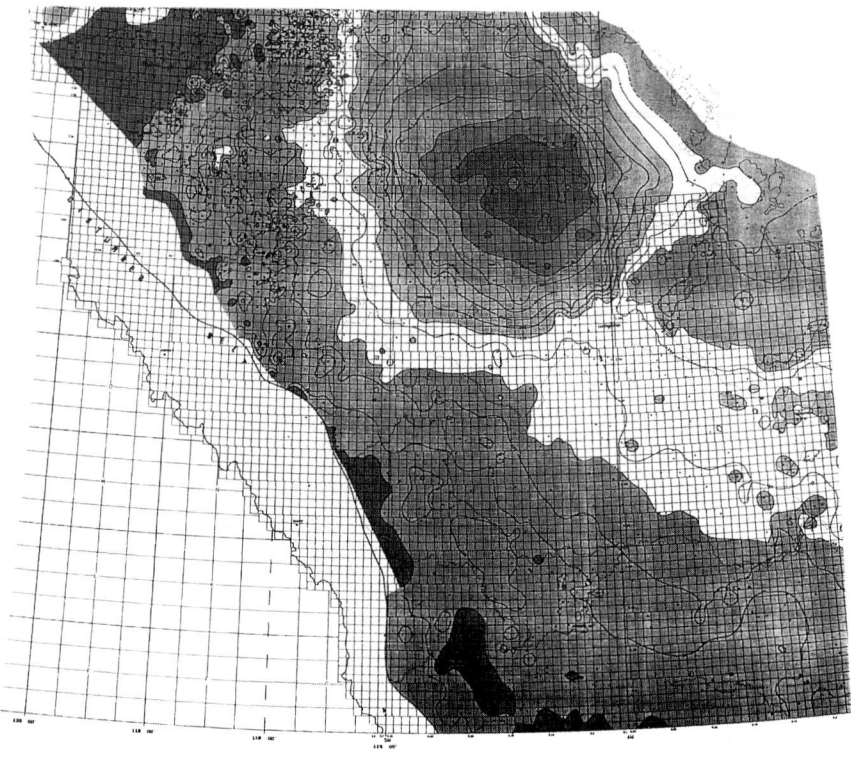

Figure 76b. Map of the relief of the pre-Devonian marker, flattened on the top of Slave Point.

Each region is geologically unique, and not all empirical principles are applicable everywhere. In the Western Canada Sedimentary Province, for instance, seismic picks of the basement top are often unreliable, because beneath the very reflective (and signal-attenuating, multiple-producing) Paleozoic succession of interbedded carbonates, evaporites and shales, to resolve the basement top is notoriously difficult. The Devonian carbonates, which lie on top of the crystalline basement, commonly have an acoustic impedance not much different from the impedance of basement rocks. This complicates their discrimination in seismic data in many areas.

Another complication arises due to velocity pull-ups in seismic sections across massive carbonate bodies (e.g., barrier reefs common in the Alberta Basin). These pull-ups create false images of non-existent basement highs, in particular under the famous, oil-rich Leduc reefs, adding to the interpretational confusion.

Basement-sourced magnetic anomalies are not always related to the basement top, and they may also be distorted by local intra-cover anomaly sources (e.g., channel deposits containing magnetic minerals, mineralized reefs and fracture zones, volcaniclastic deposits). Basement-depth determinations from gravity, magnetic and seismic data are widely known to be unreliable in the Alberta Platform in general, especially where drillhole calibration is sparse. Even drillhole rock samples from the basement can be deceptive, as weathered rocks from the top of the basement and near-basement sedimentary rocks made up of coarse, reworked basement-derived detritus often look similar, thus undermining the reliability of some basement-top picks.

The boundaries of a platform are conventionally defined by the extent of its cover. Whereas the basement of cratons can be as old as Archean to Mesoproterozoic, the platformal cover has developed in the Neoproterozoic to Recent. Some tectonic platforms, overlapping the extinct Middle/Late Proterozoic through Phanerozoic mobile megabelts, are, by definition, non-cratonic; they are younger and less stable. Examples are the southern part of the Midcontinent Platform over the Grenville-Ouachita foldbelts in the U.S., or the North European Platform between Poland and southern England where the undeformed flat-lying sedimentary cover overlies a Paleozoic Caledonian and Variscan basement (Fig. 77).

Old cratons extended into regions now occupied by mobile megabelts. Some pieces of these ancient cratons are identified geologically in miogeosynclines and median massifs. In miogeosynclines, pre-existing cratonic platforms and shields were buried to depths of 20-30 km and more (as, for example, in the Omineca Belt of the Canadian Cordillera). In median massifs (e.g., the Intermontane Belt and Yukon-Tanana province in the Cordillera in British Columbia and Alaska), the burial and reworking was much less intense. Cratonic shields and platforms that were not included into mobile megabelts

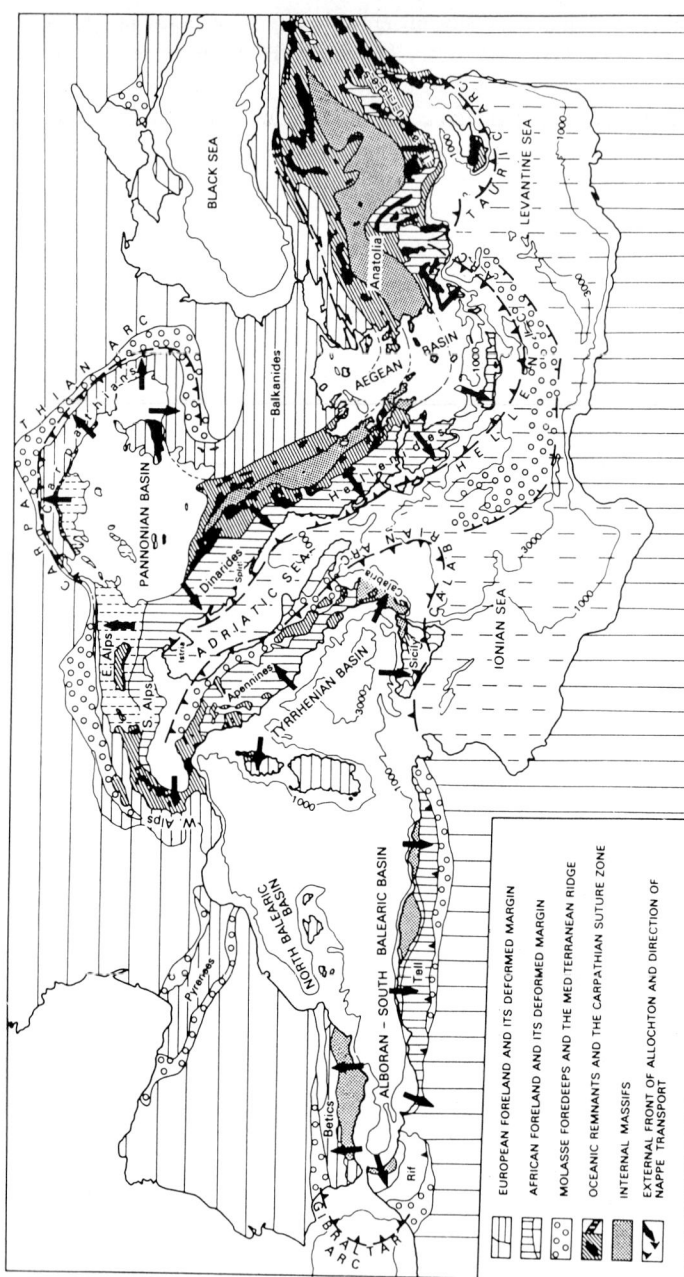

Figure 77. Tectonic map of southern Europe and Mediterranean region, showing the Alpine mobile megabelt, Pannonian Basin (lying on top of a median massif), and platformal regions to the north (modified from F. Horvath, AAPG©1988; reprinted by permission of the American Association of Petroleum Geologists).

were buried and reworked even less, as evidenced by the preservation of unmetamorphosed sedimentary cover.

Time and again, tectonic regimes typical for cratons expanded temporarily towards mobile megabelts, but not beyond the miogeosyncline-platform domains. Such episodes have been recognized on the rims of the Russian, Siberian and North American cratons.

Obvious median massifs have for decades been distinguished as comparatively stable crustal blocks atypical for mobile megabelts. They served as buttresses in relation to orogenic belts which were deflected around them. The best-studied examples of median massifs are the Pannonian massif in the Carpathians of Central Europe and the Intermontane massif of the Canadian Cordillera, whose basement and cover bear a record of strong orogenic influences in the specific types of sedimentation, magmatism, metamorphism and deformation. By contrast, ancient miogeosynclines had sedimentation of a type transitional to platforms, and are less magmatic than most other parts of the mobile megabelt; but rocks in them are nevertheless metamorphosed and deformed. Tectonic platforms, whether cratonic or not, have a sedimentary cover essentially lacking metamorphism and magmatism, and the styles of indigenous deformation in them are limited to vertical warping, high-angle faulting and block movements.

The main difference between the fully cratonic and younger tectonic platforms is that the latter exhibit more manifestations of tectonic activity inherited from the preceding orogenic stage. In young platforms, the linearity of basement uplifts (ridges, arches, etc.) and depressions (used as depocenters) is more pronounced than in fully cratonic platforms. This structural linearity is closely linked with the orogenic belts in the basement.

The transitional miogeosynclinal-platformal tectonic domain in western Canada contains the same kind of thick, extensive Devonian carbonate platforms as in the Alberta Basin. Remnants of these carbonate platforms are found in many parts of eastern Cordillera, including the Omineca Belt (Chapter 3). As the foreland evolved in this domain (since the Late Devonian and Carboniferous), the transitional zone between the platform-shield tectonic domain of the platform and the increasingly orogenized miogeosynclinal belt of the Cordillera became narrower, while the foreland basin shifted cratonward. In the Early Cretaceous, the eastern flank of this transitional zone lay as far east as the Pembina bulge in the Alberta Platform.

During the Late Cretaceous-Tertiary tectonic stage, manifestations of the powerful Laramian tectonism in the Cordillera spread into the Alberta Platform as well. Eastward advance of the Rocky Mountain fold-and-thrust belt was different in the southern, central and northern parts of the Cordilleran miogeosyncline adjacent to the Alberta Platform. The farthest-advanced thrust sheets caused the foredeep in front of them to shift to the east, as they partly overlapped the western part of the foreland sedimentary basin. But even the enormous stack of Laramian thrust sheets did not produce a noticeable bulge at the platformward edge of the Tertiary foredeep in Alberta. The thin-skinned thrust stacking was only a response to active orogenic processes which were concentrated to the west, in the Omineca miogeosynclinal belt, while deep crustal levels in the Alberta Platform remained essentially uninvolved.

Cratonic shields and platforms as contrasting features in paleogeography and tectonics

Cratonic shields and platforms are distinguished by, respectively, the absence or presence of a continuous sedimentary cover, which in platforms overlies metamorphic rocks usually of Archean to Early-Middle Proterozoic age. This division is somewhat arbitrary structurally and temporally. A bipartite basement-cover regional structure formed in some regions as early as in Late Archean, and broadened in the Early Proterozoic. In the Early and Middle Proterozoic, many volcano-sedimentary basins lay between basement arches. The extensive, continuous Late Proterozoic and Phanerozoic sedimentary cover, whose appearance reflected the formation of broad tectonic platforms, overlapped disparate metamorphic terrains of preceding epochs. The time gap between the erosionally beveled basement and the basal horizons of the cover in most cratonic platforms is huge, but variable from area to area. The age of the basement varies greatly. The onset of cover sedimentation occurred in places in the Mesoproterozoic, in places in the Neoproterozoic Ediacaran or Vendian, and most commonly in the Phanerozoic Cambrian.

Edges of the sedimentary cover, which are often used to separate the shields from the platforms, actually shifted considerably in time: during some tectonic episodes, as sedimentation spread, they advanced over the erosionally beveled cratonic terrains; at other times, they retreated, and the thin sedimentary veneer they left behind was eroded to again expose the metamorphic bedrock. The zero edge of the cover can thus be historically ephemeral and tectonically meaningless. This is particularly true of the western and southern boundaries of the present-day Canadian Shield with the Interior and Midcontinent platforms.

Tectonic shield-platform boundaries, in contrast, are apparent where big faults separate shield-confined crustal blocks from platform-confined blocks (e.g., boundaries of the narrow, elongated Lawrence Platform in southeastern Quebec or some boundaries of the round Hudson Bay Platform in northern Ontario; Fig. 78a). Erosional and tectonic boundaries can easily be confused. In Asia, west of Lake Baikal, the huge Baikal-Enissey fault separates a cratonic tectonic region, which contains the Phanerozoic platformal Irkutsk Amphitheater as well as some Precambrian crystalline shield terrains, from a mobile megabelt to the south (Fig. 78b). But some workers (e.g., Mitrofanov and Taskin, 1994) traditionally put the craton's boundary not at the fault but along the

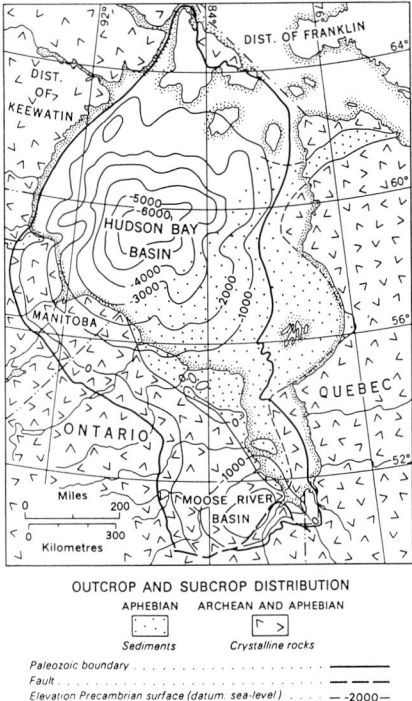

OUTCROP AND SUBCROP DISTRIBUTION

APHEBIAN   ARCHEAN AND APHEBIAN

Sediments   Crystalline rocks

Paleozoic boundary . . . . ───────
Fault . . . . ─ ─ ─ ─
Elevation Precambrian surface (datum: sea-level) . . . . . ─ -2000 ─

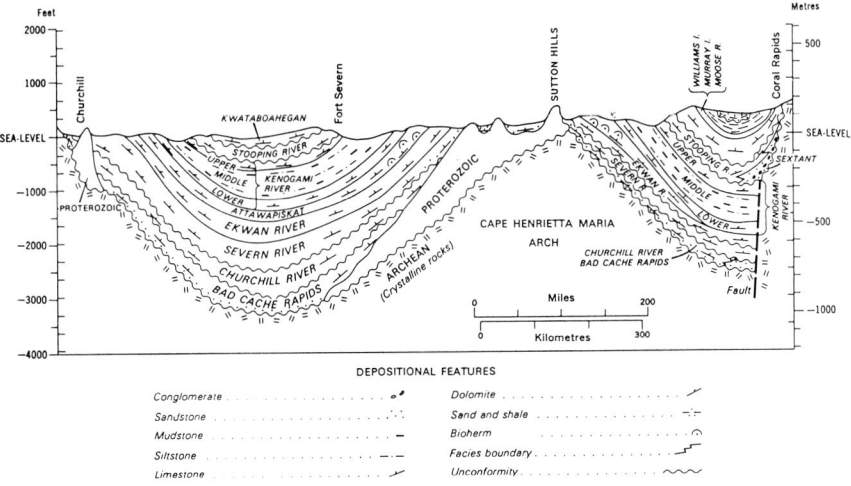

DEPOSITIONAL FEATURES

Conglomerate	. . . . ∘	Dolomite	. . . .
Sandstone	. . . .	Sand and shale	. . . .
Mudstone	. . . . ─	Bioherm	. . . .
Siltstone	. . . . ─ ─	Facies boundary	. . . .
Limestone	. . . .	Unconformity	. . . . ∼

Figure 78a. An example of relationship between steep crustal faults, erosional, and platform boundaries: southern Hudson Bay Platform (simplified from Stockwell et al., 1970). Both types of platform boundary, structural and non-structural, are seen in this region.

234

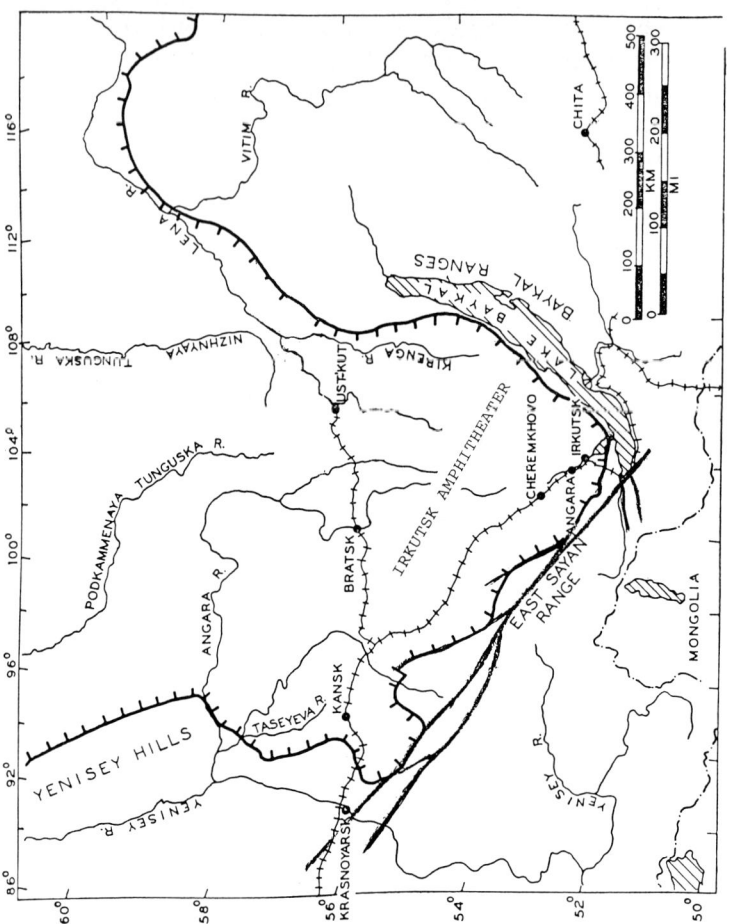

Figure 78b. Mismatch between the Baikal-Enissey fault (in the East Sayan Range) and the shield-platform boundary in the Irkutsk Amphitheater (Basin) region in southern Siberia. Barbed line represents basin edge (shield-platform boundary), whereas the Baikal-Enissey fault separates the craton from the mobile megabelt. Cratonic crystalline rocks are exposed in the zone between this fault and the basin edge, which belongs to the craton but not to the basin. Location in Fig. 4c.

erosional zero edge of the Phanerozoic platformal cover that separates the Amphitheater from the shield areas. The Fennoscandian Shield in northern Europe is also flanked by sedimentary basins, whose zero edge is erosional and not tectonic.

Erosional shield-platform boundaries depend on changeable factors such as fluctuating sea level, regional and local crustal subsidence and uplift, rates of sedimentation and of subaerial and submarine erosion. Tectonic boundaries, in contrast, are related to faults, which makes them more stationary. These faults may be of different sizes, up to lithospheric-scale, and they narrowly restricted the geographic band within which paleogeographic boundaries could shift.

On a continental scale, cratons are huge lithospheric blocks, and tectonically determined shields and platforms are their first-order constituents. Since the Neoproterozoic, tectonic shield-platform boundaries separate blocks of predominant uplift (in shields) from those of predominant subsidence (in platforms). Such persistent dynamics, for hundreds of millions of years, can only be explained by forces applied from below due to deep-seated tectonic processes. Heating, for example, makes the blocks more buoyant and causes them to rise; removal of a heat source results in sagging. Other causes of block uplift and subsidence could be density changes of rocks at their deep levels due to changing metamorphic conditions, and so on.

But as these processes fluctuate, the crust retains a memory of its reworking. This is apparent from surface geologic mapping and geochemical analysis of rocks, and is suggested by deep geophysical surveys. The crustal structure in ancient shields and Neoproterozoic-Phanerozoic platforms is now known to be different. Although pre-cratonic orogenic zones can be traced from shields into platform basement (e.g., in potential-field maps), the crustal seismic structure of contrasting cratonic tectonic blocks is not the same - and the underlying upper mantle seems to vary as well (Hajnal et al., 1997).

In North America, since the latest Precambrian and Early Cambrian, and especially since the Ordovician, tectonic subsidence expanded laterally over much of the craton. The main tendency of the Canadian Shield to rise and of the Midcontinent and other tectonic platform areas to subside has been strong throughout the Phanerozoic. The differentiation of the Proterozoic cratonic mass into the shield and platforms, established near the Proterozoic-Phanerozoic boundary, exists to this day.

The initial pericratonic subsidence was broad and uniform. Further tectonic differentiation of the platforms occurred later, separating the Midcontinent Platform from its Interior counterpart and segmenting the Interior Platform into smaller parts. The shield, meanwhile, remained less differentiated internally, rising slowly (a notable exception is the Transcontinental Arch, which subsided following its initial rise).

Different geodynamical systems functioned deep in the crust and lithosphere of the contrasting tectonic domains of the continent (cratons and mobile megabelts, shields and platforms), though some interactions occurred between them. Structural and formational characteristics of the tectonic étages in the Alberta Platform show that the initial pericratonic basin has existed there since the Middle Cambrian. The foreland basin, whose internal variations largely reflected the along-trend differences in the Antler orogenic belt in the U.S. and Canada, appeared in the Late Devonian-Early Carboniferous. The first foredeeps on the western margin of these pericratonic and foreland basins were formed in the Late Carboniferous and Permian, but mostly the foredeeps evolved later, the Late Cretaceous and Early Tertiary, as the Laramian fold-and-thrust belt prograded. As can be seem from the detailed analysis of the Alberta Platform, the boundaries of its sedimentary cover shifted through time, from one tectonic stage to another, due to restructuring episodes reflecting rearrangements in the endogenic geodynamical systems. The structural rearrangements associated with changes in the regional tectonic regime (primarily, stress field) gave each étage its own characteristic structural-formational face. Some of the platform's structural features were short-lived, others persisted through several étages and stages. Some of them were strongly localized in certain deeply rooted zones, such as the Athabasca-Peace River-Skeena zone of crustal weakness.

Southeast of the Alberta Platform (and Basin), purely cratonic restructuring in the Western Canada Sedimentary Province created the undisputably intracontinental and intracratonic Williston Basin belonging to the Midcontinent Platform. Structural-formational étages in this basin have been described in detail by Sloss (1963, 1988a). In the deepest part of the basin, in North Dakota, their total thickness is up to almost 5,000 m. Nine major unconformities developed in this basin's Paleozoic and early Mesozoic successions in response to vertical tectonic movements of the crust. The considerable thickness and round shape of this basin imply considerable bending of the crust during the long periods of subsidence, interrupted by block-like movements up and down. The upper crust responded to such bending brittly, and the patterns of many local blocks and faults is complex (Gerhard et al., 1990). From subsidence curves (e.g., Haidl, 1991), the most rapid subsidence on the site of the Williston Basin occurred in several relatively short intervals: 525 to 490 Ma, 450 to 420 Ma, 390 to 330 Ma, 270 to 250 Ma, 100 to 50 Ma. Between the episodes of rapid subsidence, there were periods of slower subsidence or upward inversion. Vertical offsets of major fault-bounded blocks reach 400-500 to 1,500 m (Redly and Hajnal, 1997). These faults were studied in detail because they control the lithology and structure of Paleozoic and Mesozoic étages. Extensive drilling confirms long-lived fault activity (e.g., in the Cedar and Nesson anticlines) which affected various étages; the Cedar fault is mappable even at the ground surface (e.g., Gerhard et al., 1990).

Round-shaped basins are known in Proterozoic terrains of the Russian (e.g., Igolkina, ed., 1981) and North American (e.g., Campbell, ed., 1981) cratons as well as in younger platforms the world over. Amadeus Basin in Australia, Michigan Basin in North America, Ordos Basin in China, Paris basin in France - all of them possess striking similarities in their structural characteristics, despite differencesin their basements. The Williston Basin in the Cambrian to Tertiary cover on an Archean to Early Proterozoic cratonic basement, and the Paris Basin in the Triassic to Tertiary cover on a Paleozoic basement, both resulted from downwarping complicated by block displacements. The Williston Basin is six times bigger than the Paris Basin (300,000 km^2 vs. only 50,000 km^2), but the >3,000-m thickness of the latter suggests even sharper crustal downwarping than in the Williston Basin.

Such intraplatformal downwarps contrast with the upwarped shields. A chain of positive and negative warps could explain the swells and necks in the outline of the Interior Platform between the Canadian Shield and the Rocky Mountains (as was discussed above). The fundamental restructuring of the entire craton near the Proterozoic-Phanerozoic boundary, which resulted in the differentiationof platformal and shield areas (though their boundaries were still modified later), was apparently associated with deep-seated intracratonic tectonic processes which were dissimilar in platforms and shields.

Like a shoreline between land and sea, an erosional edge of the sedimentary cover is very ephemeral, although in some areas (e.g., on the south side of the Sino-Korean Shield) major crustal faults control the position of erosional shield-platform boundaries. These faults may be exposed when the edge retreats towards the platform, as happened in the Great Lakes region in North America, or they may be hidden beneath the cover.

In the practice of economic geology, agriculture, environmental work, the cover blanketed platforms and bare shields are the terrains of principal importance. But for many purposes, including resource exploration, of great importance are the tectonic boundaries between the shields and platforms.

6 FINDING AN ADEQUATE REGIONAL TECTONIC INTERPRETATION CONSTRAINED BY MAPPABLE PROPERTIES OF ROCKS

Rationale in tectonic studies of shield terrains

Two harmful prejudices tend to occur in studies of cratonic regions: (a) that all cratonic tectonism is induced from outside; and (b) that all orogenies are produced by collisions of continents and/or subduction of oceanic lithosphere. Wilson (1966) postulated that continental masses are split repeatedly; that oceanic domains are opened due to this splitting; that later they close again, accompanied by collsions, subduction, orogenic compression, etc. In the so-called Wilson cycle, the Atlantic Ocean supposedly opened and closed at least twice, first time as the pre-mid-Paleozoic Iapetus Ocean. Following the same logic, a so-called Manikewan Ocean was proposed to have existed on the site of the Trans-Hudson Orogen in the Early Proterozoic (Stauffer, 1984; Symons, 1991). This ocean supposedly closed by about 1,800 Ma, and remnants of oceanic crust have been postulated along some of the assumed tectono-stratigraphic terranes (Hoffman, 1990). In this scheme, the Archean terrains in the Canadian Shield (the Superior and Slave cratons) were regarded as Archean continents with separate histories of motion. They were supposedly supplemented with accreted juvenile crust, and amalgamated together into the Laurentian craton (Hoffman, 1988, 1990). This hypothesis of tectonic evolution of the North American continent is currently dominant.

But the regional geology is more complex than that, and many facts contradict this simple scheme. The nature and extent of the so-called Trans-Hudson Orogen is not clear (e.g., Lewry and Collerson, 1990). The room for accreted oceanic crust is reduced by the recognition of many large Archean blocks there, including the recently outlined Archean Saskatchewan (Sask) craton. The southward continuation of this orogen into the U.S. is only a foldbelt separating the Superior and Wyoming cratons (Bickford, 1988; Leclair et al., 1997). In South Dakota, the N-S Trans-Hudson trends are truncated by E-W structural trends attributed to a younger Proterozoic orogen (Central Plains or Mazatzal, of about 1,790-1,630 Ma; Van Schmus et al., 1987; Van Schmus and Bickford, 1993) in the basement of the Midcontinent Platform.

Lewry et al. (1996) have recently summarized the history of the Trans-Hudson Orogen in Saskatchewan as an assumed succession of seven major Early Proterozoic events: (1) rifting of a previous Archean craton and oceanic-arc development (on the site of the Reindeer zone) around 2,100 to 1,920-1,870 Ma; (2) amalgamation of juvenile-arc and oceanic-crust fragments into a Glennie-Flin Flon proto-continent (or superterrane) from

1,870 to 1,830 Ma; (3) accretion of island arcs (La Ronge, Lynn Lake, Rottenstone belts or terranes) to the Wollaston (Hearne) continent's margin and the formation of the Wathaman-Chipewyan batholith at ~1,855 Ma; (4) development of southern-polarity subduction with numerous overriding microplates, and appearance of a continental magmatic arc over the Glennie-Flin Flon continent, with the Kisseynew Basin in its backarc; (5) collision of this continent with the Sask continent (or block) around 1,835 to 1,810 Ma; (6) collision of the Hearne-Rae-Slave craton with the Superior craton and formation of a suture in the Thompson Belt after 1,800 Ma; and (7) post-1,780 Ma whole-region and whole-crust block movements. This scheme incorporates the old postulations from the 1970s and 1980s, and competes with many other such schemes with different scenarios.

Baird et al. (1996, p. 416), for instance, argued that comparison of their E-W COCORP deep seismic profile across Montana and North Dakota with a parallel Lithoprobe profile in Saskatchewan (described by Lucas et al., 1993; White et al., 1994; Fig. 70) "documents significant along-strike variation in orogenic evolution". In reality, seismic data "document" nothing, but merely record an acoustic image of crustal geometry (not "evolution"). But more important for the present discussion are the "variations" in seismic images between these two transects. Baird et al. (1996, p. 423) stated that the band of east-dipping reflections in their data is evidence "against the continuation of the Superior province westward beneath the Trans-Hudson [Orogen]". But just to the north, in Canada, the Superior E-W structural trends expressed in potential-field maps continue much farther to the west, into southern Saskatchewan (Lyatsky et al., 1998), where Superior-age rocks have been recovered in drillhole basement samples as well (Collerson et al., 1990). Baird et al. (1996) regarded this east-dipping reflection band as "east dipping thrust faults that carried Superior province *over* the eastern margin of the orogen" (italics ours). Yet, field geology shows that shearing in the exposed Thompson boundary zone in Caaba was mostly in a dip-slip sense (Fueten and Robin, 1989); the exposed thrusts there have a westerly vergence, so that Proterozoic formations are thrusted on top of the Archean Superior rocks (Weber, 1990). Baird et al. (1996, p. 424) reported some near-horizontal lower-crustal reflections in the eastern part of the Wyoming craton, and interpreted them as "an extensional feature associated with the initial opening of the Manikewan ocean". However, subhorizontal lower-crustal reflectivity is normal in the cratonic continental crust. Because their data show "no prominent low-angle reflector on the Trans-Hudson profile", these authors rejected the idea of" underthrusting of the interior orogen by a passive [Superior] continental margin" and argued instead that "Superior province crust was thrust westward over the [Trans-Hudson] internides... at least during the final stages of the Hudsonian collision" (op. cit.). Of course, to derive such an elaborate tectonic scenario from a seismic profile amounts to overinterpretation.

The large Archean blocks along the axis of the N-S-trending South Saskatchewan crustal arch make up half the width of the province between the Superior shield and the Wyoming-craton Custer platform. The probability that the Kisseynew block is also Archean further increases the proportion of Archean blocks in the "Trans-Hudson Orogen". Thus, what has been lumped together as this "Orogen", at least in its N-S branch in Saskatchewan and the Dakotas, seems to be a series of Archean blocks formed in situ and dissected by many deep and broad rifts. These rifts were squeezed together between the unreworked Archean cratons and blocks that had been detached from them by rifting. The composite Glennie-Hanson Lake block, exposed in central Saskatchewan, is more than 110 km wide, wedging out to the north and south. The outline of this polygonal block is defined by major, steep crustal faults. The N-S-oriented Tabbernor fault zone, which runs along the entire Trans-Hudson Orogen, as well as other major faults such as Saskatoon, Thompson and others, is straight and steep (e.g., Saskatchewan Geological Survey, 1994), though with some overturned splays at shallow crustal levels (Gordon et al., 1990). It is long-lived, having been active since before 1,735 Ma and reactivated repeatedly in the Phanerozoic (Elliott, 1997).

To rely too heavily on seismic images of the deep crust in studies of regional tectonic evolution may be very misleading. These images depend on the specific parameters and techniques of data acquisition and processing used in a survey, and no geophysical image can be interpreted uniquely. It was noted recently, for example, that steep faults and dikes crossing a seismic profile at an oblique angle may produce dipping low-angle reflections cutting across other reflection bands; these and other complications are often "overlooked and misinterpreted" (Zaleski et al., 1997).

Not just geophysical but also geochemical data are not subject to unique interpretation. Even eclogites are now thought to have different possible origins (Snyder et al., 1997), and the so-called juvenile crust does not necessarily have an oceanic-plate origin. Only a broad geological visualization of different data sets makes it possible to compile an internally non-contradictory tectonic-evolution concept.

Most of the geochemically juvenile rocks in the Trans-Hudson Orogen are assigned to domains between the Glennie and Kisseynew blocks, such as Flin Flon (Thom et al., 1990). Mafic gneisses and amphibolites in it are now attributed also to the basement of the Namew Basin (Leclair et al., 1997). According to Thom et al. (1990), the Flin Flon rock assemblage has geochemical and isotopic signatures comparable to those in island arcs, whereas the metavolcanic rocks of the Glennie-Hanson Lake block are remnants of continental magmatic arcs (also Watters and Pearce, 1987). A tectonic contact separates the Hanson Lake and Glennie blocks from the large Kisseynew block, which also seems to be Archean and continental. Farther east, across the Thompson fault zone, the Superior shield (craton) is undisputably Archean and continental, as is the newly outlined Sask craton (Glennie-Hanson Lake block).

The Kisseynew block is distinctly rectangular. Presence of an Archean crustal block beneath is the most likely explanation for this shape. Its mappable extent is represented by the distribution of an 2-2.5-km-thick assemblage of paragneisses, whose protoliths were siliceous clastic sedimentary rocks (sandstone, conglomerate). Pebbles having Trans-Hudson Orogen affinities are found as well (Stauffer, 1990). This distinctive assemblage is usually denoted as the Kisseynew Basin, with the Missi Group at its bottom.

The Missi Group makes up a distinct structural-formational étage formed at ~1,800 Ma in the late Paleoproterozoic. The orientation and vergence of folds in it vary; particularly around some big granitic plutons and what seems like uplifted reworked basement rocks (Gordon et al., 1990; Zwanzig, 1990). The most intensive, isoclinal and recumbent folding is concentrated along the major block-bounding and basin-bounding faults, whose orientation is sharply discordant with the general N-S structural grain of the Trans-Hudson Orogen. Indeed, large granitic plutons found in the central part of the basin have strong S-type geochemical signatures. The boundaries of the Early Proterozoic Kisseynew gneissic basin are reliably defined by the sillimanite metamorphic isograd (Zwanzig, 1990), but this isograd may not be exactly coincident with the steep fault bounding the underlying Archean block. These boundary faults propagated upward, and are represented by several peripheral, narrow Early Proterozoic volcanic belts. The Archean Kisseynew block could have acted as a median massif within the Early Proterozoic Trans-Hudson mobile megabelt. Changes in lithofacies observed across the Kisseynew Basin are gradual (Tran et al., 1996; also Zwanzig et al., 1996). Stauffer (1990) interpreted the Missi Group as a rock succession analogous to the late- to post-orogenic continental molassic assemblage in the modern European Alps.

Stauffer (1990) and Gordon et al. (1990) took account of many structural differences in rock characteristics above and below the base-Missi unconformity: different composition, age and initial dips, disharmonic folding patterns, shearing and décollements, and so on. The unconformity at the base of this group and basin is major: it cuts 2,500-3,000 m into the basement, whose top is marked by regoliths. These observations confirm that the base-Missi unconformity separates two different tectonic, structural-formational étages.

Neither Lithoprobe nor COCORP publications have described the vertical and lateral tectonic units as they are discussed here. The significance of low-angle seismic discontinuities is greatly exaggerated in those publications. Though some geologically meaningful seismic events are observed in the deep sections, not all of them can be interpreted as thrust faults. In many areas, low-angle seismic reflections may be related to buried unconformities, such as the pre-Missi unconformity well recognized in the Kisseynew block. In other regions, low-angle discontinuities have been found to indicate crustal zones of higher porosity (Kozlovsky, ed., 1984). They may also be caused by extensive igneous sills, transitions between brittle and ductile crustal zones, and so on.

True primary reflections from acoustic-impedance contrasts in a seismic profile are easy to confuse with various forms of undesirable coherent noise: diffractions, multiples, off-line arrivals, processing artifacts. In the crystalline crust, contorted complexly by folding and faulting, subhorizontal and gently dipping seismic events are often seen to cut across the more coherent reflection fabric, but many such events are not representative of the structure along the profile. In the Superior province, for example, Zaleski et al. (1997, p. 709) found that "many inclined and subhorizontal linear events in the... seismic survey originated from subvertical dikes [which cross the profile obliquely]... We cannot distinguish between reflections from the side and reflections from depth on the basis of velocities". They reported also (op. cit., p. 707): "On the basis of forward modeling studies, the events occur at two-way traveltimes for horizontally propagating waves... A number of dipping and subhorizontal events represent refracted waves propagated along horizontal ray paths near the surface and reflected from subvertical dikes."

On the east side of the Kisseynew Basin, Hudsonian orogenic and early post-orogenic rocks are juxtaposed against the Superior shield across a zone more than 100 km wide. Shortening at the Churchill-Superior boundary (Thompson fault belt) took place between 1,820 and 1,720 Ma (Weber, 1990), i.e. mostly after the Hudsonian orogeny (which is conventionally bracketed between 1,900 and 1,800 Ma). Some sheets in this thrust zone are 15-20 km thick, and they verge towards the Superior craton. The thrust sheets, displaced laterally by up to 50-60 km, are imbricated with folds. The fact that shortening occurred only in this fairly narrow zone supports the idea that the Archean Superior craton and Kisseynew block were rigid enough to resist deformation. Most probably, at the time of thrusting the Kisseynew block stood higher than the Superior block, accounting for the vergence of thrusting away from the Kisseynew domain. Thrusting in the opposite direction, towards the Kisseynew block, has been suggested from the eastward dips of deep crustal reflections across the Thompson belt (Baird et al., 1996), but these dips may represent no more than the eastern flank of the South Saskatchewan crustal arch.

Modern seismic techniques are often unable to detect the steep faults on the flanks of Archean (and other) blocks, or to discriminate between low-angle seismic events based on their origin. A more integrated approach is needed to interpret the seismic images realistically. Rock-based control is essential for the restoration of tectonic events, which varied from region to region.

A buried southward continuation of the Churchill-Superior boundary zone (Thompson fault zone) lies under the platformal cover in southern Manitoba and North Dakota. In potential-field maps, it is marked by a set of NNE-elongated anomalies that pass west of Lake Winnipeg (Lyatsky et al., 1998). Between the Thompson and Tabbernor fault zones, differential vertical movements of blocks have been proven by geologic mapping (e.g., Weber, 1990). Conventionally, these movements are ascribed to post-Hudsonian tectonism, such as "orogen-parallel escape wedging ...probably related to collisional indentation to the northeast" which supposedly occurred "via fault-bounded southwesterly block movement and linked lower-crustal detachment" (Lewry et al., 1996, p. 174). That, of course, is not seen in seismic profiles nor by geologic field mapping. What is seen instead is evidence for differential vertical movements of crustal blocks.

The typically cratonic, post-orogenic vertical displacements of big crustal blocks are confirmed by a variety techniques: isotopic, fission-track, thermo-bathymetric. The Kisseynew block was evidently buried considerably, first to make room for deposition of the thick sedimentary succession of the Kisseynew Basin and then for amphibolite- and even granulite-grade metamorphism of the Missi Group. Such metamorphism requires high pressures of 7 to 9 kbar (700 to 900 MPa) and temperatures of 700° to 800°C, suggesting depths of 25-30 km. Such depths are consistent with the ductile deformation of Missi Group paragneisses intruded by granites of palingenic origin (Lewry et al., 1990; Zwanzig, 1990, 1997; Zwanzig et al., 1996). Later, uplift of the Kisseynew block brought the high-grade gneisses back to the surface sometime after the peak of metamorphism around 1,815 Ma (Gordon et al., 1990). Leucogranites, commonly associated with regional extensional regimes (cp. Colpron et al., 1996), yield ages between 1,820 and 1,790 Ma (Zwanzig et al., 1996), and rocks of the Missi Group cooled past the temperature of 280°C at 1,705 Ma. The vertical movements of blocks, elucidated by detailed structural analysis, often involved re-use of old zones of mylonitization by younger, brittle-fault deformation (Bleeker, 1990). As the Kisseynew block rose, its upper crust became brittle (Bleeker, 1990); offsets on these brittle faults are estimated to be 2 to 4 km on the perimeter of the Kisseynew Basin (Stauffer, 1990), whose rocks are now juxtaposed against rocks of much lower metamorphic grades. The rise of the Kisseynew block was over by late Mesoproterozoic time. Emplacement of NNW-trending mafic dikes at ~1,270 Ma marked a shift to a new tectonic regime in the reorganized craton (Heaman and Le Cheminant, 1988). These late Mesoproterozoic dike swarms in large regions of the Canadian Shield mostly ignore the older structural fabrics and cut across them. In the Reindeer Lake zone in Saskatchewan, they cross-cut the major Hudsonian-age faults without being offset by later tectonic displacements.

Rationale in tectonic studies in Alberta and other cratonic platforms

The number and interrelationships of Archean cratons (shields) on the site of the modern North American craton are still unclear. Of the many orogenies that affected the cratonic crust, the Kenoran (ca. 2,500 Ma) and Hudsonian (ca. 1,900 Ma) were the most important in Canada. Whether the Penokean, Trans-Hudson and Wopmay orogens are branches of a single Hudsonian mobile megabelt or not is not yet known. Links of the Trans-Hudson Orogen with the Penokean orogen in the east and Wopmay orogen in the west are still under debate.

The growing number of recognized remnants of Archean material (in the form of slivers, blocks, protoliths, radiometric inheritance) in the Paleoproterozoic mobile megabelt(s) contradicts the idea of their origin in exotic and remote parts of the globe, far travels and assembly in an accidental collage to make the North American continent. Translations of thousands of kilometers are also hard to reconcile with the fact that cratons and continents have deep crustal and subcrustal roots. The best-studied parts of the Trans-Hudson Orogen show it is structurally very complex, containing median massifs (e.g., the Kisseynew crustal block), local orogenic belts (e.g., Rottenstone, La Ronge) and younger platformal volcano-sedimentary basins (e.g., Namew). No grounds exist for the speculations about continental-crust collisions, oceanic-crust subduction and so on in the Alberta Platform basement, since such phenomena remain unproven even in the exposed areas of the craton. In a general sense, it is apparent that Wyoming cratonic terrains underlie the western part of the Williston Basin and the southern part of the Alberta Basin, and that the Wopmay mobile megabelt continues into the basement of the Interior Platform in the north and may be connected with the Trans-Hudson Orogen beneath the Alberta Basin.

As the recognized Archean blocks in the Trans-Hudson Orogen (Glennie "protocontinent", Saskatchewan or Sask "microcontinental craton", Dakota block) become more and more numerous, it becomes increasingly hard to support the outdated ideas about their collisions and assembly (for example, geologic similarities indicate the Glennie block has pre-Hudsonian affinities with "the Superior and/or Hearne provinces"; Chiarenzelli et al., 1998, p. 247). No evidence supports these Archean blocks' non-native origins, and new paleomagnetic results cast doubt on the reliability of earlier determinations (Dunlop, 1995; Halls, 1995). Appeals to seismic images as evidence for this or that regional tectonic evolutionary scenario (e.g., Baird et al., 1996; Clowes, 1996; Leclair et al., 1997) are also inconclusive. More probably, these Archean blocks are semi-stable pieces of native, North American Archean craton(s) which developed more or less in situ. Archean-age rock samples have been recovered from the basement of the Alberta Platform (Villeneuve et al., 1993). Their position near the axis of the Southern Alberta crustal arch (Fig. 79) is similar to the position of the Glennie block along the axis of a similar arch in Saskatchewan.

Exotic tectono-stratigraphic terranes in the Alberta Platform basement were initially postulated with just 57 age determinations of basement samples from petroleum-industry wells (Ross et al., 1988, 1991), but this unfounded scheme continues to be used as a template for Lithoprobe interpretations (Eaton et al., 1995; Clowes, 1996;

Figure 79. Southern Alberta crustal arch, as imaged in the E-W deep crustal Lithoprobe seismic reflection profile (location of seismic data in Fig. 73).

Ross et al., 1997). But the idea of exotic terranes is under growing criticism even in its birthplace, the Canadian Cordillera. Since the late 1980s, many artificially designated "terranes" have been revised in Alaska, British Columbia and conterminous U.S. (e.g., Ernst, 1988; Woodsworth et al., 1991; Dover, 1994). Now, this whole approach is undergoing deep re-evaluation (see Lyatsky, 1996 for review).

Generations of (volcano-)sedimentary basins - or, more precisely, structural-formational étages - in the platformal cover bear evidence of tectonic stages that followed cratonization. This is particularly obvious in well-studied and densely drilled platformal regions such as the Athabasca or Alberta Basins. Tectonic reconstructions based on evidence contained in rocks show that Archean rocks older than 2,500 Ma south of the southwestward projection of the Snowbird-Virgin River fault system may be correlated with the Wyoming craton. Early Proterozoic rocks in this area are generally metamorphosed to a low grade and include felsic tuff, phyllite, marble and quartzite which probably overlie the Archean basement (Frost and Burwash, 1986; Ross et al., 1991; Burwash, 1993). Gravity and magnetic trends of the Wyoming shield, which have a general NNE orientation (Dutch, 1983; Green et al., 1985a-b; Sharpton et al., 1987; Thomas et al., 1987) continue into southern Alberta at least as far as the E-W-trending Brooks (Vulcan) gravity low, north of which the anomaly pattern becomes more complex (Geological Survey of Canada, 1990).

From these and other data, as discussed above, the Wyoming craton may extend to the Snowbird-Virgin River fault system. Ross et al. (1991) attributed this fault system to an Early Proterozoic subduction zone. However, Lewry and Collerson (1990) pointed out that the idea of a suture there has no supporting field evidence in shield outcrops. No evidence exists for arc magmatism on either side of this fault system, in the Canadian Shield and the Alberta Platform basement alike. In the Shield, granitic plutons adjacent to it were shown to have resulted from melting of a continental crust (Lewry and Collerson, 1990). Hoffman(1990, p. 20) acknowledged that "no magmatic arc related to the Snowbird line has been found". The Snowbird-Virgin River fault system is very ancient, probably pre-Hudsonian: as Hoffman(1990) noted, it has no structural connections with the Trans-Hudson Orogen. Lewry and Collerson (1990) reasonably regarded it as a reactivated, very ancient intracontinental fault zone. Hanmer et al. (1994) described its western segment, the Virgin River fault, as an Archean structural zone reactivated in the Early Proterozoic.

The Snowbird-Virgin River fault zone also truncates the broad Taltson magmatic belt, which lies farther north parallel to the western margin of the Canadian Shield. The Taltson belt is conventionally thought to be correlative with the Thelon magmatic belt still farther north, at the margin of the Archean Slave shield, whose width is only one-third the width of the Taltson belt. The segment between these two belts, related to the boundary between the Slave shield and the Rae platform, is aligned along the

McDonald fault of the complex Early Proterozoic Great Slave Lake shear zone (Hanmer, 1988; Bostock et al., 1991; Bostock and van Breemen, 1994).

McDonough et al. (1995) described the Taltson belt "composite continental magmatic arc", supposedly "resembling" an Andean-type magmatic orogen. Interestingly, however, the paragneiss protoliths in both these Proterozoic belts are predominantly sedimentary, whereas orthogneisses in these belts have been linked to a probable continental-crust basement. Thus, the Taltson and Thelon magmatic belts probably lie on a continental crust, and Hudsonian partial remelting of the Archean crust at least in part accounts for their formation. The age of the two belts is similar: 1,970-1,930 Ma; no orogenic deformation younger than 1,930-1,920 Ma has been noted there. Late-stage brittle deformation there took place between 1,840 and 1,735 Ma, and produced visible offsets on many faults which were also rejuvenated later (Hanmer and Connelly, 1986; Burwash et al., 1994). Hoffman (1990) hypothesized the Slave shield collided with the Rae platform around 1,970 Ma, before the assumed Hearne-Superior and Rae-Hearne collisions assembled the Laurentian craton between 1,950 and 1,850 Ma. But Hoffman himself (op. cit., p. 21) added a caveat: "if they [these structural provinces] were ever separated".

The Great Slave Lake shear zone of crustal weakness is considered to be similar to the Snowbird zone (Hanmer, 1988). The basement block between them contains highly metamorphosed rocks of pre-Hudsonian Proterozoic age (2,320 to 2,000 Ma; Villeneuve et al., 1993). The rocks with Archean affinities were later overprinted by intense metamorphism, including strong K-metasomatism of the Athabasca polymetamorphic province (Fig. 54; Burwash et al., 1994). Lewry and Collerson (1990, p. 7) were also cautious in their interpretations of the tectonic nature of the "Rae-Hearne platform", which they suggested is still "largely a matter of speculations". Stressing how uncertain the current interpretations are, they added that "we are not even sure how many independent Archean microcontinents were actually involved in the Early Proterozoic assembly of the North American craton" (op. cit., p. 12): seven as postulated by Hoffman (1990), or only four (North Atlantic, Superior, Slave and Rae-Hearne-Wyoming). A more commonly assumed cratonic combination is Slave-Rae-Hearne. Discussing some instances of adventurous interpretation of the Slave shield area in terms of accretion, subduction and assembly, King et al. (1989; vs. Kusky, 1989) called for caution in postulating fancy tectonic models without "accurate, supportable data".

Some prominent structural trends are traced across the Canadian interior Platform, from the Shield into the Cordillera. The most apparent such crustal discontinuities are Snowbird-Virgin River and Great Slave (MacDonald). But other structural trends which played a fundamental role in the development of the platform are discordant with those typical in the Shield. These are the E-W trends in the Athabasca and Peace River regions, and many others.

To study these fault trends with seismic data alone is difficult, because the basement in Alberta is overlain by a strongly reflective sedimentary package of interbedded carbonates, shales and evaporites which considerably degrades the seismic images of the basement-cover interface. More useful regionally are potential-field data which, when interpreted properly, permit to discriminate networks of faults related to various, pre-cratonic and cratonic, tectonic stages. In places, it is possible to distinguish incipient NNW-SSE Cordilleran structural patterns (Kimiwan; Fig. 56; Chacko et al., 1995) predating the appearance of the Cordilleran mobile megabelt itself. In other platformal areas in the Alberta and Williston basins, NNW-SSE structural trends were active in the Phanerozoic and influenced several structural-formational étages in the cover (Lyatsky et al., 1998).

Many of the modern hydrocarbon-exploration targets in the Western Canada Sedimentary Province lie in the upper tectonic étages of the platformal sedimentary cover, and often the oil-industry seismic surveys are not designed to image the deeper cover horizons. Deep seismic profiles, such as those acquired by the Lithoprobe program, are aimed at deep levels of the sub-cover, crystalline crust and mantle. They do not depict the basement-cover contact, either.

The specific acquisition and processing parameters of the Lithoprobe surveys make it possible to resolve some seismic characteristics of the middle and lower crustal levels, which are of great interest. Their geologic interpretation, of course, remains non-unique, because there is no drillhole control on the interpretations. But removing co-herent and incoherent noise by data processing cleans up the deep seismic images, and these data are able to provide invaluable information for the regional tectonic studies of the Alberta Platform.

Besides the limitations inherent in any geophysical method, there are several alarming points in the way the deep seismic reflection data are often interpreted. One is the prac-tice to publish the data early, before detailed analysis could be undertaken to separate primary reflections from various types of coherent noise. Particular difficulties are en-countered with vertical to moderately dipping faults and dikes running parallel to the seismic profiles or crossing them obliquely: they produce low-angle reflections and dif-fractions that cut the main reflection fabric and are often misinterpreted as low-angle faults and sills. In Alberta, such events have been linked to an assumed huge sill suite (Ross and Eaton, 1997), having been modeled as "tabular mafic intrusions", but Zaleski et al. (1997, p. 710) cautioned that such "models alone do not provide a means" for a definitive assessment.

If overlooked, the poor ability of existing seismic techniques to detect steep faults may easily cause misinterpretations and exaggeration of the importance and role of low-angle events. In the past, subhorizontal seismic discontinuities were boldly interpreted as unconformities; now they are just as boldly interpreted as low-angle thrusts. These huge pitfalls in interpretation are often ignored, as easy interpretations reduce a geophysicist's work load. But deep seismic surveys are only aimed to assist regional tectonic interpretation, and they are no substitute for rock-based tectonic investigations of a region. To interpret seismic images properly, one has to consider them together with the results of other studies. In any case, thorough analysis of structural-formational étages the most reliable control on geophysical interpretations and modeling.

During the last decade, once-innovative and -attractive supermobilistic speculations have stopped being satisfactory to many geologists (Struik, 1987; Henderson, 1989; Richards, 1989; Aitken, 1993b-c). The hypothesized structural connections of western North America with Eurasia or Australia (Sears and Price, 1978; Eisbacher, 1983) have been found to be very tenuous, and arbitrary timing of the assumed continental-separation rifting event - 1,500 Ma (Sears and Price, 1978), 780 Ma (Aitken et al., 1993a-b) or 575 Ma (Bond and Kominz, 1984) - raises doubts on its very reality. After many years of searching, neither a hinge line nor a breakup unconformity have been found in the well-mapped and heavily-drilled Alberta Platform (Henderson et al., 1993; Richards et al., 1993). Morrow (1991) reports that he found no fixed position of such a hinge line in the Paleozoic, when the passive margin supposedly existed (see also Cecile et al., 1997). Henderson et al. (1993) questioned the applicability of Bond and Kominz's (1984) scheme (see also Aitken, 1993a-c).

Characteristics of the structural-formational étages in the Canadian Shield and the Alberta Platform provide a solid ground for distinguishing real evolutionary tectonic stages. The sedimentary cover of the North American craton as a whole is apparently a product of pericratonic subsidence which at different times affected the whole or parts of the Interior, Midcontinent, Hudson Bay platforms. In most of Alberta, this subsidence took place since the Middle Cambrian. Since the Ordovician, pericratonic subsidence spread over much of the continental interior, in Midcontinent and Hudson Bay platforms as well as the platforms in the east along the Appalachians and in the west along the Cordillera (Sloss, ed., 1988; Stott and Aitken, eds., 1993). In the Devonian, western Interior Platform was segmented, and the Alberta Platform appeared. Since the Late Devonian-Mississippian, different segments of the Interior Platform acquired the characteristics of foreland subsidence and locally of foredeeps. Some tectonic settings typical of the Alberta Platform in many places continued laterally far west into the Canadian Cordillera, into the Rocky Mountain and Omineca belts, during the entire Paleozoic and earliest Mesozoic. During several short time intervals, platformal settings prevailed even in the miogeosynclinal zone of the eastern Cordillera. Pericratonic and foreland subsidence of the Alberta Platform was later supplemented by the subsidence of foredeeps, first in the Carboniferous and Permian and more prominently in the Late Cretaceous and Tertiary.

Factual evidence from rock-made tectonic étages is inconsistent with many current speculative concepts. Ricketts (1989, p. 7) argued that the Jurassic and younger foreland basin was "superimposed on the cratonic succession" in Alberta. In fact, the Alberta Basin has from the beginning been cratonic, and the cratonic foreland and foredeeps appeared there as early as in the Paleozoic. Podruski et al. (1988, p. 14) attributed the later, Jurassic to Tertiary stage in the Alberta Basin to "the rising thrust sheets [which] provided both the detritus to fill the foredeep and the tectonic load to tilt the earlier sediments towards the continental margin." To link the foredeep(s) with a continental margin is unreasonable. Even more unreasonable is the idea to relate the Jurassic-Tertiary evolution of the Alberta Basin to plate and terrane interactions at the distant western Canadian continental margin (as was done by some authors in Macqueen and Leckie, eds., 1992; or in Mossop and Shetsen, comps., 1994).

Deficiencies of models constructed without adequate geological control have been demonstrated worldwide, particularly by drilling superdeep wells in the Russian part of the Fennoscandian Shield, the Paris Basin in France, southern Germany (KTB), Arizona, and so on. One of the recent findings is that the continental crust has variable stress conditions in different areas and at different depths (e.g., KTB drilling results; Borm et al., 1996). Emmermann and Lauterjung (1997) and Huenges et al. (1997) noted, in particular, that many predictions of porosity and seismic velocity in the upper crust from the assumed mineral composition were not confirmed, and that in the penetrated rock units the rock fabrics, pressure temperature, water content and other properties were not as expected, which illustrates how significant misinterpretations of geophysical data can arise from incorrect assumptions. All this underlines the need for more restraint in making unverified predictions, and for cautious calibration of geophysical and geophysical interpretations for each particular region and geologic time. The available geological and geochemical data suggest an apparent specificity of two distinct mega-stages of the Earth's evolution: Archean and Early-Middle Proterozoic, and Late Proterozoic and Phanerozoic (e.g., Hamilton, 1993; Rudnick, 1995).

These considerations should caution against oversimplifications caused, in particular, by misapplication of the popular terrane-tectonic and similar models. Some interpreters are all too quick to interpret Lithoprobe seismic images simply and unambiguously. For example, Cook et al. (1997) reported some east-dipping reflection bands in a newly-shot seismic profile along the Slave-Northern Cordillera Lithoprobe transect across the northern Interior Platform, which they interpreted as a relic from subduction around 1,900-1,860 Ma. A dipping reflection band in the Superior province has been interpreted in terms of Archean subduction (Calvert et al., 1995), and Clowes (1996, p. 113) claimed "these data and their interpretation provide the first *direct evidence* for modern-style plate tectonics being active in Late Archean time!" (italics ours). This is incorrect, because the geometry of seismic reflections is not direct evidence for any geologic process.

It would be more practical, of course, to discuss all the many other alternatives. If, for example, the supposedly subduction-related seismic band recorded in the Superior province was linked to the tectonic event that created the Kapuskasing Structural Zone, its age would be Proterozoic. A growing volume of geologic evidence indicates that even cratonized crust continued to be active during the Proterozoic and Phanerozoic, in specifically cratonic forms. Seismic reflection geometries and signatures created by various processes may be similar or dissimilar, causing non-uniqueness in geological interpretations of geophysical data. Repeated appearance of new configurations of cratons and mobile megabelts around them suggests that cessation of pre-cratonic, orogenic processes was not necessarily accompanied by drastic changes of thermo-tectonic conditions deep in the crust and upper mantle. The Kapuskasing horst, where the Proterozoic lower crust is exposed, is one of the visible results of tectonism in cratons unrelated to subduction. A steep reverse fault has been mapped at this horst's flank.

During the cratonic tectonic megastage, the cratonized crust was affected by episodic extension, and each such event provided conditions for vertical crustal movements and magmatism. Several craton-scale dike suites record these episodes. This suggests that at great depths, processes of deep reworking never stopped. Rocks there could flow in a ductile manner, and low-angle reflection bands in deep seismic images could mark such juvenile flowage patterns. At the upper crustal-levels, meanwhile, the manifestations of cratonic tectonism are brittle deformation and some specific styles of magmatism (including the Mesozoic and Tertiary kimberlite-pipe emplacement in the North American craton in Canada).

The appeal to subduction as the only possible explanation for dipping seismic reflections is unrealistically simplistic. More probably, in regions like the Canadian Shield, such reflection bands resulted from later deep-seated cratonic processes. Magnificent surface expression of these processes are great phenomena like dike magmatism in the Canadian Shield, Midcontinent felsic magmatism, Kapuskasing upturn, etc. They are much more widely represented deeper in the cratonic crust (for example, in the form of delamination and recycling of lower-crustal material into the mantle, or intrusion of mantle melts into the lower crust, with a resulting formation of seismically visible lamellae; e.g., Kay and Kay, 1993; Rudnick, 1995).

An international team of 29 authors (BABEL Working Group, 1991, p. 77) has stated that in deep seismic sections "...it is inappropriate to distinguish genetically between Phanerozoic and Precambrian reflectivity patterns". Geophysical methods are indeed insensitive to the age of geologic features they image. Physical characteristics of geologic bodies may be similar despite the differences in their age and tectonic essence. Estimates suggest that the modern rise and shape of Fennoscandia and Hudson Bay regions, and even the average elevation of entire cratons, cannot easily be explained just by passive effects such as post-glacial rebound (e.g., Peltier, 1985; Forte et al., 1993a; Wu, 1993), suggesting driving forces in the crust or mantle unrelated to subduction.

Though simplistic old ideas that the continental crust is made up of the "basaltic" and "granitic" layers have been refuted by superdeep drilling (Kozlovsky, ed., 1984), the general notion that the lower crust is more mafic than the upper is accepted widely (Anderson, 1995; Rudnick and Fountain, 1995). If high-grade metamorphism and mafic intrusions in the lower crust produce significant changes in rock density, they may control the buoyancy and subsidence of the crust. Eclogitization, for example, can make the lower crust denser, whereas retrograde metamorphism or delamination of denser lower-crustal layers into the mantle may restore the crust's lower density. Repetition of these processes is suggested by geochemical (e.g., Rudnick, 1995) and petrological (e.g., Snyder et al., 1997) studies, and it may cause upward and downward movements of blocks. Eclogitization of lower-crustal horizons close to the Moho (Kay and Kay, 1986, 1993) is indicated by various lines of evidence, including studies of xenoliths in deep-sourced intrusions. Some workers argue it does not appear to have been massive (Pearson and O'Reilly, 1991; Downes, 1993), but granulitization of rocks in the lower-crustal conditions is broadly acknowledged to be common, and "eclogitization of crustal rocks is likely" (Fountain et al., 1994, p. 411-412).

The nature and origin of the crust-mantle boundary and reflectivity of the lower crust remain unclear (e.g., Griffin and O'Reilly, 1987). There can be variations in pressure-temperature conditions and metamorphic state of rocks, mechanical delamination, layered intrusions, phase transformations, and so on. Though normally seismic P-wave velocities in the continental crust are between about 6 and 7.5 km/s (Christensen and Mooney, 1995), higher values often occur in the lower crust. In many areas, lower-crustal horizons are found with seismic velocities of 7.7-7.0 km/s or more. Velocities up to 8 km/s are rather common in the lower crust in the Canadian Cordillera (Clowes et al., 1995) In active tectonic zones like the East African rift in the African craton, thickness of a high-velocity and high-density lower-crustal layer has been found to be from 2 to 9 km (Mooney and Christensen, 1994), possibly indicating high-grade metamorphism and an abundance of mafic intrusions. Whatever the cause, high velocity usually goes hand-in-glove with high density. If thermo-mechanical conditions were to change, the lower-crustal layer of high-density rocks may either grow or decay, inducing upward or downward vertical movements of the entire crust.

Removal of lower-crustal material by delamination and recycling into the mantle may also involve partial melting, increasing the amount of light felsic material in the crust and decreasing the crust's density. Lateral flowage of the lower crust may also change the density of the crustal column. Besides, subsidence and uplift may be induced by mantle convection under the continent (Forte et al., 1993a). Rudnick (1995) has argued that delamination played a leading role in the chemical evolution of the lower crust. In this process, the dense part of the lower crust should be mechanically detached and re-cycled into the mantle. Such detached and descending slices could produce dipping reflection bands in seismic images, similar to those at modern subduction zones.

Craton's boundaries and how to define them

King (1977) noted the uncertainties in delineation of the North American craton's western boundary in the vast U.S. Rocky Mountain province, including the boundaries of the Colorado block. Sloss (1988a, p. 28) regretted that "the lessening of faith in the inboard margin of a "miogeosyncline" as the limiting boundary of a craton leaves tectonists without a broadly accepted definition of craton margins". In the modern North American craton, this arises often on its eastern and western margins, where huge inboard thrust sheets induced by mobile megabelts override and imbricate the platformal cover. This necessitates clarification of how to define the boundaries of cratons.

Because cratons of each particular generation reveal themselves as fundamental tectonic entities contrasting with mobile megabelts, definition of a craton depends on its boundaries with the megabelts. As a new megabelt arises or an old one becomes extinct, the craton changes. The number of generations of cratons thus depends on the number of generations of megabelts.

During the Phanerozoic, three main orogenies occurred in the Appalachians: Taconian, Acadian and Alleghenian. Four orogenies took place in the Cordillera: Antler, Early Nevadan, Late Nevadan-Columbian, and Laramian. Not all of them affected the corresponding miogeosynclines and the marginal parts of the cratonic mass. But when and where they did, they changed that craton's boundary - for example, by shifting it cratonward in front of the advancing fold-and-thrust belts.

Mobile megabelts begin with the appearance of precursors - whole-crust rift zones - which involve dramatic differential subsidence. Subduction-related phenomena make these zones fully orogenic (eugeosynclinal, in earlier classifications). Kay (1951) described North American eugeosynclines and miogeosynclines as lying next to each other in pairs, though the time of their development may not be the same.

Dietz and Holden (1974) proposed that a modern example of Kay's geosyncline pairs is evolving at the Atlantic continental margin of the North American continent. The submerged shelf and slope were interpreted as a new miogeosyncline, and the continental rise was assigned to a future eugeosyncline. This picture ignores the distinction between eustatic and geocratic (i.e. tectonic) shelves, which have different relationships with the marginal crustal blocks (Lyatsky, 1969, 1974). Dietz and Holden's (1974) interpretation of the shelf-slope-rise system at the Atlantic continental margin of North America was in fact rather sedimentological. Sloss (1988a) included paleoshelves

(Appalachian Shelf, Cordilleran Shelf) into the North American craton, and he defined these shelves mostly paleogeographically. By contrast, in Lyatsky's (1974) classification of the extensive shelves of northern Eurasia, the paired submerged platformal shelves and adjacent emergent lowlands were grouped together as belonging to the same crustal block, whose very existence consigns them to a single geodynamical system. Sloss (1988a) made no distinction between different types of shelf recognized in the geologic past and lumped them together, because the paleogeographical and sedimentological approaches prevailed in his work. Despite their differences, Sloss and Bally pictured miogeosynclines from a paleogeographical and sedimentological viewpoint. Bally (1989) favored the continued use of the term *miogeosyncline*, but only to denote ancient passive continental margins; he recommended that *eugeosyncline* be abandoned. Bally rightly rejected the attempts to replace the word *miogeosyncline* with a shortened and more meaningless word *miocline*: indeed, miogeosynclines are structurally neither mioclinal nor monoclinal.

Though it is a misnomer, many authors in North America continue to use the word *miocline*, quite often loosely. However, in the current usage, *miogeosyncline* or *miocline* has lost its important tectonic meaning as crustal (or lithospheric) zone whose deep orogenic reworking created geologic manifestations inherently different from those on cratons. Sloss (1988a) complained that because the tectonic notion of miogeosyncline was sterilized, it is no longer clearly formalized. It is evident that to equate miogeosynclines/mioclines with continental margins, at least with regards to the Canadian Cordilleran miogeosyncline, is wrong. Sedimentary basins which formed along the eastern Cordilleran miogeosyncline in late Precambrian and Paleozoic time were not west-facing, open to an assumed proto-Pacific ocean. Continental cratonic crust lay in their basement. They were two-sided structural features which developed on the site of the miogeosyncline during particular tectonic stages, and sedimentary basins there has continental-crust provenance areas on their eastern and western flanks (Struik, 1987; Henderson, 1989; Richards, 1989; McMillan, 1991; and many others).

In the eastern Cordilleran miogeosyncline in Canada, the rift-related Late Proterozoic Windermere Basin and the orogen-related Paleozoic Prophet/Ishbel Basin reflected deep crustal reworking in the evolving miogeosyncline. They were orogenized (magmatized, metamorphosed, deformed) to turn into typical foldbelts. Complex reworking masked the earlier boundaries of the craton and the megabelt (at least, in its Omineca miogeosyncline and Intermontane median massif; cp. Cook, 1995a-c). But remnants of these boundaries continued to reveal themselves as rather stationary zones of crustal weakness with a NNW-SSE orientation. Strong faulting gave them many magma conduits, and tectonic manifestations there were considerable. Strongly reworked remnants of a cratonic basement have been recognized in the miogeosyncline, as seen in rock samples from the metamorphic core complexes in the Omineca Belt. But the precursor rift can still be delineated, and it marks the boundary between the less reworked cratonic remnants to the east and more reworked ones to the west.

The later cratonward shifts of the axes of foreland and foredeep basins in the broad miogeosyncline-platform transition zone were associated with crustal reorganizations and new fault systems. The newly formed sedimentary piles lie away from the initial craton-megabelt boundary, and the axes of the new basins reflect cratonward shifts of the western boundary of the cratonic mass. This is certainly the case in the western Alberta Platform from late Paleozoic time onward, as foreland and foredeep sedimentary sequences with total thicknesses of ~6,000 m prograded eastward. During Late Cretaceous and Early Tertiary time, the Laramian thrust sheets advanced spectacularly platformward. In the modern Rocky Mountain Belt, these thrust sheets partly overlapped the existing basins and partly localized new ones in the sedimentary cover. The thin-skinned fold-and-thrust belt formed almost entirely above the west-tilted basement top of the pre-modern North American craton. As a result, the western boundary of the present-day Alberta Platform, marked by the Tertiary deformation front, is completely separate from the initial craton-miogeosyncline boundary, which was related to a major, steep, whole-crust Proterozoic fault zone.

The present-day deformed belt, including the Foothills, which almost reaches Calgary, lies far to the east from the original craton-mobile megabelt boundary. Hundreds of kilometers separate the modern North American craton boundary from the initial, Grenvillian-age boundary between the Laurentian craton and the evolving Cordilleran megabelt. Yet, influences of the powerful Laramian tectonism were felt far to the east. Taylor et al. (1964) thought that in the Alberta Platform, the most distant feature to respond to Laramian mountain building was the Northern Sweetgrass uplift of Tertiary age, though others have noted cratonic inversion tectonics in that region. These inversions are often overlooked by modelers of basin evolution in miogeosynclinal-platformal domains, and acceptance that crustal inversion may occur without tectonic influences from mobile megabelts requires a new set of parameters (Coward, 1991; Buchanan and Buchanan, eds., 1995; Beratan, ed., 1996). Causes of inversion are still poorly understood, but phenomenologically these vertical movements are too clearly manifested in the geologic record in both cratons and mobile megabelts to be ignored.

When discussing the position of a craton-megabelt boundary, the time must be specified because ancient and modern boundaries are not in the same place. Thin-skinned fold-and-thrust belts involve parts of miogeosynclines and cratonic platforms alike, masking deep boundaries between them. By definition, sedimentary cover of a platform lies flat, and deformed rocks in the foothills regions should be included into non-cratonic zones. But sedimentologically, especially in the outer foothills, some of these deformed rocks are of cratonic platformal origin. Farther towards the megabelt hinterland, even in the inner foothills, some of the thrust-displaced rocks are of non-platformal miogeosynclinal origin. Still farther into the mobile megabelt, in the main mountain belts, rocks of platformal and non-platformal origin are intercalated structurally in very complex, multiple-faulted and -folded stacks, and closer to the megabelt interior, miogeosynclinal rocks become predominant. But these thrust sheets, being detached from the miogeosynclinal hinterland and from deep crustal levels, do not represent the position of the miogeosyncline's deep roots.

By rock composition alone, the marginal fold-and-thrust belts are partly cratonic and partly miogeosynclinal. By their deformation, they are completely non-cratonic. Massive folding and thrusting are typical of orogenic provinces, but craton-facing thrust belts (sometimes called foreland belts, which unfortunately confuses the nomenclature) are more difficult to classify. They are no in-situ parts of mobile megabelts, but are induced by a mature megabelt's dynamics. In this sense, inner parts of these belts should be assigned to orogenically-induced mountain zones, but only their small parts are truly connected with in-situ orogenic processes at deep crustal levels beneath.

These deep-seated processes are the most vividly manifested in the most active continental tectonic zones, which we customarily call orogens (sometimes meaning a whole mobile megabelt, sometimes just its most-mobilized parts excluding median massifs). But studies of continental tectonic history show that deep reworking - orogenic by definition - occur not only in orogenic zones but also in median massifs and, surprisingly, even in the lower levels of cratonized crust of the cratonic masses.

Folding and thrusting are the most common visible forms of structuring in orogenic zones, median massifs and marginal fold-and-thrust belts. In cratonized regions, they are occurring at deep crustal levels not subject to direct observation. Block-like and warp-like deformation in the crust occur in both cratons and mobile megabelts, but in the megabelts they are overwhelmed by folding and thrusting and therefore often overlooked. Vertical crustal movements such as doming, sagging, arching and block displacements are not restricted to cratons. They occur all across continents, though in orogens they are overshadowed by greatly manifested folding and thrusting.

In the modern geological literature, folding and thrusting are related much too readily to subduction. For ancient times in the geologic past especially, this is often done on no better grounds than the need to explain some inferred horizontal force. Bally (1981) even proposed two types of subduction: B, related to the Benioff zones, and A, linked to fold-and-thrust zones along mobile megabelts. Later, he (Bally, 1989) reasonably stepped away from that idea (only B-subduction, in different variants, is real). Far-field influences from distant subduction zones have been called upon to explain any compressional deformation in large rock bodies. Whole-crust warping has been linked to plate-wide compression and extension due to plate interactions at continental margins (e.g., Ziegler, ed., 1987; Coward et al., eds., 1989). Such an explanation, applied sometimes to the Rocky Mountain fold-and-thrust belt, is hard to justify, as has been noted time after time (e.g., Stock and Molnar, 1988). Externally supplied stresses do not explain how light blocks rich in granitic material were able to subside to depths of many kilometers, permitting the development of many sedimentary basins in cratons and megabelts (Greenwood et al., 1991).

Broadly speaking, forces that drive vertical tectonics are gravity, which pulls the crust down, or effects of heating, which push it up. These fundamental forces, in combination, drive even the mantle convection responsible for movements of lithospheric plates. Vertical crustal movements vary in type: epeirogenic rise and fall of whole continents or their very large parts, regional crustal upwarping and downwarping, block displacements.

The broad Laurentian craton on the North American continent was cut on the east and west by the Mesoproterozoic Grenville-Appalachian and Neoproterozoic Cordilleran mobile megabelts. A craton of the next generation, Grenvillian, appeared later, during the Appalachian-Ouachitan cycle of the Grenville-Appalachian megabelt. Fold-and-thrust belts advancing from the mature mobile megabelts in the Appalachian, Ouachita-Marathon and Cordilleran provinces shrank the extent of craton. Only after the powerful Laramian orogeny in the Cordilleran megabelt, accompanied by the formation of marginal fold-and-thrust belt, did the modern North American craton finally acquire its familiar boundaries.

Proper definition and delineation of the cratons of different generations on the North American continent is not just an academic matter. It has direct applications to the predictions required by mineral and petroleum exploration as well as by the environmental needs. Miogeosynclinal development, which is not accompanied by complete reworking of the pre-existing crust but involved recurrent extensional tectonic regimes, included magmatism, development of dikes and subhorizontal sills, formation of huge exhalite Ba-Zn-Pb deposits (Fig. 80; Gordey et al., 1987; Nelson, 1991; MacDonald et al., 1992; Goodfellow et al., 1995). Mature miogeosynclines produced bilateral fold-belts rimmed by fold-and-thrust zones that contain many oil and gas traps fields. Trapping mechanisms in the miogeosyncline-platform domains depend heavily on the tectonic history and patterns of structural deformation in each particular zone.

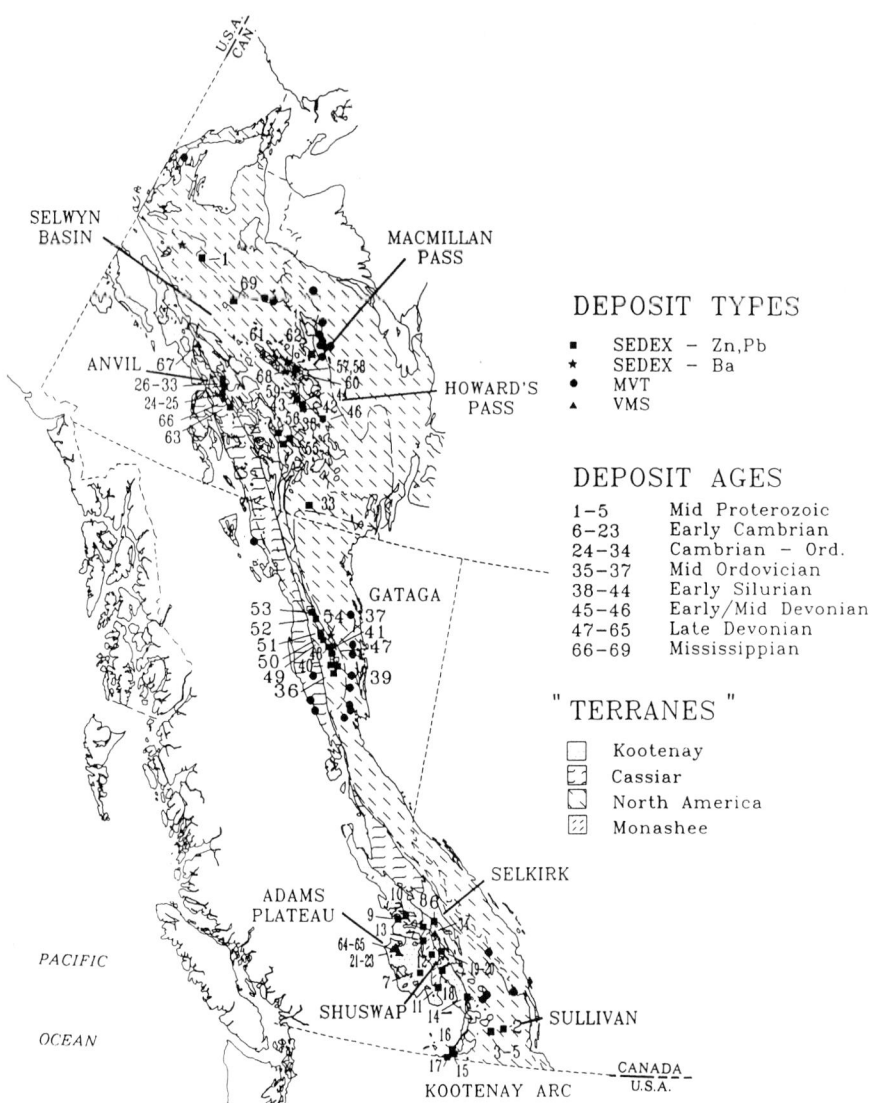

Figure 80. Distribution of sediment-hosted exhalative (SEDEX) mineral deposits in the Canadian Cordillera (modified from MacIntyre, 1991). Most of these deposits are confined to the miogeosynclinal Omineca Belt. Other abbreviations of mineral-deposit types: MVT - Mississippi-Valley-type; VMS - volcanogenic massive sulfide.

7   PLACE OF NEOTECTONIC AND CURRENT CRUSTAL MOVEMENTS IN REGIONAL TECTONIC STUDIES

Manifestations of neotectonic and current crustal movements, and their significance

Past crustal restructuring events reveal themselves in the structural peculiarities of particular tectonic étages. The latest such restructuring is still unfinished, and also unfinished is the latest tectonic étage. Current tectonic movements, which have affectedthe crust since some time in the recent past, will continue into the future till the next reorganization. These recent crustal movements are seen in the tectonic topographic relief, but tectonic fluctuations are reflected in the physiography only partly. Most physiographic features are affectedmy massive surficial denudation (especially river erosion), which greatly modifies the tectonically-induced relief.

Even the relief components directly produced by crustal movements are not all reflective of ongoing tectonism: many of them resulted from tectonism in the past. But it is fairly recent tectonic crustal movements that created a tectonic topography that is still preserved in the landscape. For this reason, such movements are called *neotectonic*. Most of them occurred in the Tertiary, some are late Mesozoic; very occasionally, tectonic topography was shaped by movements dating back to earlier Mesozoic or even late Paleozoic.

A special field of geological and geographical knowledge, called geomorphology, deals with the visible topographic relief. Its job is to separate erosional (denudational) relief from tectonic, and neotectonic relief from current-tectonic. In the last job, geomorphology relies on geodesy, as well as such modern tools as planetary gravimetric studies and mantle geodynamic modeling.

Geodetic and gravity-derived estimates of the shape of the entire Earth, its various spheres (including perisphere) and outer surface provide valuable information. They show that tectonic relief is affected variously by extraterrestrial bodies (e.g., the Moon), dynamics of the mantle, and self-driven dynamics of the crust itself. To separate their results is extremely difficult. To consider the modern tectonic relief, tectonics mainly takes into account the data provided by geomorphology, because geomorphology deals with the natural, visible relief of the subaerial (physiographic) and subaqueous (bathymetric) surface of the Earth's perisphere.

Geomorphology studies the physiographic and/or bathymetric relief, its genesis and evolution, including the effects of tectonic (endogenic) or non-tectonic (exogenic: e.g., deposition, denudation and erosion) factors on land and on the sea-floor surface. Erosional topographic surfaces are the norm in mountainous terrains, accumulative surfaces in many lowlands and abyssal plains, abrasion surfaces on some shelves, glacially carved surfaces on land and sea floor at high latitudes. But the general outline of most broad areas is shaped in large part by crustal tectonic deformation induced by endogenic forces such as subduction and non-subduction lithospheric crustal warping and block movements (e.g., Bowin, 1983; Forte et al., 1993a).

Ancient zones of weathering and paleosols are good indicators of old surfaces generally unchanged since the time of their origin. In topography, such surfaces occur as uncharacteristically beveled terrains (plateaus, highlands) surrounded by mountainous areas. Some erosional (denudational) low-relief surfaces in the mountains can be correlated with low-energy surfaces in the plains nearby. Broad beveled geomorphological provinces are usually produced by a combination of erosion and accumulation. If such a surface is elevated and affected by a new pulse of erosion, its remnants may still be preserved between incised river valleys. Such remnants of old beveled surfaces of different ages are found in the Alberta Plains and in the mountains of the Canadian Cordillera.

Some of the preserved neotectonic surfaces in western Canada are as old as late Mesozoic, others are Tertiary, still others post-glacial. To identify them is not always easy, because they rarely contain marine or other distinctive faunal assemblages (the resolution obtainable from the continental Cenozoic fauna and flora is usually not sufficiently high). In the Ancestral Rocky Mountains region of the U.S. and Canada, some old beveled surfaces have been raised to altitudes of 500 m, 1,000 m (in Alberta) and ~2,500 m (in the Colorado Plateau). In the intracontinental Tien Shan Mountains in Asia, proven Tertiary beveled surfaces are found at an astonishing altitude of ~5,000 m. Movements of this magnitude must have been tectonic. They are purely intracontinental and, as in the western North American craton, intracratonic.

It has been estimated that many continental cratonic areas stand either much higher (e.g., Fennoscandia) or much lower (e.g., Hudson Bay area) than can be accounted for by post-glacial rebound given the common understanding of mantle viscosity and flow. To reach their present-day elevation, these areas had to be affected by additional mantle-driven tectonic factors that created neotectonic anomalies in Eurasia and North America. Some very deep-seated causes related to the dynamics of the mantle have been proposed to account for such huge continental warping, having no relationship to plate boundaries. Rather, Forte et al. (1993a) linked such warping to non-subduction mantle currents (also Forte et al., 1993b vs. Gurnis, 1993).

Neotectonic geomorphologic features permit to restore vertical crustal movements using the preserved dated erosional surfaces (unconformities) as markers. In the Cordillera, a number of horsts and grabens were superimposed on the Laramian Rocky Mountain thrust sheets after thrusting had ended in the Oligocene. In Nevada and Utah, variably tilted crustal blocks formed the spectacular Basin and Range province (Beratan, ed., 1996). Some chains of tilted horsts and half-grabens stretches from Nevada and Utah to western Montana, Idaho and eastern British Columbia (Janecke, 1994; Constenius, 1996). In the Omineca Belt of the Canadian Cordillera, post-Eocene elevation exhumed a pre-mobile megabelt basement in metamorphic core complexes. This basement-rock material was raised from the Tertiary lower crustal levels at ~30 km depth to the surface in just 10 m.y. (Parrish, 1995). An estimated 1 to 3 km of rocks (Kalkreuth and McMechan, 1984; Issler et al., 1990) were removed from the Rocky Mountains, and partly redeposited in the Western Canada Sedimentary Province including the Alberta and Williston basins (Edwards et al., 1994). Yet, a later Tertiary and Quaternary uplift affected the Plains as well.

King (1977) reasonably suggested that in North America, more attention should be paid to the observable variations in the timing and amplitude of vertical crustal movements in the Cenozoic. Sloss (1988a) supposed that along the eastern Cordillera, equilibrium between crustal compression and extension was reached near the Paleocene-Eocene boundary. Carr (1992) timed this event at 58±1 Ma (latest Paleocene). Rapid cooling due to regional uplift in the Rocky Mountains is thought to have occurred around 45 Ma (late Middle Eocene; Constenius, 1996). The wide distribution of pre-glacial gravels in Alberta, with boulders and cobbles of quartzite sourced from the Cordilleran Omineca Belt (Edwards et al., 1994), resulted from this neotectonic uplift.

The end of Laramian crustal shortening in the eastern Cordillera did not produce a tectonic collapse: no such structures are seen there. The idea that the horst-graben pattern resulted from passive downdropping of some crustal pieces that lost their support lacks factual substantiation. The horsts and grabens were formed by differential vertical tectonic movements of blocks, both up and down. These movements continued till the Early Miocene Burdigalian, ca. 20 Ma (Constenius, 1996). Two distinct episodes of normal faulting have been identified from studies of volcano-sedimentary successions in the grabens (Eisbacher, 1977; Lamerson, 1982). In the Columbia River basalt province, volcanics are younger than 17 Ma.

The listric and steep planar faults originated variously from the base of thrust sheets or from the ground surface, and in places they cut through the entire thrust-sheet stack 6-8 km thick (Dahlstrom, 1970). The 150-km-long Kishenehn graben is related to some of these major faults. In the Southern Rocky Mountain Trench, some 50 km to the west from this graben, two separate Paleocene and Miocene clastic successions contain boulders and cobbles of Omineca-Belt rocks (McMechan and Thompson, 1993; Constenius,

1996). The maximum extension on the faults bounding the Trench is estimated to be only ~10 km (Van der Velden and Cook, 1994), on the faults bounding the Kishenehn Basin ~15 km (Constenius, 1996). Most commonly, this extension was accommodated by a multitude of small faults mapped in the region. Not a collapse but a normal-fault network characterized the extended brittle upper crust.

The up and down crustal movements were the largest in the inner part of the eastern Cordilleran miogeosynclinal Omineca Belt. In seismic profiles across the Interior Platform and eastern Rocky Mountains, the basement seems to be uninvolved in Laramian thrusting (Bally et al., 1966). Seismic reflections attributed to the crystalline basement in many areas suggest the basement is essentially undeformed (e.g., Yoos et al., 1991). In a 150-km-wide zone from the Flathead fault to near the Moyie fault, depth to the top of a pre-modern cratonic basement that underlies the Rocky Mountain and eastern Omineca belts changes from 5 to 20 km. It is difficult to accept that it was not affected by the neotectonic block movements. Vertical faults certainly affected the miogeosyncline-platform tectonic domain, and the throw amplitude on these faults was greater in the miogeosyncline.

Insights into neotectonics from the distribution of the latest Cenozoic deposits and drainage systems

Pre-glacial late Cenozoic gravels and sands, widespread in the Alberta Platform, were sourced from both the Cordillera and the Canadian Shield, which at that time experienced substantial general uplift. Accounting for the abundance of Shield clasts in the Plains, the regional drainage system contained tributaries flowing from the east. The non-uniformity of neotectonics across Alberta in the Tertiary is well recorded. In the Paleocene, as central Alberta relatively subsided and pebbles were laid down unconformably over the Upper Cretaceous rocks, in a broad area from Montana to Swan Hills and from the Foothills to Edmonton uplifts took place. The uplifted areas broadened in the Eocene, involving much of Alberta except for its southern and eastern parts. The Oligocene uplift, in contrast, affected most of southern Alberta, while in some areas in the north sedimentation took place (Edwards et al., 1994).

Neotectonic movements were different in the southern surroundings of the Canadian Shield. A brief marine incursion is recorded in the Paleocene in northeastern North Dakota. In the Turtle Mountains, near the Manitoba border, the corresponding marine beds lie at an elevation of some 600 m above the modern sea level. Regional neotectonic uplift of the south-central part of the Canadian Shield began in the Late Paleocene and continued through the Miocene. It was evidently subdued later, and the area was depressed by the weight of the Quaternary glaciers. The regional uplift was, however, renewed after deglaciation around 11,000-9,000 BP. The huge Lake Agassiz appeared, covering much of southern Manitoba and nearby parts of the Midcontinent Platform. It left a record of ancient shorelines, which permit to recognize and estimate the upwarping. The present-day elevation of the Shield in this region is usually 500-700 m above sea level, and its ongoing rise is recorded by leveling surveys.

In the Midcontinent region since the Paleozoic, local uplifts in areas such as the Ozark Dome reached a magnitude of 4,000-4,500 m (see reviews by Friedman, 1987a-c). Pre-Grenville-age rocks of the Adirondack Mountains in New York state are still rising today; as the uplift is thought to have begun only at 10 to 15 Ma, these mountains are young (Isachsen, 1992). The general rise of the U.S. Great Plains, since the Neogene, has in places reached up to ~1,500 m (Sloss, 1988a). This recent uplift of the Plains is also recorded in many parts of the North American craton by the continuing incision of rivers.

Local variations in the current crustal vertical movements in the Alberta Platform are also apparent. The modern Lake Athabasca lies at slightly more than 200 m above sea level, whereas the surrounding rolling hills stand above 500-600 m. Beveled Tertiary surfaces now lie at the tops of the Caribou Mountains, Birch Mountains, Stony Mountains and other block-like highlands at an elevation of 800 to 1,000 m (Fig. 40). Incised rivers have cut deep valleys into the Alberta Plains, partly reusing pre-Quaternary drainage channels (Fenton et al., 1994, 1996).

During the Pleistocene glaciation, a large area in North America was temporarily lowered by continental-scale ice sheets many kilometers thick. On the rims of this general depression, tilts towards the center of the glacier were expressed in the topography along and south of the Canada-U.S. border. In the early Holocene, as post-glacial rebound began to restore the continent to its natural isostatic level, streams from the rising parts of the Canadian Shield flowed away from the Shield center towards the Plains, except for the anomalously lowered area of Hudson Bay. The modern drainage patterns, which formed in the last ~5,000 years, show their sensitivity to fault-related lineaments and to curves of the growing crustal warps. The present-day Arctic-Atlantic continental divide in central Alberta separates the modern Alberta Platform into northern and southern domains. This E-W to NE-SW zonation is highlighted also by a geochemical trend of elevated values of Co, Cr and Li content in the Quaternary till. This trend is traced across the northern part of the Birch Mountains and Buffalo Hills areas, roughly along the latitude 58°N (Fenton et al., 1996). The distribution of shallow heavy oil and tar sands all over Alberta (Fig. 81) illustrates links with this neotectonic warping: many occurrences follow the neotectonic arch responsible for the Arctic-Atlantic divide. Current crustal warps are also highlighted by the modern drainage system, especially in the area of confluence of the North and South Saskatchewan rivers. The festoon shape of the western edge of the Canadian Shield, with its promontories and re-entrants, was apparently caused by current upward and downward regional warping in the crust.

Figure 81. Distribution of heavy oil deposits in the Alberta Basin (simplified from Jardine, 1974). 200 miles = 322 km.

The appearance and disappearance of the continental ice load distorted the normal manifestations of gradual epeirogenic uplift of the North American continent. The on-going crustal movements of very long wavelengths are reflected in particular in the general tilt of the Midcontinent towards the Gulf of Mexico, as well as in the high stand of the continental interiors. Crustal warping of shorter wavelength is recorded elsewhere. The curved, sinuous modern outline of the Canadian Shield and surrounding mountainous provinces (Fig. 82) and of the edge of the platformal cover are in part products of this warping. Where epeirogenic movements and warping have the same sign, the Shield protrudes south and west. Significantly, in the Alberta Platform, platformward protrusions of the Canadian Shield and of the Cordilleran mountains face each other, creating a set of juvenile swells and necks in the shape of the Interior Platform between the Canada-U.S. border and the Arctic Ocean. The continental divide between the Arctic (Mackenzie) and Atlantic (Mississippi-Missouri) drainage systems lies in central Alberta, perhaps influenced by an E-W to NE-SW neotectonic arch seen as a topographic high. The position and orientation of this continental divide are apparently congruent with the regional swell-and-neck pattern of promontories and re-entrants.

Thickness of lacustrine glacial deposits is known to increase northward along the Mackenzie River (Barton et al., 1964). To the south, deposits from several Wisconsinan-age lakes have also been identified. The late-stage glacial deposits are progressively younger to the north. This suggests regional tilting of the northern part of the Interior Platform towards the Arctic Ocean during and after the Wisconsinan glaciation. The modern Arctic-Atlantic continental divide (and the neotectonic upwarp) in central Alberta has risen only since the Pleistocene. Pleistocene lacustrine deposits are clayey and fairly uniform; their mineralogy is fairly consistent. They mostly developed from the Upper Cretaceous shale in the bedrock, and additional material was brought in by glaciers as erratic clasts. Unlike the northward-tilted platform in the north, the platform in southern Alberta was flat during the Late Cenozoic, as it generally is today. The remarkable regional monotony and flatness of the western Canadian Plains is a result.

A zone of topographic lows in the modern physiography of these plains lies south of the North Saskatchewan River, from Fort Vermillion to Lloydminster. The Wabasca, Athabasca and North Saskatchewan rivers and their tributaries often use these depressions. In its narrowest part, the low topographic zone is about 20-30 km wide, but reaches some 70 km where the Wabasca River flows through it. The average depth of incision by river currents is ~50 m. Southeast of the Arctic-Atlantic continental divide, in an area between the cities of Athabasca, Saskatoon and Regina, a belt of unconsolidated sand and gravel deposits follows a linear structural trend marked by a very gentle gradient, from just over 600 m above sea level in central Alberta to about 500 m near the Saskatchewan-Manitoba-North Dakota border junction. This trend coincides with a band of topographic lineaments seen in satellite remote-sensing images of the Alberta Plains (Figs. 41, 42; Misra et al., 1991). Roughly equidistant between the edge of the

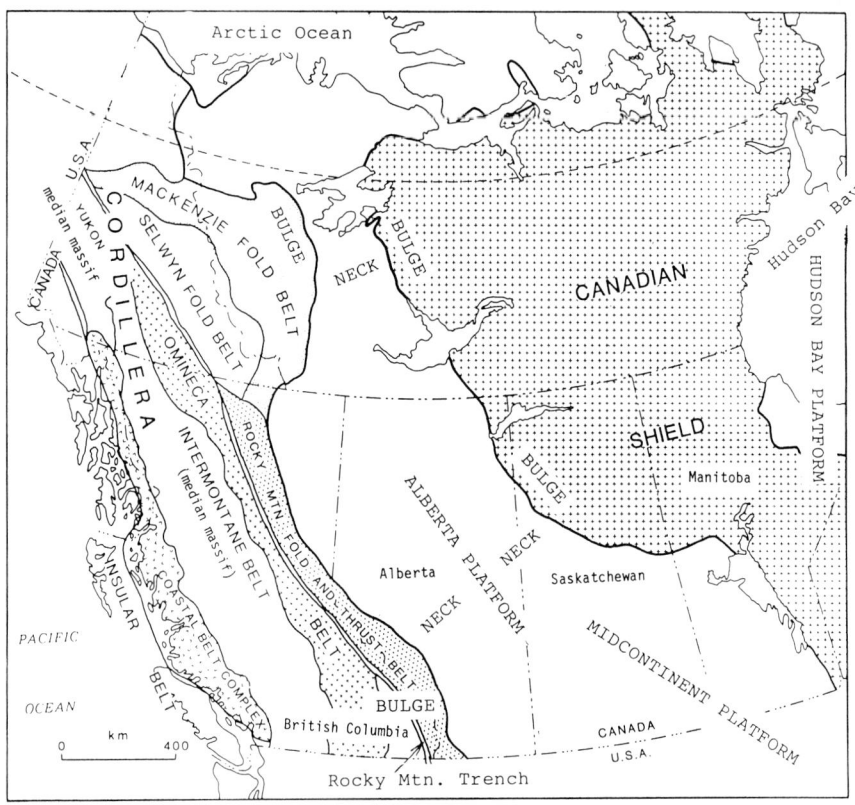

Figure 82. Bulges and necks in the outline of the Western Canada Sedimentary Province. Facing platformward protrusions of the Cordillera and the Canadian Shield (marked "bulge") create narrow zones in the sedimentary province (marked "neck"). These bulge-neck zones probably represent broad uplifts of the continental crust.

Canadian Shield and the Rocky Mountains, the Median neotectonic structural zone marks the western boundary of a chain of relative topographic lows; to the east of it lies another band of NNW-trending lineaments. The Median zone is neotectonic and recent, representing a subtle flexural topograhic bend. But the two principal tectonic domains on the two sides of the young Median zone have been recognized in many old structural-formational étages in the Alberta Platform. There passed the boundary between the miogeosyncline-platform domain to the west and the platform-shield domain to the east. This indicates the neotectonic and ongoing tectonic crustal movements in the Alberta Platform are in part inherited from the past, possibly due to continuing evolution of the Cordilleran mobile megabelt.

Cenozoic uplift likely affected the fluid-flow patterns and thermal equilibrium in aquifers in this region (Hitchon et al., 1990, their Figs. 2, 9, 15): eastward and westward flows of subsurface water in the Alberta Basin seem to run into each other near the Median zone. Attempts have been made to assess the role of tectonic disturbances in defining hydraulic systems in the stratified platformal cover across Alberta. The terrestrial heat flow on the Alberta Platform is estimated to be 55 mW/m^2 (Bachu, 1985). But the downward geothermal gradient varies: 25°C/km in southern Alberta, over 35°C/km in the north of the province (Hitchon et al., 1990). Evidence that a variety of hydraulic systems exist in the Precambrian crystalline rocks was discussed earlier with regards to the Canadian Shield (Fritz and Frape, eds., 1987). In the basement of the Alberta Platform, such systems may greatly complicate the modeling of regional thermo-tectonic events (Issler et al., 1990). Variations in the geothermal properties and fluid-flow patterns are indeed found all through the sedimentary cover, including in the Quaternary till (see also Tóth, 1978; Fenton et al., 1994).

Fracture fabrics in the Western Canada Sedimentary Province: their identification and classification

Linearity of some physiographic features - ridges, river valleys, shorelines - was related to geology long ago. Hobbs (1904, 1911) termed such features *lineaments*, and noted that they are often connected with faults and fractures. The term *lineament* has since been expanded to all linear features detected in landform studies, geologic mapping and geophysical surveys at the surface and at depth. Locke et al. (1940) termed linear geomorphologic zones as *lanes*, as did Maughan and Perry (1986), but the term *lineament* still predominates in the geological literature.

That many oil and gas fields in the Interior Platform, particularly in Alberta, are aligned has also been noted for decades, attracting much interest to detailed analysis of airphoto and remote-sensing images. Unfortunately, over-optimistic expectations led to disappointments, and the initial excessive optimism has faded, putting some sound ideas to oblivion. The current, more sober approach has shown the real usefulness of combined analysis of lineaments.

The Interior Platform in the U.S. and Canada offers many examples where important geologic features and potential-field trends are correlated with fault-block tectonics (Thomas, 1974; Maughan and Perry; 1986; Baars, 1988; Penner and Mollard, 1991; Gregor, 1997; Edwards et al., 1998). Prominent NW-SE-trending lineaments, such as Walker-Texas, Lewis and Clark, Olympic-Wallowa, and NE-SW-trending ones like Colorado, Snake River, Snowbird-Virgin River, Great Slave Lake, are of great inter-regional significance (Fig. 83; Warner, 1978; Maughan and Perry, 1986; Baars, 1988; Hoffman, 1990; Lyatsky, 1996). These lineaments are traced for thousands of kilometers as alignments of geologic and physiographic features. They often ignore visible geologic boundaries, such as those between the North American craton and the adjacent megabelts. The network of these continental-scale lineaments in North America is thus fundamental. The pattern of these lineaments is orthogonal, though their distribution is uneven. Major NW-SE lineaments are common in the southwestern U.S., but big NE-SW lineaments are more typical in western Canada. These are continental-scale structures of the highest order.

Lineaments of lower orders are now recognized in many regions. In different combinations, they are characteristic of different parts of the Western Canada Sedimentary Province (Stauffer and Gendzwill, 1987; Penner and Mollard, 1991; Edwards et al., 1998). It is possible to distinguish lineament-pattern domains in the Alberta Platform. To distinguish these domains is easier than to identify different generations of linear structural trends. Still, it is conventionally accepted that there are at least three classes of fault-related lineaments: (a) preserved from the past but now dead, (b) rejuvenated, or (c) newly formed. But although bands of surface lineaments trending NE-SW, NW-SE, N-S and E-W have been found in abundance all over Alberta and Saskatchewan (Babcock, 1973, 1974; Stauffer and Gendzwill, 1987; Mollard, 1988), it has been shown that their relationships with geologic structures are not always straightforward.

Importantly, detailed analysis of structural-formational étages in the sedimentary cover of platforms makes it possible to classify lineaments in relation to the regional tectonic stages. Without this information, the lineament mosaic recognizable in landforms or potential-field anomalies may be very confusing. Some meaningful lineaments can be misinterpreted, others overlooked. They can be classified by their relationship to particular tectonic stages only through detailed local analysis.

The prominent NE lineaments in Alberta are very old. They seem to predate even the Early Proterozoic Hudsonian orogeny (Hanmer, 1988; Hoffman, 1990). Geologic field evidence suggests segments of the Snowbird-Virgin River lineament in the western Canadian Shield may be Late Archean (Hanmer et al., 1994). Also old are the N-S pre-

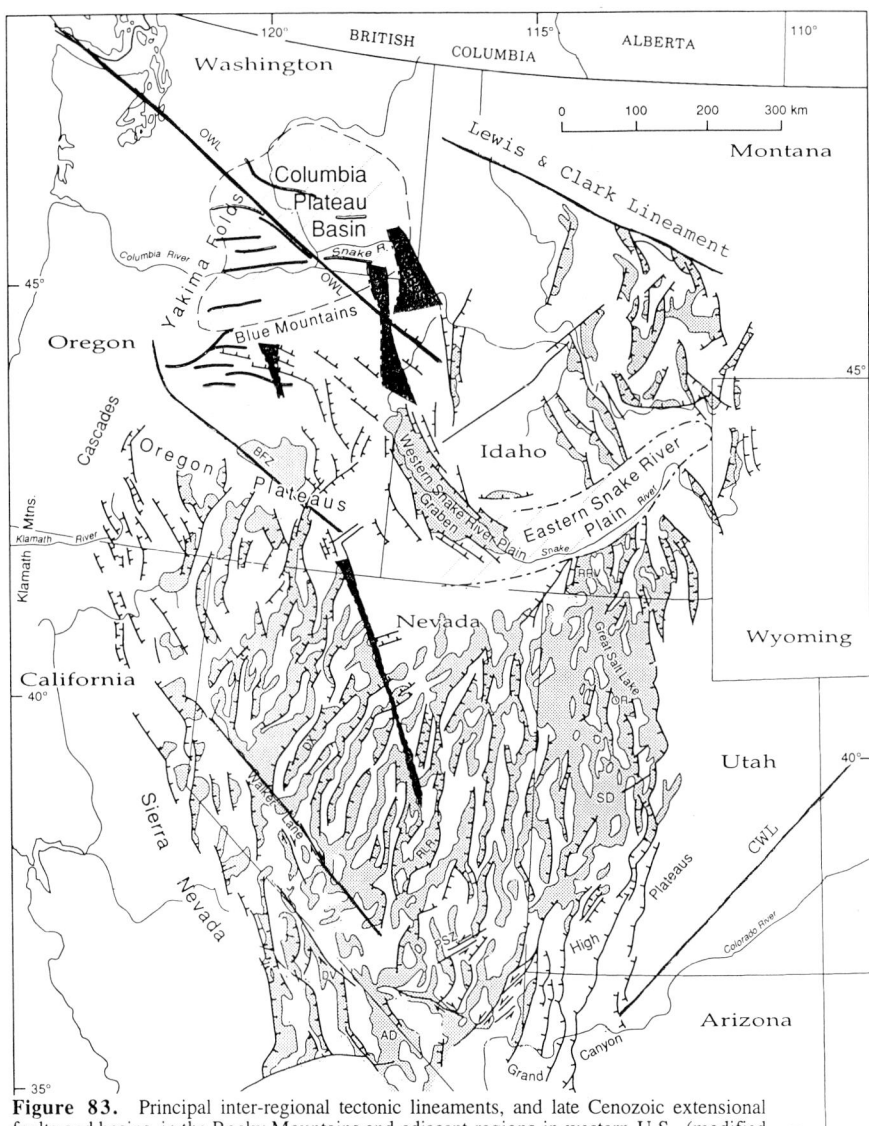

Figure 83. Principal inter-regional tectonic lineaments, and late Cenozoic extensional faults and basins, in the Rocky Mountains and adjacent regions in western U.S. (modified from Christiansen and Yeats, 1992; for detailed local descriptions, see also Beratan, ed., 1996). Dark fields and bands indicate major dike swarms. AD - Amargosa Desert; BFZ - Brothers structural zone; CDS - Cornucopia dike swarm; CWL - Colorado-Wyoming structural zone; DV - Death Valley; DX - Dixie Valley; FCFZ - Furnace Creek fault; GRDS - Grande Ronde dike swarm; OR - Oquirrh Range; OWL - Olympic-Wallowa structural zone (see also Lyatsky, 1996); PSZ - Pahranagat shear zone; RLR - Railroad Valley; RRV - Raft River Valley; SD - Sevier Desert; WF - Wasatch fault.

Hudsonian trends controlling the boundaries of the Archean Glennie and Hanson Lake blocks and some Early Proterozoic basins. Examples of such faults are Tabbernor and Saskatoon.

Steep brittle faults in the cratonic upper crust were formed under conditions of tectonic extension. With detailed mapping in the western Canadian Shield in Alberta, Langenberg (1983) found that several generations of joints there had an extensional origin. Low-angle joints have no strong expression in topographic lineaments, though they are also common. Rose diagrams of steep joints show that the main orientations change from one tectonic domain to another, and the joint patterns are not orthogonal in all of them. On the whole, the vertical fractures are clustered into two main orthogonal pair-sets: NE/NW and N-S/E-W, which corresponds to the principal regional fault patterns. Post-orogenic, truly cratonic brittle fractures cut the older, orogen-related structural features such as ductile recumbent folds, shear zones, foliation fabrics.

Orientations of numerous joints on the ground surface of the sedimentary cover in Alberta and Saskatchewan are also paired into NE-SW/NW-SE and N-S/E-W orthogonal sets (Babcock, 1973, 1974; Babcock and Sheldon, 1979; Mollard, 1988). Such a similarity of brittle lineament patterns in basement and in the sedimentary cover confirms their genetic connection.

Different fracture zones were reactivated or formed at different times during the cratonic stage, in response to changing stress conditions. Studies of tectonic étages and stages permit to differentiate these fault sets and relate them to concrete tectonic stress regimes more precisely. The orogenic fault network is expressed more strongly in gravity and magnetic maps, largely due to the abundance of igneous rocks along these faults, whereas post-orogenic lineament patterns are less explicit (unless they were inherited from pre-cratonic time). Only some of the faults, of whatever age, correspond directly to linear potential-field anomalies (Sprenke et al., 1986). Younger generations of faults can be inferred from offsets and terminations in other potential-field anomalies, and from linear geologic features in the cover or in the topography (Fig. 84). But in the history of the craton, the brittle fractures tend to be increasingly barren of magmatic intrusions, especially during the formation of the sedimentary cover. The observable lineament network also includes fractures active during the neotectonic and current periods of crustal movements. Lumping all the lineaments together is not helpful to the practical petroleum-exploration efforts.

Mollard (1988, p. 750) stated that "the patterns of photolineaments in drift-covered southern Saskatchewan [northern Williston Basin] were similar to fracture networks in the Canadian Shield and in exposed Phanerozoic sedimentary rock terranes south of the Shield". By detailed mapping, he proved the correspondence of some structural and stratigraphic features in sedimentary-cover tectonic étages to older faults. General simi-

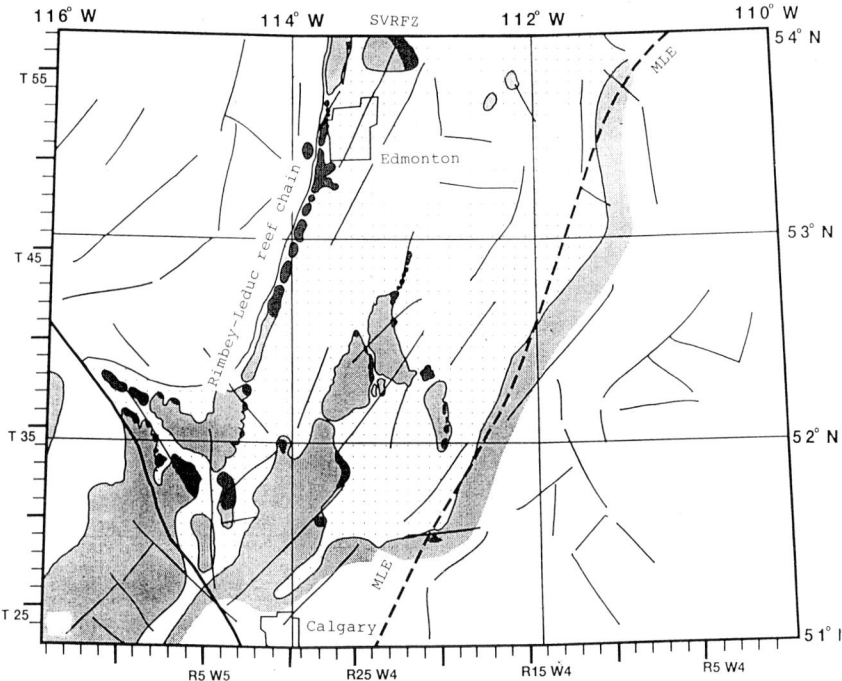

Figure 84. Relationship of the Devonian Woodbend Group reef complexes in central Alberta (cp. Figs. 20, 22) and major lineaments in Bouguer gravity horizontal-gradient vector maps (Fig. 74). Note the similarity of the Rimbey-Leduc reef chain with gravity lineaments along the Snowbird-Virgin River fault system. In the Rocky Mountain fold-and-thrust belt, reef position is restored palinspastically (Andrews, 1987). T - township; R - range; MLE - Meadow Lake escarpment; SVRFZ - Snowbird-Virgin River fault zone. Shaded areas indicate reefs, dark areas mark oil and gas pools, dotted area is the Cooking Lake carbonate platform (modified from Edwards et al., 1998).

larity of the studied Cretaceous and modern lineaments across the Western Canada Sedimentary Province illustrates the persistence of the fracture sets during the neotectonic stage (Stauffer and Gendzwill, 1987; Mollard, 1988; Misra et al., 1991). Various fault systems were active and affected the sedimentation settings during the Phanerozoic stages. But during each stage, at least some of these faults did manifest themselves. Osadetz (1989) emphasized that all of the Alberta cover was deformed in some fashion at one time or another. He noted folds of orogenic origin (on the western limb of the Alberta Syncline and on the southern limb of the Northern Sweetgrass Arch); disturbances in the strata over the dissolution edges of the 100-150-m-thick Devonian salt layers (including the Punnichy "Arch" that marks the dissolution edge on the southwestern edge of the Devonian Elk Point Basin); astroblemes; and faults of different generations (Fig. 85).

In southern Saskatchewan and adjacent parts of the U.S. Midcontinent Platform (Stauffer and Gendzwill, 1987; Mollard, 1988; Shurr, 1994), the most common is the NE/NW fracture pairset. In Alberta, the most typical trends are NE-SW, NW-SE and N-S, with E-W trends appearing in places (Babcock, 1973, 1974; Babcock and Sheldon, 1979; Lyatsky et al., 1992; Edwards et al., 1998). In central Alberta, the pattern of fractures in the cover is especially complex (also Jones, 1980; Greggs and Greggs, 1989) along the broad zone of long-lived contact between the platform-shield and miogeosyncline-platform domains. The NNW-SSE trends are particularly common in western Alberta parallel to the Cordilleran deformation front, and the NE-SW/NW-SE pairset, along with distinct NW- to NNW-trending bands, is widespread. Recent satellite images reveal this regional segmentation even more clearly. In those images, some major NE-SW-oriented faults (e.g., MacDonald) are expressed well, whereas the Snowbird fault zone's expression is subdued (Misra et al., 1991). This is consistent with higher activity in the northern Interior Platform during the neotectonic stage of regional crustal movements. The satellite images also highlight the NW-trending Median zone of dense recent fracturing zone that transects central Alberta obliquely to the NNW-SSE Cordilleran structural trend.

Structural fractures are induced by episodic regional or local restructuring. Sparsely or densely spaced brittle fractures in the cover and basement may be faults, joints or cracks of different ages. In a changing stress field, some old steep faults in the crystalline basement and cover were rejuvenated, and new ones formed. Some faults propagated upward, others downward. Some families of fractures (joints, cracks) formed during lithification, diagenesis, dewatering and compaction of sedimentary deposits. To discriminate between different classes of fractures is one of the hardest jobs in exploration.

Cratonic fracture fabrics are commonly characterized by orthogonal pairsets (e.g., Stille, 1924; de Sitter, 1956; Gay, 1973; O'Driscoll, 1982; Wellman, 1985; Maughan and Perry, 1986; Anfiloff, 1988; Baars, 1988; Cohen et al., 1990; Ameen, 1992; Lyatsky,

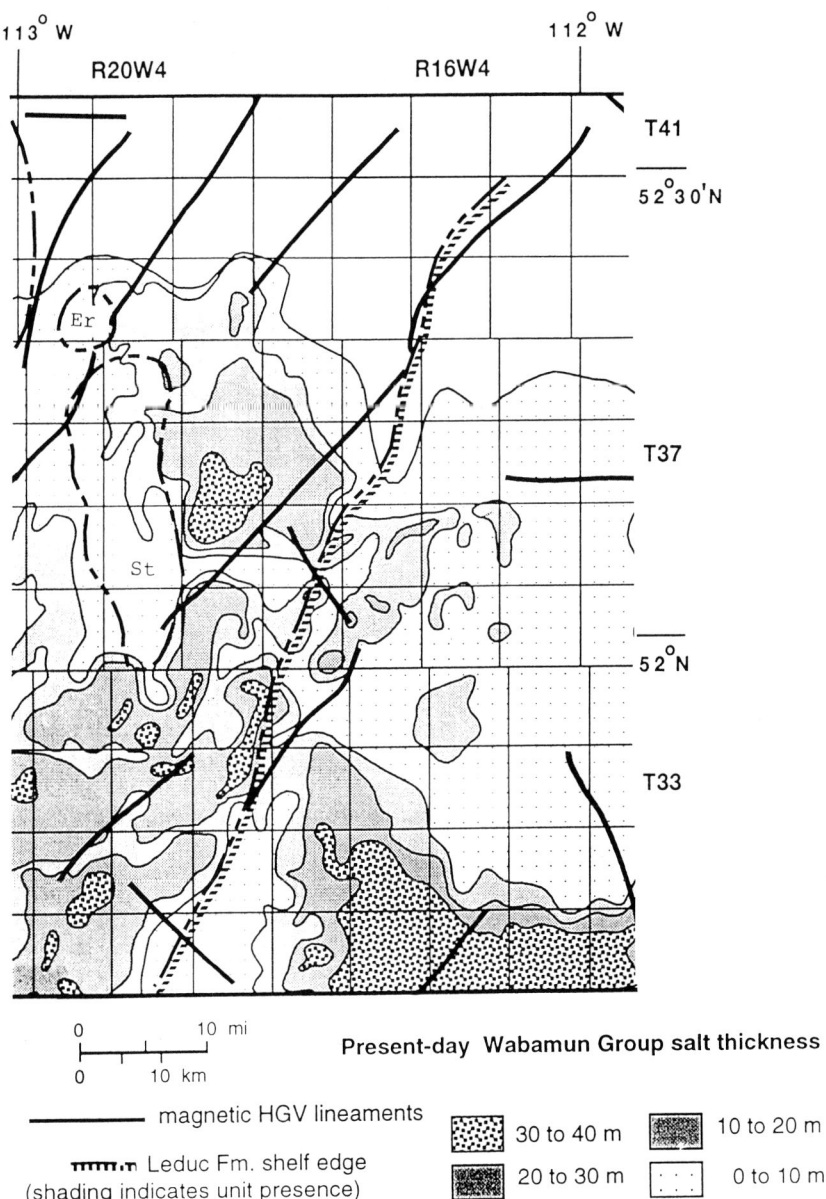

Present-day Wabamun Group salt thickness

———— magnetic HGV lineaments

ттттт.т Leduc Fm. shelf edge
(shading indicates unit presence)

30 to 40 m   10 to 20 m

20 to 30 m   0 to 10 m

Figure 85. Present-day thickness variations (in meters) of Devonian Wabamun Group salt in eastern Alberta, in relation to lineaments in total-field aeromagnetic horizontal-gradient vector maps (Fig. 75). Underlying earlier Devonian reefs: Er - Erskine; St - Stettler. T- township; R - range. Reduced salt thickness over the reefs suggests control on salt dissolution by water conducted through permeable reefs. The common orientations of salt-dissolution fronts and magnetic lineaments - NW-SE, NE-SW, E-W - suggest that water flowing along steep faults likewise controls salt dissolution (modified from Edwards et al., 1998).

1993, 1996). The extent of upward propagation of faults from great depths depends on the history of regional, continental-scale or planetary stress regimes. If pre-existing zones of lithospheric or crustal weakness are persistent, the oldest of them have had a chance to be reactivated repeatedly, in different stress fields (e.g., Isachsen et al., 1983; Stauffer and Gendzwill, 1987), though their exact manifestations can differ from one epoch to the next. These phenomena are well known all over the Interior Platform, in New Mexico, Colorado, eastern Utah, Montana and Alberta (Jones, 1980; Peterson and Smith, 1986; Sloss, ed., 1988). Downward-propagating faults are also recognized, especially in areas of flexuring and tectonic inversion. Nur (1982) analyzed some formation mechanisms of tensile fractures and estimated the depth of their downward penetration to be more than 3 km. Despite their genetic differences, all these phenomena are sometimes referred to together as "basement control".

In the late Precambrian, south of the U.S.-Canada border, three large E-W troughs (Central Montana, Uinta, Grand Canyon) separate four major crustal blocks: from south to north, Arizona, Utah, Wyoming, Alberta. Some of these blocks did not always act independently in the Phanerozoic, but they were reactivated several times. In the resulting distribution of basins and arches, the predominance of E-W, N-S, NW-SE and NE-SW trends is evident in these regions, in geology and topography (Maughan and Perry, 1986; Peterson and Smith, 1986). In the Alberta Platform, persistence of some structural trends, especially NE-SW ones, through time is obvious (e.g., Cecile et al., 1997). The challenge is to determine the timing of these fracture families, which is done the most reliably through analysis of structural-formational étages. At times, the difficulties in classifying local joints and regional lineament patterns, as well as separating them into distinct fabrics, have given rise to skepticism and quirks about "linesmanship and the practice of linear geo-art" (Wise, 1982, p. 886). But only poor-quality lineament work could provoke such dismissive skepticism.

Gravity and magnetic anomaly lineaments, as well as topographic lineaments, can highlight some fault-related structural features. The prominent Precambrian Great Slave Lake-MacDonald and Saskatoon faults are expressed at the surface in the alignment of river valleys, landscape breaks and bands of closely spaced topographic joints. Surface fracture fabrics are expressed well in modern drainage patterns (Figs. 6b, 86). The neotectonic structure of the Williston, Alberta and Northern Alberta (Hay River) basins, as well as of the Sweetgrass and Peace River arches, is revealed with such techniques in considerable detail (Mollard, 1988; Misra et al., 1991). Trajectories of the maximum horizontal compressional stress, as determined from studies of oil-well breakouts in Alberta, are in places roughly congruent with or slightly oblique to the deformation front in the Cordilleran Foothills (Fig. 87; Bell et al., 1994). These trajectories are clustered differently, being subparallel to the Median fracture zone in the heavily drilled areas in central and northeastern Alberta (also Bell and Babcock, 1986). The neotectonic Median fracture zone is almost equidistant between the Rocky Mountain Belt of

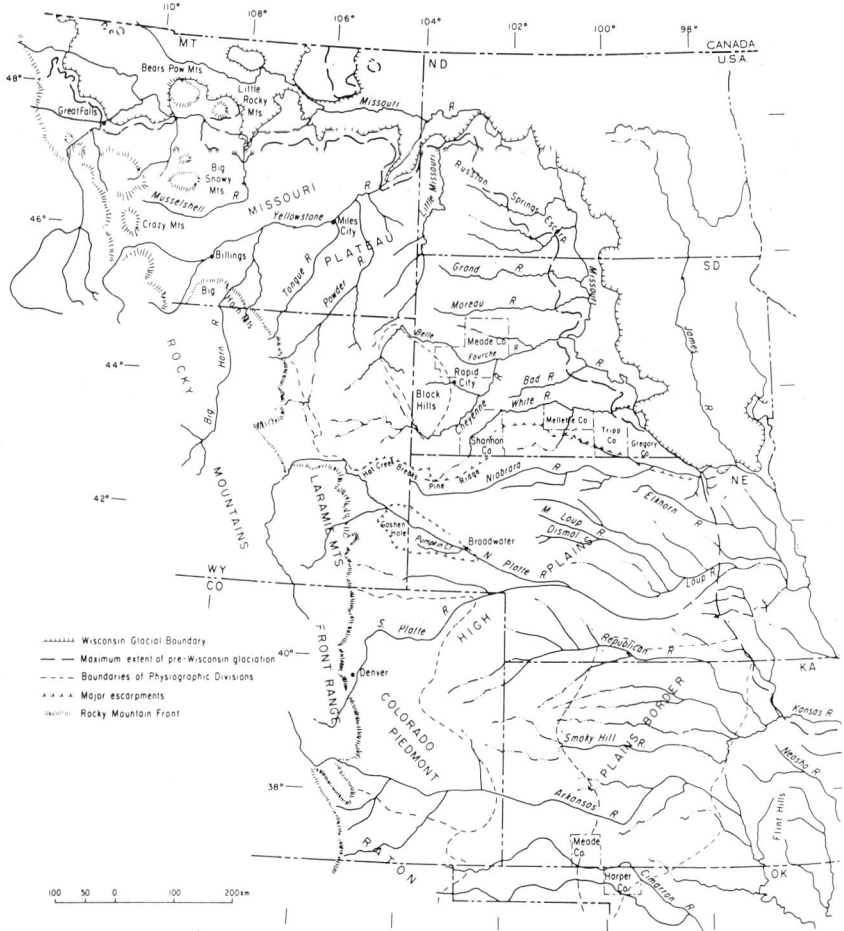

Figure 86. Regional drainage pattern in the northern U.S. Great Plains (simplified from Wayne et al., 1991). Note the predominance of E-W trends in the southern Interior Platform, as well as of the NE-SW trends in the northern Interior Platform that begins in Montana and continues into Canada. A rectilinear NE-SW/NW-SE drainage pattern is common also in the cratonic platforms in western Canada (Fig. 6b).

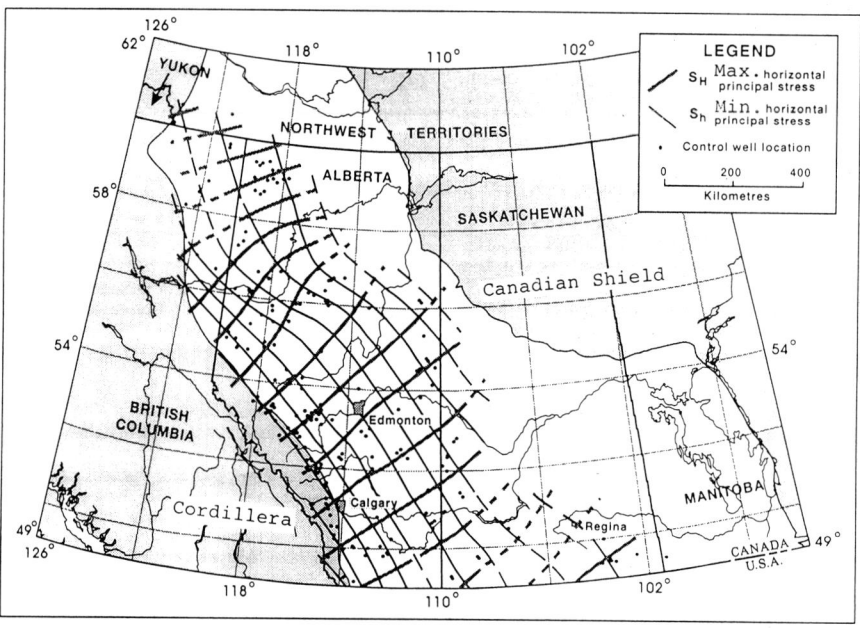

Figure 87. Horizontal-stress orientations in the Phanerozoic sedimentary cover of the Alberta Platform (modified from Bell and McLellan, 1995). The same stress pattern may or may not persist deeper, into the crystalline crust, as deep continental drilling elsewhere has shown that stress orientations change with depth (Emmermann and Lauterjung, 1997).

the Cordillera and the edge of the Canadian Shield. A number of Mesozoic petroleum fields in the sedimentary cover of the Alberta Platform are concentrated along this zone (Podruski et al., 1988; cp. Penner et al., 1987). To analyze all these data is one of the jobs involved in ongoing hydrocarbon exploration in a mature region like the Western Canada Sedimentary Province.

Seismic techniques cannot resolve small fault offsets, less than 10-15 m, in most structural-formational étages of the Alberta Platform sedimentary cover. The cratonic-stage fracture fabrics, not associated with magmatism, are expressed in potential-field data only weakly. Detection of faults, crucial in petroleum exploration, requires very detailed integrated analysis of subtle changes in seismic profiles, cross-well log correlations and potential-field anomalies. The temptation to interpret potential-field data simplistically in terms of structural trends prospective for mineral and hydrocarbon exploration is unjustified and may lead to disappointments. Wholly incorrect but common is a mistake to treat gravity and magnetic anomaly highs as basement structural highs and anomaly lows as basement lows. Even along proven petroleum-producing trends in Alberta, drilling success of later wells has been mixed (Greggs and Greggs, 1989). The distribution of oil and gas deposits does not simply follow structural trends, being also a function of syn-depositional paleoenvironments, subsequent restructuring, burial, fluid migration, etc.

Another usual mistake, discussed above, is to put the boundaries of potential-field anomaly domains along the zero anomaly contours, whose position is arbitrary and depends on specific assumptions made in the data processing. It may also be affected by secondary regional tectonic effects, such as basement tilt. The boundaries of potential-field domains corresponding to tectonic units commonly cross the zero anomaly contours (Hood and Teskey, 1989; Lyatsky et al., 1990, 1992, 1998). Gravity and magnetic lineaments related to steep crustal faults are independent of the zero anomaly-amplitude contours. Besides, only those faults are important in exploration that affected the sedimentary cover. Without proper local studies, overly optimistic promises that lineament analysis is the long-sought silver bullet in exploration have led to disappointments.

Compartmentalization of an area, interpreted from domains with visibly consistent anomaly patterns, is an important first step. Pieces of the crust of similar or dissimilar composition and structure can be distinguished in gravity and magnetic maps, although different rock assemblages and rock-body relationships commonly produce very similar potential-field anomaly signatures. To distinguish even potential-field anomaly domains is thus not always easy. Much more difficult is to correlate these assumed potential-field domains with possible tectonic units. There is a lot of room for subjective conclusions, depending on a particular individual's biases and experience. To reduce subjectivity, direct geological observations of anomaly-producing rocks are indispensable. From experience, some empirical principles have arisen: anomalies caused by igneous bodies are often round or curved (e.g., Sprenke et al., 1986), whereas tectonic blocks in the crust are expressed in angular or polygonal anomaly geometries (e.g., Lyatsky et al., 1990, 1998). Correlation of potential-field anomaly lineaments with tectonic boundaries, and of the latter with proven geologic boundaries and topographic lineaments observed at the surface, makes the interpretation more reliable.

Drillhole data show a vast variety of ways in which fault activity can be read from the distribution of cover lithofacies, rock alteration, local structures, and so on (Jones, 1980; Plint et al., 1993; Bergman and Walker, 1995; Dietrich and Palmer, 1996; Gregor, 1997; Edwards et al., 1998). In a recent summary of a decade-long Lithoprobe effort in the Western Canada Sedimentary Province, Eaton et al. (1995) and Clowes (1996, p. 118) reduced the great variety of basement control to only three "styles": (a) passive effects of the basement topography on the lower cover horizons; (b) faulting affecting the Cambrian strata; (c) "abrupt lateral facies changes of uncertain origin". In reality, the number of ways the basement could have influenced the cover is much greater, and some pre-cratonic and pre-cover structural characteristics are more significant than others. A more cautious approach is needed to defining the modes of basement fault and block influence on the sedimentary cover and its economic-resource potential. Of maximum importance are whole-crust, combined basement-cover movements during the development of the cover.

Distribution of Mesozoic oil and gas fields all over the Western Canada Sedimentary Province (e.g., Penner and Mollard, 1991) has been shown to be very sensitive to the position of crustal flexures and fracture zones that were active during various tectonic stages. Most of this tectonic activity affected both the basement and the cover. Patterns of lithofacies distribution, thickness and depth-to-top variations of some late Mesozoic structural-formational étages are also correlative with the regional patterns of surface lineaments mapped from remote-sensing images and potential-field data (Cant and Abrahamson, 1996; Gregor, 1997; Edwards et al., 1998). Reactivation of NE- and NW-trending faults across the Alberta Platform affected the localization of some Cretaceous sand bodies containing oil and gas in Alberta (Podruski, ed., 1988; Walker and Eyles, 1991). Late Cretaceous-Tertiary tectonism created favorable conditions for the localization of sand bodies in the Laramian foredeep basin (e.g., Bergman and Walker, 1995).

In the Late Albian-Maastrichtian Western Interior Seaway, conventionally ascribed to a worldwide highstand of sea level (Caldwell and Kauffman, eds., 1993), sedimentation occurred under extracratonic (Cordilleran) and intracratonic influences. According to Plint et al. (1993), wedges of Cretaceous thick marine shales were laid down on the Alberta Platform during periods of higher tectonic activity in the adjacent Cordillera, and coarser sediments were deposited during periods of Cordilleran relative tectonic quiescence. The latter periods saw a higher cratonic crustal instability in this platform, which affected the basement and the cover. During Late Cretaceous-Early Tertiary tectonic stage, the structural pattern of the Alberta Platform was complicated by whole-crust movements caused by cratonic warping as well as by orogen-induced subsidence (in the foredeep). The origins and inception of the neotectonic structural fabrics in the Alberta Platform were neither simple nor uniform, but these fabrics are of considerable economic significance. Coal deposits of Late Cretaceous-Early Tertiary age, mined in many parts of Alberta, show evidence of syn-sedimentary fault control and disruption by younger faults. Intermittent precursors of the modern NE-SW continental-divide arch between Arctic and Atlantic drainage systems (Fig. 40) probably appeared north of Edmonton in the Albian and gradually became a permanent structural feature of the modern Alberta Platform.

The modern structure of the Alberta Platform is a product of its entire Phanerozoic history, but it took its final shape in the Late Cretaceous and Cenozoic. In the Cordilleran miogeosyncline, reorganized once more by the Laramian orogeny, post-compressional uplift generally began in the Late Paleocene. But the modern uplifts in the Northern and Southern Sweetgrass arches had originated earlier, in the mid-Paleocene (Jerzykiewicz and Norris, 1992). The neotectonic and Recent structural framework is of different ages, generally dating back to Miocene-Pliocene, is also whole-crust, involving the basement and the cover inseparably.

Second look at geophysical data with the benefit of information about neotectonic and current tectonic activity

Geophysical data cannot be interpreted in geological terms uniquely, as differentrock-body compositions and configurations can produce similar anomalies and images. Geophysical methods respond to variations in some physical characteristics of rocks but are unable to reveal many other geologically meaningful variations in the crust. Geophysical surveys say nothing about rock genesis and age, environments and settings of their origin. Even the structure and composition of rock bodies can be interpreted from geophysical-anomaly parameters in many differentways. The interpreter must thus be aware of differentpossible interpretation options and the specific geology of the study area. The specificity of each particular lateral tectonic unit requires geologic calibration of the interpretations from one region to another. The ultimate solution in the tectonic interpretation of geophysical data should fall to experienced geologists, who are mindful of geological facts derived from observation of rocks and rock-body relationships. Speculations must be cautious, and checked with data available from other relevant fields of knowledge including other geophysical surveys.

Steep crustal faults, missed in many modern seismic surveys, should be interpreted from mapped geology and gravity and magnetic maps. Even then, separation of lateral tectonic crustal units is often still debatable. The knowledge about sources of geophysical anomalies makes this separation less arbitrary. Linear anomalies are produced by lateral changes in rock density and magnetization which may indicate (a) primary lithologic variations, (b) secondary variations in the metamorphic grade, (c) presence of discordant intrusive igneous bodies, (d) structural offsets, or (e) a combination of the above. Straight linear anomalies across which the anomaly character changes sharply can be interpreted with some confidence as major faults. Their locations cannot be always pinpointed precisely. But in a regional view, they provide investigators with a network useful for the partition of anomaly domains in potential-field maps if the anomaly pattern on both sides of a lineament is significantly different (in terms of their shape, amplitude, wavelength spectrum, etc.). If anomalies are similar but offset across a lineament, a major crustal fault could be inferred as well.

Various anomaly-enhancement techniques are used to reveal details of particular anomalies and domains not clearly seen in conventional gravity and magnetic maps. Anomaly wavelength filtering sometimes helps highlight the crustal and lithospheric structure in western Canada (e.g., Stephenson et al., 1989; Sweeney et al., 1991). The horizontal-gradient vector method helps to highlight subtle anomalies and delineate fault-related boundaries in various parts of the Western Canada Sedimentary Province (e.g., Lyatsky, 1990, 1992, 1998; Edwards et al., 1998).

Along major faults, reworking of rocks usually affects their density and magnetic-mineral content. The more intense the fault-related reworking in a region, the more chance that a detected fault zone might be an important tectonic boundary. Igneous intrusions strengthen the potential-field signatures of domain-bounding fault zones. Where offsets of anomaly patterns across these zones are appreciable, they are more readily interpreted as tectonic displacements. In the Canadian Interior Platform, for instance, the biggest NE-SW-trending fault zones are traceable from the Shield across the Platform into the Cordillera. The Hay River and MacDonald faults, the Great Slave Lake and Snowbird-Virgin River shear zones separate crustal blocks with different potential-field signatures, which facilitates their interpretation as first-order boundaries on the scale of the platform or the whole craton.

The NE/NW lineament pairset is known in many other parts of North America (King, 1977; Shurr et al., 1989; Adams and Bell, 1991; Kent and Christopher, 1994) and on other continents (Cloos, 1948; Anfiloff, 1988). This similarity has previously led to the ideas about a global system of regmatic fractures (e.g., Sonder, 1947; Vening Meinesz, 1947). The major structural features of the North American craton, such as the very prominent Paleozoic Transcontinental Arch, fit well into the orthogonal NE/NW pattern. Conspicuously, a similar pattern is also expressed in the modern stress-field maps on the North American continent (Weimer, 1980; Zoback and Zoback, 1989).

Discontinuities and anomaly alignments at NE-SW-trending faults zones are evident from other geophysical observations as well. South of the Snowbird-Virgin River lineament, most geoelectric strike directions are NE (N35°E-N50°E), consistent with the NE-SW structural trends. North of this lineament, electrical-conductivity trends are generally oriented more northerly than in the south; these anomalies are correlative with some families of joints reported by Langenberg (1983) from the Canadian Shield in northeastern Alberta. An anomalous local area lies near the city of Red Deer north of Calgary, where some geoelectric strikes are ~N20°W, similar to a local but strong magnetic high there. Models indicate that multiple electrical conductors in Alberta lack deep roots, lying commonly in the upper 5-10 km of the crystalline basement (Boerner et al., 1995, 1996). A family of low-angle joints mapped in western Canadian Shield areas (Langenberg, 1983) may play a role in defining the conductivity structure of the upper part of the basement in the Alberta Platform. Conductors may be related to faults containing brines. In places, some vertical conductors coincide with magnetic anomalies, but typically, they are discordant with the magnetic-anomaly trends and run across them. Keeping in mind that magnetic anomalies in this region are sourced by crystalline rocks in the basement, presence of discordant conductors suggests the inferred causative basement faults are relatively juvenile. Brines, deriving their salt from Devonian evaporite horizons in the sedimentary cover and flowing along faults in the basement, make these faults detectable in conductivity maps.

There used to be a general tendency to link crustal electrical conductors with marine salt water trapped in rocks in assumed subduction zones (cp. Kurtz et al., 1986). But this is not always the case (e.g., Jones and Gough, 1995), and in Alberta conductivity zones in the crystalline basement may simply occur along brine-conducting fractures. In zones of salt dissolution, fractures have indeed been found to act as brine conduits, even if they have no structural offsets (e.g., Lyatsky et al., 1998). The porosity of rocks trapping fluids in the crystalline basement, though small, may be enough to affect the reflectivity and signal-attenuation patterns in seismic surveys in such localities, creating a pitfall in seismic interpretation.

Even the prominent NACP electrical-conductivity anomaly zone, trending generally N-S in southern in Saskatchewan and North Dakota, seems to consist of many straight and curved segments. As a whole, it has been modeled to be some 50 km wide, with separate components lying at the middle and upper crustal levels (Jones et al., 1993). Contrary to earlier ideas, the NACP anomaly does not represent an old suture that still retains oceanic water (as was supposed by Thomas et al., 1987), but is more likely caused by faults containing graphite or sulfide concentrations (Camfield et al., 1989; Jones et al., 1993). Discontinuous sets of strands along the prominent Tabbernor and Saskatoon fault zones are good candidates for such concentrations. The relationship between the electrical and potential-field anomalies is not direct, and the NACP anomaly zone transects some coherent regional potential-field anomaly domains (Lyatsky et al., 1998). A relatively young origin of this anomaly seems probable.

In the north, the NACP anomaly corresponds to a part of the Tabbernor fault zone. Elongated magnetic anomalies there indicate this fault zone's position, trending N-S along the eastern edge of the Archean Glennie block. It corresponds to the part of the NACP anomaly which in the north coincides with the Tabbernor fault zone. To the west, the magnetic quiet domain between the Tabbernor and Saskatoon faults is related to a little-reworked remnant of the Early Proterozoic Custer platform. In the Dakotas, N-S-trending magnetic and Bouguer gravity anomaly trends are observed in a 30-40-km-wide knot where the Tabbernor zone meets the Thompson zone (Zietz et al., 1982; Geological Survey of Canada, 1990).

Spatial coincidence of some segments of the NACP anomaly with the long-lived Tabbernor fault may suggest these segments of the anomaly were caused by the rejuvenated strands of this fault, whose reactivations in Phanerozoic time are reflected in the sedimentary cover of the Midcontinent Platform.

A late Precambrian tectonic reactivation might and craton-wide restructuring is suggested by the appearance of the great Midcontinent rift system (Figs. 60, 88). It consists of regional faults and volcano-sedimentary basins, and is traced for over 1,000 km with a NE-SW trend, from Lake Superior to at least Kansas. Aeromagnetic data suggest this system may continue as far south as Oklahoma (Yarger et al., 1981). Though almost completely hidden under the sedimentary cover of the North American craton, it is well expressed in gravity and magnetic maps and transected by several COCORP seismic profiles. The rift fill produces strong potential-field signatures in many areas. Linear gravity anomalies mark the flanks of the rift, some 60 km apart. The rift is filled with Late Proterozoic low-grade metamorphosed sedimentary and mafic volcanic rocks and mantle-derived ultramafic intrusions. The total thickness of these rocks is ~8 km (Serpa et al., 1984). A tectonic feature so huge could not have formed without a great part of the craton being tectonically disturbed.

Post-Proterozoic rifting, though less intense, also occurred in the North American craton, at least in the U.S. Big linear structural trends are identified by mapping and drilling in some structural-formational étages of the cratonic cover (e.g., Sloss, ed., 1988; Friedman et al., 1992), and the New Madrid zone of considerable seismicity is manifested by several historical earthquakes. Rifting is a common upper-crustal manifestation of reworking at deep levels in the crust. These processes have been indigenously cratonic. Continuing tectonic activity in the North American craton makes it possible that the causes of the NACP anomaly are also reworking-related.

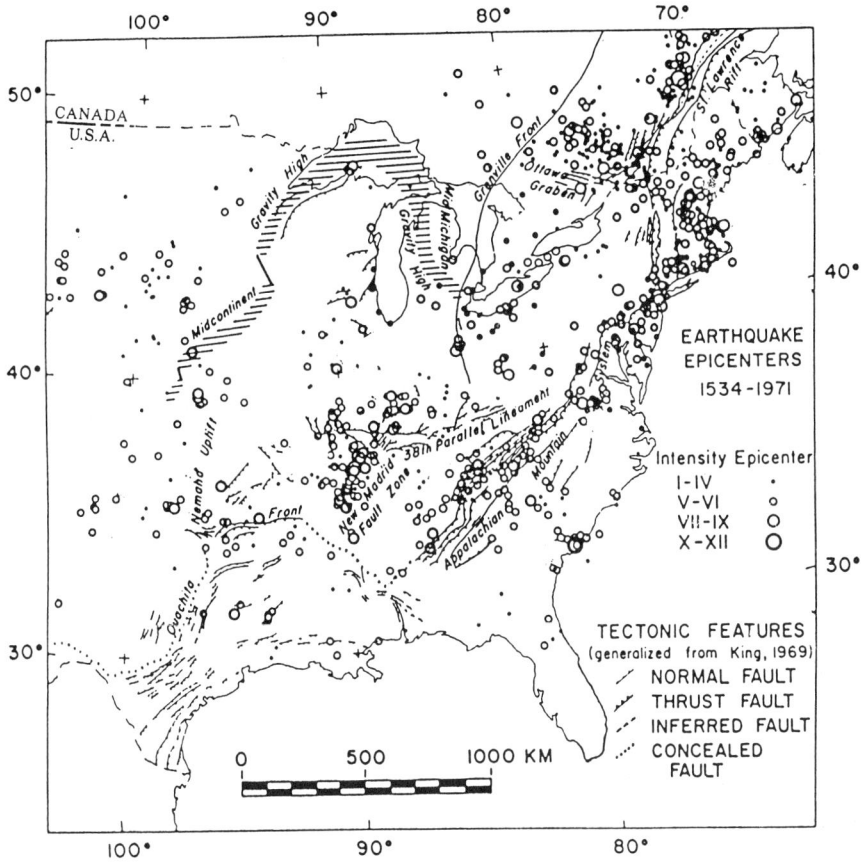

Figure 88. Major tectonic zones and seismicity clusters in eastern North America (modified from Hinze and Braile, 1988). The Midcontinent gravity high marks the Proterozoic Midcontinent rift system (Figs. 48a, 60).

A revised view can be taken of some enigmatic seismic events in COCORP and Litho-probe deep reflection profiles from various parts of the North American craton. Unfortunately, all too often these data are interpreted based on some currently fashionable biases, without sufficient consideration of possible alternatives. The common exaggeration of the role of detached thrust slices resulted from the deficiency of modern seismic techniques in detecting steep faults. Another misconception is to casually interpret subhorizontal seismic events cutting the predominant dipping reflections as igneous-rock sills. A general error is to try to interpret from seismic images the genesis and age of the supposed geologic bodies, though seismic discontinuities of similar shape can arise in many different geologic situations.

The prejudice that the continental crust consists of thrust slices stacked in an unpredictable fashion is hampering seismic interpretation of these very valuable data, particularly in the western craton and the Cordillera (e.g., White et al., 1994; Cook, 1995a-c). Another biased tendency, in western Canada, ignoring many other reasonable alternatives, is to simply ascribe low-angle seismic events deep in the crust to Proterozoic igneous sills (e.g., Mandler and Clowes, 1997a-b; Ross and Eaton, 1997). Indeed, the Lithoprobe seismic reflection profiles across Saskatchewan contain several variously continuous, bright, low-angle reflections in the crystalline crust. In the conventional geophysical jargon, such high-amplitude reflections are called "bright spots" (in sedimentary rocks, they are sometimes associated with gas concentrations). Mandler and Clowes (1997a-b) named such seismic bright spots deeper in the Saskatchewan crust the Wollaston Lake Reflector. They traced the associated seismic reflection band for "hundreds of kilometers" from the Wathaman-Chipewyan Batholith northward into the sub-cover crustal levels under the Athabasca Basin, and attributed it to the Mackenzie igneous suite. Swarms of Mackenzie diabase dikes, ~1,267 Ma in age (Hulbert et al., 1993), indeed cross many parts of the Canadian Shield. They are easy to recognize in the topography, as they form linear ridges up to tens of meters wide. Many dike swarms of this scale, Mackenzie and others, often have associated sills deeper in the crust.

Low-angle seismic events cutting the principal reflection fabric have been interpreted as sills in other regions, in Fennoscandian and the North American Cordillera. Large, thick mafic sills have been found by mapping in the extremely thick Belt-Purcell Basin; in places, these sills in the upper crust are up to 6 km thick. Sills have been penetrated by a superdeep well in Sweden. But not all near-horizontal reflection bands are caused by sills, and extensive analysis is required in each particular case to justify such an interpretation. In the central Cordillera in the U.S., Litak and Hauser (1992) correlated some seismic reflections with mafic sills whose exposures have been mapped in that area. Similar-looking reflections could be from shear zones, fronts of regional metamorphism, water-saturated porous horizons, low-angle faults, and so on.

The reflections registered by Mandler and Clowes (1997a-b) lie in the upper crystalline crust at depths of 6 to >13 km. The estimated thickness of their causative bodies varies usually between 50 and 150 m. They have no relation to the granitic Wathaman-Chipewyan Batholith, which itself is cut by small undeformed bodies of leucogranite and pegmatite dated at ~1,770 Ma (Meyer et al., 1992). Proterozoic platforms to the south and west contain anorogenic granites created at ~1,760 Ma and later (e.g., Collerson et al., 1990). Numerous younger tectono-magmatic and tectono-deformational episodes are recorded in the Athabasca Basin (see above). This offers a variety of possible ages of dike/sill magmatism (cp. Halls and Fahrig, eds., 1987). More importantly, there are many alternatives to sill origin of bright seismic events in the crystalline crust.

Massive sulfide bodies in the crystalline crust have been imaged seismically in the mineral-deposit-rich Sudbury region of the North American craton (Milkereit et al., 1996). These authors have modeled an upper-crustal section in that region to depths of ~5 km. But surface seismic reflection methods are designed to detect subhorizontal and low-angle discontinuities, and features dipping up to 60° have been imaged only in some circumstances and with difficulty. Normally, for steep boundaries in the crust, borehole seismic methods are more suitable (Milkereit et al., 1996), but the interpretation of such data is constrained by direct geologic evidence from the drillhole. With deep crustal seismic data shot at the surface, no such direct control is available.

Other subhorizontal seismic events cutting the principal reflection fabrics have been shown by modeling to possibly be off-line arrivals from vertical faults and dikes parallel to the seismic profile or crossing it obliquely (Zaleski et al., 1997). Such events are often interpreted as shear zones, thrust faults, intrusive sheets (e.g., Geis et al., 1990; Ross and Eaton; 1997; Ross et al., 1997), but they could also be arrivals reflected or refracted from such steep faults and dikes. Zaleski et al. (1997) noted specifically that although in some areas such "subhorizontal reflections have been modeled as tabular mafic intrusions (Mandler and Clowes, 1997; Ross and Eaton, 1997)", many of the same parameters (physical properties, sheet widths) "could apply to both dikes and sills, and the models alone do not provide a means of assessing intrusion orientation". The same uncertainties apply to the interpretation of deep crustal seismic bright spots in Alberta and Saskatchewan.

In Alberta south of the Peace River Arch, Ross and Eaton (1997) speculated that such bright spots (each between 60 and 80 m thick, and no broader than 40 km), lying at depths from 3.5 to 18.5 km, may be discordant igneous sheets rather than sills, which they gave them a joint name "Winagami reflection sequence". In Saskatchewan, with more exhaustive modeling (Mandler and Clowes, 1997a-b), similar reflectors have been modeled to vary in thickness from 8 to 216 m, while lying in a much narrower crustal depth interval. These authors reported up to 7 high-velocity layers in that reflective band, and found "...extreme variation of the internal structure of the features often within only a hundred meters of lateral offset." The latter observation does not seem to match the generally uniform composition and style of mafic dike magmatism of various generations in the Canadian Shield (Halls and Fahrig, eds., 1987). Such lateral variability is more consistent with fossil metamorphic fronts or brittle-ductile transition zones, perhaps associated with porous zones saturated with water or gas (cp. Bailey, 1990).

The estimated emplacement depth of the Wathaman-Chipewyan Batholith is ~17 km. Metamorphic grades of its rocks require depths exceeding ~20 km (Meyer et al., 1992). The transition between the brittle and ductile parts of the crust, in Proterozoic as now, lay at these mid-crustal depths (cp. Colpron et al., 1996). Rheology of rocks changed greatly with time, largely as a function of their position in the crust. The proven vertical crustal movements in the Canadian Shield and the entire craton from time to time shifted the position of the brittle-ductile transition zone. When the fossil transition zones from previous tectonic stages were pushed into the lower crust, they were obliterated by new metamorphism. But when they were placed into the upper crust, parts of them could be preserved, and their depth may now differ from one block to another. As described and modeled by Mandler and Clowes (1997a-b), the low-angle reflectors in Saskatchewan cross the regional seismically imaged crustal geometry, have different (but narrow) thickness, and lie at differentdepths in the modern upper crust. If they correspond to former brittle-ductile transition zones, they may serve as markers deserving more detailed analysis for evaluating the amplitude of subsequent vertical crustal movements. Their age may not be the same, as post-orogenic, i.e. purely cratonic, vertical movements of the crust in southern Saskatchewan have been taking place since as early as 1,800-1,700 Ma (Gordon et al., 1990; Ansdell and Norman, 1995). Other recorded pulses of Proterozoic crustal deformation took place later: in the Kisseynew Basin area, a rearrangement of the tectonic regime was marked by intrusion of the Mackenzie dikes at ~1,267 Ma; in the Athabasca Basin, vertical movements of the crust happened sometime after ~1,300 Ma; and so on.

Fossilized Late Proterozoic and Early Paleozoic brittle-ductile transition zones are recognized in outcrops in several localities in the miogeosynclinal Omineca Belt (e.g., Tempelman-Kluit et al., 1991). Detailed mapping has shown these zones to be marked by intense low-angle shearing and brittle faulting, and metamorphism is these zones is imposed across the broader structural fabrics (also Simony, 1995). Encouragingly, despite associating their deep reflections in Saskatchewan with "extensive tabular intrusions", Mandler and Clowes (1997a; 1997b, p. 46) stressed also that "the pronounced lateral variability of the reflections suggests that rheological inhomogeneities are an important controlling factor..."

Fossil brittle-ductile transition zones have the best chance of preservation if they are pushed up, away from the reworking in the lower crust. This suggests they should be common in shield areas, where the long-lift predominance of uplift has caused elevation of the lower levels of the Proterozoic crust. No strong, regional post-Hudsonian metamorphic overprint affected the Canadian Shield in Saskatchewan (Gordon et al., 1990; Ansdell and Norman, 1995).

Main classes of vertical tectonic crustal movements as manifested in the neotectonic and current stages

The term *epeirogeny* was introduced by Gilbert (1890), who used it to describe crustal movements in the area of Lake Bonneville (a predecessor of the Great Salt Lake) in Utah. From gentle bending of old shoreline strands, he concluded that the crust there was disturbed by broad warping distinct from the short-wavelength regional structure caused by orogenic events. The essential meaning of this term was preserved for years, though with deep modifications. Epeirogenic movements are now often thought to be those which produce large sedimentary basins and arches (such as the Michigan Basin or the Ozark Dome; Fig. 89). The Colorado Plateau and the much smaller Black Hills Dome have also been described as epeirogenic uplifts. From various techniques of study, including fluid-inclusion homogenization temperatures, $\partial^{18}O$, and vitrinite reflectance, Friedman (1987a-c, 1988b) has shown that in the eastern North American craton, Ordovician strata in the Ozark Dome and Ordovician to Devonian sedimentary formations in the Northern Appalachian Basin had former burial depths on the order of 4.5 to 7 km. Though the exact amplitude of burial and subsequent uplift is a subject of discussion (Levine, 1986), it was very large. Friedman (1987a-c) argued that some epeirogenic up and down movements could have been repetitive (see also Friedman, 1988b vs. Miall, 1987). The post-orogenic, epeirogenic Mesozoic uplift in New England and adjacent parts of eastern Canada is estimated to have exceeded 4 km (Crough, 1981).

The present state of knowledge permits to classify vertical crustal movements in more detail, particularly based on the lateral wavelength of upward and downward displacements. The largest of them are epeirogenic (etymologically, continent-generating), smaller ones are warps, and the smallest are usually block-related.

Vertical crustal movements of the greatest wavelength affect vast territories, cratonic and orogenic alike, and sometimes involve entire continents. Such, purely epeirogenic movements account for the modern high stand of the whole African continent, long-lived uplift and subsidence of shield and platform areas in Eurasia and North America, a pronounced tilt of the Eurasian continent away from the modern Mesozoic-Cenozoic Eurasian mobile megabelt and towards the Arctic (Lyatsky, 1969), and the southward tilt of the North American continent towards the Gulf of Mexico.

Epeirogenic movements are warp-like, and they involve crustal buckling with wavelengths on the order of thousands of kilometers. Warps proper, as a distinct class of

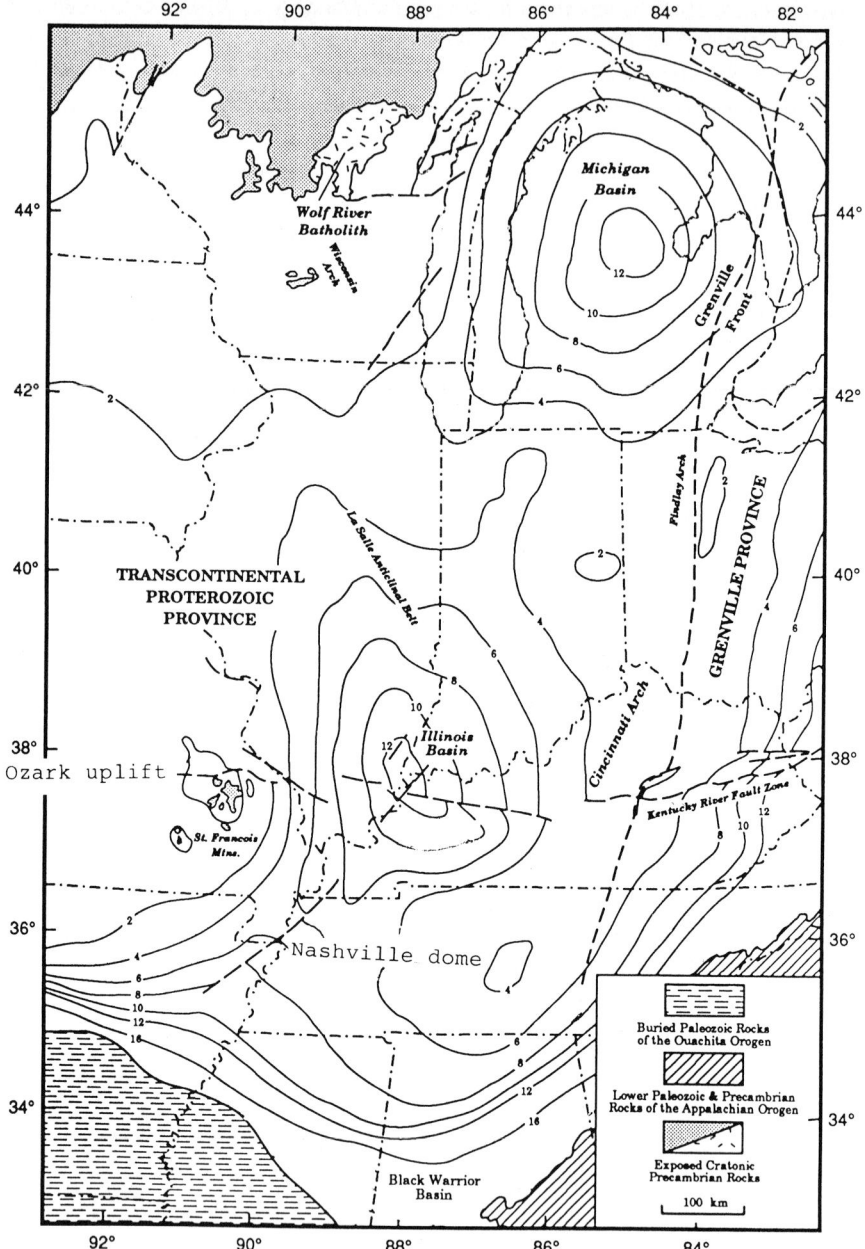

Figure 89. Round basins, uplifts and domes in the U.S. Midcontinent Platform (modified from Van Schmus et al., 1993). Contours represent crystalline-basement depth below sea level, in thousands of feet (1,000 feet = 305 m).

vertical movements, affect areas only hundreds of kilometers in wavelength: neotectonic warps in the modern Alberta Platform are good examples. Both epeirogenic buckling and more-local warping differ from block movements, which are produced by uplift or subsidence of brittly behaving crust dissected by deep, steep faults. Block movements, widely recognized under the name Germanotype, are well exhibited in mountainous topographic features, such as Harz and Schwarzwald (Black Forest), and in depressions like the Rhine and Oslo grabens. In western Canada, they are expressed in the arch/embayment and horst/graben regional structure of the Peace River and other areas in the Alberta platform. Surficial expressions of warps can be sedimentary basins, arches and domes (Michigan, Williston and Alberta basins; Nashville and Ozark domes; Northern and Southern Sweetgrass arches). Rift-related basins (e.g., Artyush-kov and Baer, 1989) may actually be a subclass of this class.

Vertical movements of continental crust occur all the time and everywhere, and they are reliably recorded by numerous structural-formational étages in both cratons and mobile megabelts. King (1977) and Rodgers (1987) took notice of chains of basement uplifts in marginal parts of the North American craton parallel to the Cordilleran and Appalachian mobile megabelts. In the Rocky Mountains in the U.S., from the Colorado Plateau to the continental interiors, vertical movements several kilometers in amplitude occurred during just ~10 m.y. in the Late Paleocene and Early Eocene (e.g., Sloss, 1988a). In Eurasia, such tectonic movements, also without extracratonic or extracontinental influences, created even the much larger, monumental northern Tien Shan Mountains, where Tertiary marine deposits now lie at altitudes of ~5 km.

The Neogene to Recent 1.0-km to 1.5-km crustal rise of the Great Plains largely created the modern topography of much of the North American continent. Sloss (1988a, p. 49) stated: "Prominent among unresolved neotectonic problems is the matter of the prevailing extraordinary elevation of much of the North American Craton above sea level." He observed that "the anomalous hypsography of Holocene time is not confined to active continental margins and adjacent forelands but is pervasive on shields, platforms, and revived ancient mountain belts on several cratons (except, perhaps, Australia) to an extent unattainable by glacially imposed eustasy. There exists no supportable body of theory to rationalize this phenomenon."

Motion of lithospheric plates fails to explain such phenomena in continental regions. It also fails to explain the apparent directional evolution of the cratons, which have progressively passed through several irreversible stages before reaching their present conditions. Directional evolution of cratons occurs, with repetition of some processes, from one tectonic stage to another. This evolutionary history is best read from structural-formational étages, which provide a realistic basis for and control on tectonic generalizations.

8   CONCLUSIONS: ADVANTAGES OF PRACTICAL TECTONICS

Historical pragmatic character of geology as a field of human activity and knowledge

Practical tectonics, like geology itself, is a rock-based field of knowledge about the surrounding nature which is the background for human activity. Rocks and rock-made bodies have always been the principal objects, information sources and evaluation criteria for geological ideas and efforts. Geology appeared from the need to use stones, and the stone age in the human history was very long. Only in the 18th and 19th centuries was geology firmly institutionalized as a distinct science. At present, it has branched out into many directly interrelated disciplines: mineralogy, sedimentology, igneous and metamorphic petrology, paleontology, geologically applied geochemistry and geophysics, etc. On the economic side, relevant industries like mining and petroleum benefit directly from geology as a field of knowledge (which is more than just a science). Environmental geology is still being defined in practice, but its place is becoming increasingly important as the impact of mankind's activity on the natural environment increases.

Theory and practice evolve hand-in-hand: practice increases comprehension, and comprehension amends the practice. The ability of modern geology to be usefully applied in human activity is an important achievement. But theoretical geology, and tectonics, can be fruitful only if constrained by observations from geologic and tectonic practice. The modern geologic and tectonic practice is mostly mapping (field work and compilation of maps of visible geologic phenomena), environmental studies and rock-related industrial activity. Thought-processed, newly established things (e.g., new minerals or fossils) and relationships (e.g., contacts between rock bodies), i.e. newly gathered facts, are essential for constructive formulation of a scientific theory. Such a theory is generalized (based on more than just some incidental local findings) and included into internally consistent concepts which do not contradict the fundamental principles of logical reasoning.

Geology, as a science, is observational and descriptive. It generalizes the facts obtained through field work, physical observation and laboratory analysis of rocks and of their mineral and chemical composition. Modern technology, with its microscopy, chemical-analysis and experimental-petrology capability, remote sensing, various geophysical methods, sea-floor sampling, drilling, expands the scope of observations greatly. Description of rock bodies, their identification and determination of their interrelationships, derived from field observations, may be extrapolated in time and space. Application of exact-science methods, especially those adapted from mathematics, coupled with improvements in computer technology, enables rapid processing of large amounts of data and their beautiful displays. But, to be useful, these images must be kept rational and realistic. Separation of geologically meaningful parts of an image is, in modern times, very much a part of tectonic studies.

The main task of natural sciences is to grasp the causal and genetic relationships between observable phenomena, and the predictions that can be derived from the understood genetic systems. Predictions often fail because the abstraction is incorrect, the geologic phenomena are oversimplified, or the initial assumptions are chosen wrongly. Extremely important are the accuracy and purity in the gathering of factual material. In geology, this includes correct definition of rocks, rock-made bodies and their observable relationships in the rock record (not vague, assumption-driven speculations about their genesis). The second requirement is to put together all the facts in an unbiased way, especially taking note of those facts which seem to contradict the existing or newly-formed concepts. Finally, an internally consistent conclusion derived from all these facts must be matched to the available relevant observations, mostly derived from geologic mapping. Evaluation of the new conclusions in comparison with other concepts promulgated previously for the same region or phenomenon, and examination of the roots of any discrepancies between them, are necessary components of scientific work. A description of the new data and their preliminary, speculative explanation does not constitute a scientific work - it is merely a collection of information. Science arises from analysis and conceptualization of the facts in observation. A review of all contradictory facts should be part of a scientific publication, along with a discussion of previous ideas and mistakes. The mistakes, especially, must be examined to derive positive lessons (it is important to remember that your own new word is also not final, and more updates and corrections will follow). These steps are compulsory if new useful knowledge is to be obtained.

This practice of thorough scientific thinking took centuries to work out. A modern tendency to substitute the meticulous data collection and analysis with faddish, assumption-based hypothesizing (called, in a current fashion, modeling) is without usefulness or future. Taking into account only a small number of parameters which do not necessarily define the system in question, and relying on some superficial coincidence of modeling results with some selected aspects of the observed phenomena, is not a scientific contribution. This tendency is, of course, temporary: being unproductive and sterile, it will pass - but for now, it remains a dangerous distraction of science from its normal work. Facile reliance on simplistic (if momentarily fashionable) postulations rather than on labor-intensive objective analysis of factual data restricts the natural evolution of free thought and condemns the results to quick oblivion.

Since the beginning of the 20th century, it has been recognized that continental crust consists of large crustal/lithospheric blocks: cratons and mobile megabelts, orogens and median massifs, horsts and grabens, and so on. This knowledge is reflected in tectonic maps in the former U.S.S.R. and Europe, compiled by N.S. Shatsky, A.A. Bogdanov and T.N. Spizharsky. In North America, a similar approach was adopted by King. The purposes of making these maps were quite practical: to create regional summaries that would form the base for understanding the continental evolution, and to offer a background for predicting mineral and petroleum concentrations.

More recent tectonic maps tend to be different. Rather than showing blocks and megablocks delineated by mapping, they arbitrarily demarcate tectonic domains derived from models derived from pre-conceived tectonic assumptions. Even the very existence of these domains is in many cases just presumed, not derived from or verified by visible geological facts. Even the best of such maps, such as the U.S.-made *Plate-Tectonic Map of the Circum-Pacific Region*, show plate boundaries, transform faults and subduction zones as well as oceanic magnetic anomaly lineations - but not the observable rock-based information. In continental regions, this information is reduced to postulated extinct rifts and major intraplate faults, supplemented with mapped thrusts. Maps of this sort, though often called tectonic (e.g., Nokleberg et al., 1994), rely too heavily on the ideas about wandering terranes, plate reconstructions, and the perception that continental tectonism is only a reflection of events at faraway plate boundaries. Presumed kinematics and, sometimes, dynamics of some crustal blocks may be of interest in specific areas, but without rock-based factual information, such maps do not replace the classical summaries of regional geology. Such an approach also wastes society's resources on unproductive or misleading research whose products lack scientific or commercial value.

But geology's historical pragmatism remains unchanged, and as before, geology must attempt to meet practical societal needs.

Social needs for rock-based regional tectonics

Behind grand speculative constructions, grass-roots geology and tectonics continue to develop. Symptomatically, a new term appeared recently in the English-language literature - *lithotectonics*. This composite word is a tautology, because tectonics, as a derivative of geology, must deal with rock-made, i.e. lithic, bodies. But the fact that this new term appeared is evidence of an acute need for truly rock-based tectonics. The awkward word *lithotectonics* denotes an alternative to speculations that are conventionally placed under the rubrics of "plate-tectonic reconstructions", "quantitative tectonic models", and so on. It signals a desire of practical geologists to operate with rock-based tectonics rather than idle speculations, because the challenge is to understand the real tectonic history and structure of an area, in order to enable practical applications of this knowledge to the search for mineral and petroleum resources, reduction of environmental risks, civilian and military construction, and prediction of earthquakes, volcanic eruptions, landslides and other natural hazards.

One reason the original term *tectonics* is now abused so much is that, in North America at least, it has not been sufficiently well defined. A standard definition of Gary et al. (1972) held it to be a branch of geology (mostly, structural geology) that is regional and general. It studies the structure of the upper part of the crust in large regions. The popular textbook of Press and Siever (1978, p. 38) expands tectonics to include "the structure of the crust in general". Bates and Jackson (eds., 1987, p. 675, italics theirs) define tectonics loosely as "a branch of geology dealing with the broad architecture of the outer part of the Earth, that is, the regional assembling of structural and deformational features, a study of their mutual relations, origin, and historical evolution. It is closely related to *structural geology*, with which the distinctions are blurred, but tectonics generally deals with larger features." But in reality, tectonics is much more than just regional structural geology which deals with only the fault and fold styles. Structural evaluations of a region are very important parts of the information considered in regional tectonic studies, but tectonics is more comprehensive than that. It includes the whole of endogenic geology, studying all compositional and structural properties of the crust that are linked with lithospheric geodynamics. To reduce the tectonics of a region to just crustal structure means limiting the scope of tectonics artificially. This narrow definition leaves out most of the rock-made perisphere of the Earth - its lithosphere. In such an emasculated form, tectonics is unwisely deprived of its rich essence: multi-aspect character and predictive power.

Objects of tectonics are rock-made bodies as big as the lithosphere or as small as the smallest self-evolving block. Kinematic and geodynamic evolution of vertical and lateral tectonic units, their origin, alteration (reworking) and destruction, all of these fall into the scope of tectonics. Tectonics is revealed in its four main interrelated aspects: sedimentological, magmatic, metamorphic and deformational. The job of tectonics is to unify all the relevant data and facts available from various disciplines into an internally consistent concept of the regional history induced by endogenic geologic processes.

The Geological Society of London has long been alarmed that "the vast amount of data" accumulated on continents during the previous 150 years of purposeful geological studies and mapping "...has not been playing an appropriate part in concepts of world tectonics and history" (Kent et al., eds., 1969, p. vi). Thirty years after these words were written, there are vastly more data, from the ocean bottom as well as from land areas, and their underuse has only increased. An unsound tendency to work chiefly with newly obtained data is understandable: it is easier than reviewing great volumes of data and their previous interpretations, and it avoids arguments and conflicts. Contradictory data and their explanations often remain in the shadow, or used cosmetically to confer the legitimacy of a pretended discussion of options.

The unexpected recent discovery of quartz-rich felsic rocks on the surface of Mars brings to light a sudden puzzle: on that compositionally differentiatedplanet, similar to the Earth, there seem to be no traces of lithospheric-plate interactions. What does it mean for tectonics on Earth? Is plate tectonics (in its modern understanding) unique to our planet? How do we reconcile this uniqueness with the fact that on Earth "one of the major quandaries in geodynamics concerns the relation between plate tectonics and mantle convection..." (Bercovici, 1993, p. 35). These fundamental uncertainties continue to bedevil what is considered a cornerstone for innumerable speculations about the assumed tectonic history of the earth's lithosphere, continents and so on.

As well, geochemists reason that basaltic magmatism, common along subduction zones of downgoing oceanic-crust plates, cannot produce the andesitic bulk composition of the Earth's continental crust. They also argue that much, if not most, of this crust was probably created in the Archean (Armstrong, 1991), when tectonic processes occurred differently than today (Hamilton, 1993). The present span of Archean terrains is 14% of the Earth's surface, and 60% of continents (e.g., Cogley, 1984; Rudnick, 1995). Discrepancies between the overall amount of continental crust formed in the Archean and the amount of Archean crust presumably inherited in Proterozoic rocks are large. But they become reduced if we keep in mind that ancient crust could be reworked beyond recognition. Archean rocks are known to be more mafic than the average composition of the post-Archean andesitic crust. Felsification had to occur to give the crust its modern composition. A possible mechanism of such transformation has been suggested to be repeated delamination and recycling into the mantle of lower, more-mafic crustal layers during the evolution of the crust from its original state (Rudnick, 1995).

Recent seismic tomographic studies have yielded improved determinations of the Earth's seismic velocity structure to large mantle depths, negating some previous models and confirmed the existence of deep (up to 600 km) roots under the worldwide system of mid-ocean ridges and continental interiors. Su et al. (1994, p. 6977) stated that "the largest velocity perturbations are at the top of the mantle", and that only "the pattern of the perturbations at shallow depths is well correlated with the tectonics of the Earth's surface". Forte and Woodward (1997) have modeled the structure of deep phase-transition zones, particularly the one inferred around 670 km depth. They found that although this zone has a relief, its lateral continuity inhibits the convective vertical flow at that level in the mantle, and the updated flow models match the dynamic topography of the Earth surface and global-scale free-air gravity anomalies. If the mantle flow is layered in this fashion, major global physiographic features and gravity anomalies (see also Bowin, 1983) might be rooted to this depth in the mantle.

The correlation is weaker at greater depths, as it should be if the crust is more than just a passive component of an inert lithosphere. Very deep roots under the North American and other continents have been confirmed by various lines of evidence, including seismic, geochemical and petrological, to reach below 400 km (cp. Jordan, 1975; Anderson and Dziewonski, 1984; Gossler and Kind, 1996). Under the Archean Superior craton, seismic velocity anisotropy matches the surface geologic grain down to at least 200 km (Silver and Chan, 1988). Late Cretaceous and Tertiary kimberlite pipes, probably rooted hundreds of kilometers deep in the mantle (see review in Lyatsky, 1994a), are recognized in various parts of the Interior Platform in Canada (e.g., Saskatchewan Geological Survey, 1994; Pell, 1997).

Ability of practical tectonics to meet the demands of the 21st century

Pre-modern geology for centuries and pre-modern tectonics for decades were able to aid the discovery of a great many mineral and hydrocarbon occurrences. This suggests the old conceptual systems had useful elements. The challenge is to find these useful grains among the obsolete chaff, and include them into a new theory that would accommodate all the new data. As this challenge is felt increasingly strongly, "lithotectonics" appeared in the past decade as a practical alternative to fruitless speculations.

Significantly, *lithotectonics* grew from geologic mapping in continental regions, where direct geological observations make it possible to restrain speculations if they conflict with the rock evidence. Directly observable facts have, on land, always been the principal source of theoretical generalizations in geology. The geosyncline theory of the 19th century (Hall, 1859; Dana, 1873) relied principally on mapping, as did its very remote descendants in the 20th century. Stille (1924, 1941), Shatsky (1956), Beloussov (1962), Aubouin (1965) and others tried to modernize this theory (see, e.g., review by Rast, 1969), with variable success but with the addition of the important theory of cratons. The geosyncline theory was abandoned in the late 1960s due to its apparent inability to accommodate the discovery of lithospheric plates and its failure to respond adequately to the new concept of plate motions. But the theory of cratons has not been reviewed and analyzed properly.

Geosynclines were originally understood to be shallow-seated furrows. With time, they came to be regarded (notably, since Stille) as whole-crust features. Even this modernized theory was undermined by Stille himself, who wrongly advocated his scheme of worldwide orogenies, and by V. Beloussov, who made things worse by insisting that crustal movements are universally vertical. But useful parts of the old ideas, related to analysis of rock characteristics, rock-body arrangements and time succession, still survive.

Practical tectonics, like geology itself, is in essence an observational and descriptive field of knowledge. Both are based on principles, i.e. empirical generalizations derived from direct observations (not from just logical constructions like in mathematics). Being part of the very old geological theory and practice, practical tectonics has incorporated all the previous achievements of geology and the new approaches, data and ideas. But the principal output of practical tectonics remains the same: regional tectonic maps, where rock units are defined and classified on the basis of their tectonic meaning in the four aspects of tectonism, and predictions about the distribution of natural resources based on the geologic properties of a region. Tectono-sedimentological, tectono-magmatic, tectono-metamorphic and tectono-deformational information must be combined to understand the regional tectonic kinematics of the crust. To compile such maps is difficult, but they are necessary as guides for users who explore for mineral and petroleum deposits or are concerned with environmental matters (e.g., storage of radioactive waste and so on). Such practical tectonic maps, in combination with specific geologic maps depicting in detail some of the noted aspects of regional geology, are able to provide reliable prognostic conclusions.

Mineral and petroleum deposits have long been understood to be products of very complex geologic processes. They are intrinsic parts of geological systems, including the initial chemicals, their interrelations with each other and with the surrounding country rocks under changing geologic conditions, and the final accumulation of fluids, gases, clasts and chemical precipitates and their preservation in suitable reservoirs and traps. In mineral and petroleum systems alike, there are source rocks or melts from which the desirable chemicals are derived, pathways and conduits for their transport through the inhomogeneous rock mass, and traps and reservoirs where they can accumulate. All this is determined by the evolution of the particular area of interest, and the mineral and petroleum systems vary from place to place (Fig. 90).

Many of these questions are just the ones practical tectonics aims to answer from the results of regional tectonic studies based on rock history. Metallic deposits are of great variety, but they are all made up of components that were transported from other localities in gaseous, liquid or solid form. Most hydrothermal occurrences are linked to deep sources, usually magma chambers, in the crust and upper mantle. Some of them (e.g., skarns) require some sort of metasomatic substitution. Diamonds form at great depths up to 300 km, in the cratonic lithosphere or deeper, and are brought upwards in a solid state (Fig. 91). Minerals in sedimentary placer occurrences were produced by separation of clasts during weathering in the provenance areas and their subsequent transport to areas of deposition. Petroleum systems require source rocks, which are usually organic-rich shale sequences; reservoirs, usually porous sandstones or carbonates; and impermeable, shaly or evaporitic, caprock to form the traps. When the organic-rich source rocks are buried to appropriate depths and windows of thermal maturation, they generate oil and gas. Migration of the resulting fluids through the rock mass strongly depends on

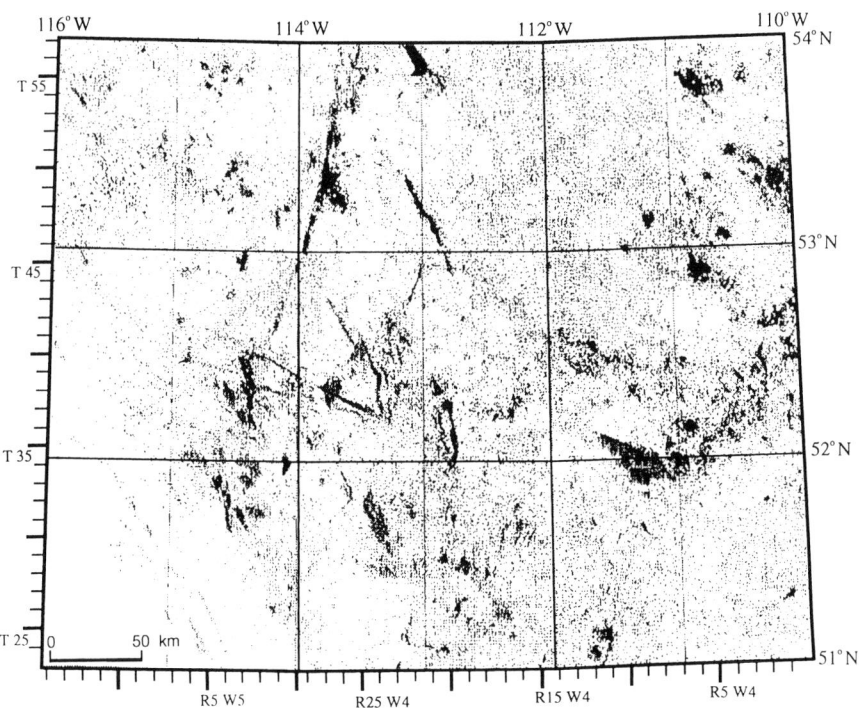

Figure 90. Distribution of oil and gas exploration wells in central Alberta. T - township; R - range. Linear alignments of wells represent alignments of exploration targets, as hydrocarbon reservoirs in the Ètages of the sedimentary cover often tend to concentrate along steep faults.

Figure 91. Distribution of kimberlites in North America (modified from Fipke et al., 1995). Recent discoveries of kimberlite pipes in Canadian cratonic regions in the Northwest Territories and Saskatchewan have attracted considerable commercial interest.

the regional and local tilts, unconformities, fault patterns. Entrapment of hydrocarbons requires unbreached structural and stratigraphic traps. Preservation of hydrocarbon occurs under the right tectonic conditions (stability and lack of big uplift, deep burial or strong faulting).

Coal deposits also depend strongly on regional tectonic conditions. The economic quality of coal increases with burial, to the anthracite rank. On the other hand, fault displacements of coal seams hamper mining. Other immobile occurrences, such as salt (though it can flow slowly at times), also require certain post-depositional tectonic conditions to be economic: too much uplift and faulting can disrupt the deposit or destroy it completely (Fig. 37).

Tectonic classifications of sedimentary basins are many. Most of the ones put forward in the last quarter-century were based on the basins' presumed position in the lithospheric plates. Bally (1975) assumed that one kind of basin is associated with tectonic sutures. Dickinson (1976), following the island-arc concept (Wyllie, 1971), classified sedimentary basins formally based on their position in relation to such arcs: forearc (in front), backarc (behind). Induced by buckling of lithospheric plates at subduction zones, forearc basins do exist (e.g., in Indonesia or in Cook Inlet in Alaska). But the definition of backarc basins is still too loose to be useful: the inland extent of so-called backarc basins has never been defined, and their internal characteristics are not determined. Galloway (1974) used more lithological criteria to classify sedimentary basins (e.g., Tertiary ones along the western North American continental margin).

Importantly, many basins which developed at continental margins may seem similar based on their geographical position, but they are different geologically (in terms of their rock types) and tectonically (in terms of their history). They are often correlated in the literature, but they are dissimilar in their essence and hydrocarbon potential. Sedimentary basins along the Cascadia subduction zone in Oregon and Washington are different from the Queen Charlotte Basin at the western Canadian continental margin. The latter is tectonically intracontinental and not related to subduction (Lyatsky, 1993, 1996; Lyatsky and Haggart, 1993).

Of little use are some recent classifications of mineral deposits also formally related to assumed episodes of terrane accretion. Terrane partition of regions is subjective and inconclusive: the grouping of deposits into pre-, syn- and post-accretion classes is arbitrary. These classes are meaningful only if they unintentionally happen to coincide with pre-, syn- and post-orogenic groups of previous classifications. Metallogeny of the early stages in orogenic zones, especially where subduction and suturing were involved, is commonly associated with mafic, ophiolitic rocks (e.g., in the Franciscan or Teslin belts of the Cordillera) and typified by concentrations of Cu, Cr, Ni and Pl-group elements. Later, in the syn-orogenic (or syn-kinematic) stage, tectonic conditions are favorable for batholithic granitization and accompanying polymetallic accumulations (with Cu, Zn, Pb, Mo, Au in porphyry, skarn and other ores). Post-orogenic crustal extension and uplift in each particular orogen are associated with epithermal deposits of Au, Ag, Sb, Hg. Median massifs, possessing a pre-mobile-megabelt crystalline basement, are sites of abundant polyphase volcanism and formation of massive stratiform sulfide deposits. Miogeosynclinal belts, bearing memory of platformal conditions as well as deep reworking of rocks and orogenic magmatism, host exhalite W-Mo-Su and Pb-Zn mineralization in stratal carbonate complexes. Of defining importance for metallogeny in mobile megabelts is the degree of reworking of the old basement as well as the type of rock succession in the orogenic grain of each particular zone (Fig. 80).

Only rock-based tectonic maps may serve as a reliable basis to reduce the risk in predicting possible locations of mineral and petroleum occurrences. Because these occurrences are intrinsic parts of geological systems controlled by the region's tectonic history, their prediction often directly depends on correct reconstructions of the history of the relevant rock systems. Emphasis on lithologic assemblages in a temporal succession and on lithofacies zones, as well as on the distribution of specific rock types in these successions and facies zones within tectonic étages, is needed to guide the exploration effort to the path of minimal risk. For environmental studies, other rock characteristics also included in tectonic maps may be more important: fracture and fault patterns (both extinct and active), permeable and impermeable rock units in various terrains, etc. Properly compiled, rock-based tectonic maps help to understand the causes and affinities of crustal warping, seismicity, volcanic disturbances and so on. They may serve as a background for different specialized maps.

Significance of rock-based tectonics in the development of sciences about the Earth

Since the mid-20th century, technological advances have made it possible to expand geological observations to the entire Earth, including its vast parts covered by sea. This turned geology into a truly global science. But the observation of rocks in submerged areas still remains very limited: the drilling is sparse, the dredging of seabed local and surficial. Horizons of the oceanic crust below the basaltic and sheeted-dike layers are unsampled. Most regional data have been derived from geophysical, chiefly gravity and magnetic, surveys. Since the early 1960s, they have provided a huge amount of information, which in many cases was unexpected, intriguing and puzzling. The classical, on-land tectonics was unable to explain it. Geophysics, even physics, became the main source of new data; soon these experts deviated from their field to begin drawing conclusions about geology and tectonics, of which often they knew little. This created a climate in which speculative hypothesizing became more popular than fact-based conclusions. Ingenious suppositions took the place of exhaustive studies of rocks, subverting the principal scientific method of geology in general and tectonics in particular. Geophysicists of the country that in the 1960s was the most technologically advanced - the U.S. - brought their own methods, the methods of exact sciences, and began to "improve" on the geological ways of reasoning. Yet at first, old-school geophysicists remained cautious in their use of physical and numerical models to constrain geological interpretation options, and in that application the new methods provided new constraints for speculative fantasy (Isacks et al., 1968; Morgan, 1968). But soon after, the tendency to subvert geology with non-geological methods prevailed, although the new arbiters of tectonic solutions were right in their criticism of the original geosyncline theory. Unfortunately, this theory was perceived only in its obviously outdated 19th-century version, and it was discarded indiscriminately along with the useful elements of its later updates.

A new tectonics, operating mainly with marine geophysical data and physical principles, arose without the benefit of traditional constraints derived from rocks. This new tectonics, incorrectly called "plate tectonics", has never been conclusively defined, as the exact meaning of *lithospheric plate* remains unclear to this day. Moreover, lithospheric plates have from the beginning been treated as mechanically stiff, undeformed internally, with all deformation confined to the proximity of their boundaries with other plates. "Plate tectonics" has never been able to explain cratonization of continental crust, nor its subsequent directional 2,000-Ma evolution with many episodes of internally induced tectonic reworking, nor many cratonic structural features (e.g., domes and round basins), nor its specific tectonic regimes, in-situ stresses, and so on.

Globalization broadened tectonics geographically, and new techniques have permitted to include into its consideration the entire rock-made lithosphere. This improved vision of tectonics did not change its essence: as in the past, it considers its study objects in the tectono-sedimentological, -magmatic, -metamorphic and -deformational aspects. The classical, rock-based approach cannot be replaced with abstract models. Besides, many of the current models' assumptions are derived from little more than the stripe-like magnetic anomaly lineations in oceanic areas that are not seen on continents. Strangely, too, so-called "plate tectonics" and "terrane tectonics" focus their attention on mobile megabelts, which include only relatively small parts of the continental masses, while the vast cratons are ignored. In North America, since the first appearance of cratonized-crust patches in the Archean, cratons have expanded to include most of the modern continent; in the past, cratons included most of the modern-day Appalachians and Cordillera. All this is largely overlooked in the fashionable "tectonic models".

The evolution of the cratons and continents worldwide has much in common. All of them possess geodynamical energy of their own, sufficient to maintain their state as self-developing systems. Vertical tectonic movements with up and down fluctuations produced many generations of arches and volcano-sedimentary basins. The highest-grade Archean and Early Proterozoic metamorphic rocks must have at one time been downdropped to depths of 30-40 km before being exposed at the ground surface. The impressive and long-lived tectonic differentiation of the cratons since the Neoproterozoic and early Phanerozoic has resulted in a shield-platform partition that has persisted through time (mobile megabelts, by contrast, have developed much more variably). The continuous sedimentary cover in North America was preceded by an extensive, predominantly magmatic rock assemblage linked to indigenous non-orogenic tectono-magmatic activity. The dissimilar modern seismic velocity structure of the mostly rising shields and mostly subsiding platforms reflects a long-lived tectonic partition of cratons, whereas most second-order features such as arches and basins were inverted repeatedly.

The huge Mesoproterozoic Belt-Purcell Basin in western North America was intracontinental and intracratonic, with provenance areas someplace in the future western Cordillera. This basin's elongation was defined by the very old NW-SE and E-W structural trends, which predate (Winston, 1986) the familiar NNW-SSE Cordilleran grain. The total thickness of the intracratonic Belt-Purcell Basin, reconstructed from exposed slices, was up to 20 km, with sills up to 6 km thick intruded into the basin sediments. The lifespan of Belt-Purcell subsidence was enormous, some 600 m.y., beginning around 1,500 Ma (Obradovich et al., 1984), but it is not much different from the lifespan of the Alberta and Williston basins in the Phanerozoic.

Most of the Proterozoic volcano-sedimentary basins in Canada and the U.S. are apparently linked to the prominent NE-SW and NW-SE fault systems typical for the North American as well as other continents. This implies a common cause, perhaps some variations in the global stress regime. A global ~200-m.y. periodicity of chrono-mineralogenic events is established for all of the Proterozoic and Phanerozoic (e.g., Lyatsky, 1965), ignoring the assumed plate movements and reorganizations. This near-200 m.y. periodicity is used in the newly approved standard worldwide time scale for all of the Proterozoic from 2,500 to 650 Ma.

Manifestations of tectonism such as grandiose folds and thrust faults, majestic mountains and so on, are typical for mobile megabelts. In areas of cratonized crust, now or in the past, extensional stress regime has been more common. Huge tectono-magmatic episodes, such as several generations of craton-scale dike swarms mapped in the Canadian Shield or the Midcontinent province of anorogenic felsic intrusions and lava flows, are spectacular manifestations of such extension. It was induced by inherent sporadic changes of physical conditions at deep levels of the cratonized crust and lithosphere.

Some obvious deficiencies of physical models that ignore rock-based tectonic constraints

That the Earth's perisphere is broken up into a small number (>20) of large lithospheric plates is a proven fact. That these plates interact with one another in some fashion is also a fact. Most probably, this activity is induced by a convecting mantle. This set of ideas (e.g., Morgan, 1968) was a revolutionary breakthrough in tectonics, because for the first time it truly regarded the crust and lithosphere of continents and oceans together. The related physical models were combined under the name "plate tectonics". But all these ideas are still substantially hypothetical. The exact mechanism of plate motions is still not clear. Thermodynamic currents in the sublithospheric mantle can form a complex set of cells with upgoing and downgoing limbs, but the dimensions of these cells are not yet known.

Unconstrained by evidence from practical tectonics, speculations about the motions of plates, continents and terranes help little in understanding the Earth's geology. The dynamics of plate motions is still unclear, movements of deep-rooted continents around the globe are unexplained, terrane-tectonic postulations are questioned. Neither age calibration of linear magnetic anomalies in the oceanic crust nor studies of paleomagnetic properties of rocks on land provide conclusive data for reconstructions of supercontinents and major continent displacements. Even paleomagnetic determinations of the Early Proterozoic position of the Archean Slave, Superior and Wyoming cratons in North America are still debatable, and probably it was much the same as at present.

Lithospheric plates, as they are conventionally pictured, are thick, stiff slabs floating on top of the underlying convecting, viscous mantle. Boundaries between plates are commonly marked by zones of higher seismicity, but the bases of plates are often still unclear from the known physical parameters of the mantle. In the general model, plates move away (diverge) from spreading centers marked by mid-ocean ridges. Along these zones, new oceanic lithosphere is formed, filling the space between the diverging plates. Belts of this new lithosphere are systematically accreted to older parts of the lithospheric plate. The resulting lateral enlargement of plates is compensated by their subduction beneath other plates at zone of convergence. The subducting plates, still rigid at shallow depths, are heated, bent downward and absorbed into the mantle. Many phenomena are logically linked with this general model: mid-ocean ridges; deep bathymetric trenches; island arcs and volcanic chains at plate boundaries where underthrusting is taking place; probable crustal buckling along some subduction zones and their characteristic continentward-dipping Benioff-Wadati zones of seismicity. But contrary to the physical models, it is incorrect to treat continents as just dead pieces of continental crust in the lithosphere. Continents have their own energy sources, and they are not inert masses merely reacting to forces applied from the mantle and the plate boundaries.

The term *plate tectonics* is a misnomer, because of the fundamental assumption in the physical models about an essentially rigid and internally undeformed (i.e. untectonized) character of plates. All these models quantify the supposed plate behavior ignoring in-plate deformation. But in-plate tectonism does exist; it is not just deformational, and on continents at least it is driven by indigenously sourced energy.

The theory of lithospheric plates triumphed because of its ability to explain these facts in a simple, internally consistent way, and deviations from this elegant early model did not come to light till later. The fundamentals of plate-motion reconstructions are also hypothetical, and rely on incomplete knowledge of geophysical phenomena. By mapping onshore, it was found that the direction of remanent magnetization in igneous rocks varies from normal, as at present, to reverse. These polarity changes, correlated with the age of rocks, suggest sharp changes in the polarity of the Earth's magnetic field. The causes of these reversals are unclear. Magnetic anomaly lineations over the oceanic crust were logically explained by these apparent reversals. The oceanic magnetic stripes were linked to the eruption of basaltic lava at spreading centers. It was supposed that these basalts cooled during the alternating periods of normal and reverse polarity of the ambient geomagnetic field. This kind of magnetized oceanic crust has been compared to a huge tape recorder (Vine and Matthews, 1963).

These natural magnetic records at the ocean bottom were interpreted as being strongly systematic, each anomaly band marking a reversal in the geomagnetic field; the bands were surmised to become older systematically in the direction perpendicular to the mid-ocean ridge. This logic was employed in a great number of models of oceanic-crust formation and plate motions. Assuming that each linear magnetic anomaly is part of an age progression from older to younger, anomaly lineations were numbered and cataloged, and each stripe was assigned an age. To support this procedure, it was noted that the age of the deepest sediments on top of the ocean-floor basalts increases away from the mid-ocean ridges.

On this data base, the universal magnetic-polarity time scale was created (Heirtzler et al., 1968). This scale (also Cande et al., 1989) provides the key to the reconstruction of assumed motions of plates and their episodic rearrangements, from which was derived a template of tectonic influences of oceanic plates on the inert continental-crust masses (Atwater, 1970). This purely geophysical model is still used as the defining criterion to check the validity of regional tectonic generalizations (e.g., Atwater, 1989; Cook, 1995a-c; Hyndman, 1995), but its very essence is questionable. Each significant force applied to the crust or lithosphere produces a stress that results in a huge number of fractures, cracks and faults. Faults have indeed been found to be numerous in the oceanic crust (e.g., Macdonald et al., 1991 and references therein). All of them are potential conduits for magma injection under favorable thermo-mechanical conditions. This weakens the assumption that the oceanic crystalline crust consists only of bands arranged in age succession, reflected in magnetic lineations without significant distortion.

Juvenile anomaly sources could be formed along crustal fractures, whose orientation may or may not follow the age zonation of the crust. Straight chains of basaltic seamounts are know to be younger than the ocean floor on which they stand, and basaltic flows from these localized eruption centers have in some cases been found to spread laterally for tens of kilometers (e.g., Davis, 1982). As a result, the pattern of magnetic anomalies is complicated (e.g., Lyatsky, 1996). In a complex fracture pattern, magma injection of the same age could have a distribution far more intricate than on a ribbon of a tape recorder.

After decades of uncritical fascination, discussions and questions (see, e.g., Cande, 1976), it is impossible to believe that variations in magnetization of the oceanic crust are simply correlated with the crust's age. The fact remains that sediments of the oceanic-crust layer 1 lying directly above the basalts of layer 2 tend to be the oldest on the oceanic basins' periphery and the youngest near the spreading centers. But clastic deposits have their provenance areas on the continents around the ocean basins, and anyway, the age of rocks above the unconformity is not enough to define the age of rocks below. No precise ages of ocean-crust basalts are available, except in a few drilled and dredged localities, because of extreme difficulties of their isotopic analysis (e.g., in Berggren et al., eds., 1995).

The younging of oceanic lithosphere towards the spreading centers is a good first approximation, but it does not define the age of rocks responsible for the pronounced zonation of oceanic magnetic anomalies. Vine and Matthews (1963) attributed this zonation to a zoned arrangement of lava flows. This was later proved to be incorrect: other contributions are also needed to account for the observed anomalies. Thermoremanent magnetization of layer 2 is not enough, and other forms of remanence in other layers (e.g., sheeted dikes) of the oceanic crust and upper mantle are needed to explain the local magnetic stripes and the non-striped long-wavelength magnetic anomalies over oceanic-crust regions (e.g., Yañez and LaBrecque, 1997 and references therein). Bulk magnetization of the oceanic lithosphere has been found to differ from region to region.

The simple mechanism of Vine and Matthews (1963) envisions only thermoremanent magnetization acquired by mafic rocks cooling in an ambient geomagnetic field, but the geologic reality has turned out to be far more complex. Even some of the veterans of oceanic magnetic studies remain unsatisfied. No simple crustal fissuring took place at spreading centers, and eruptions there are neither systematic nor orderly. Along the mid-ocean ridges, magma chambers are discontinuous and lie at intersections of variably oriented faults. Age variations of ocean-floor basalts are complex, being controlled by various sets of active faults. During and after eruption, when the lavas first come into contact with seawater and thereafter, basalts undergo chemical alterations which strongly affects their remanent magnetization (see, e.g., Macdonald et al., 1991). After three decades of active involvement in studies of the ocean bottom, Pitman (1997, p. 122) remained perplexed: "Did the ridges push? Did the trenches pull? Both or neither? All of the above or none of the above?" More strongly, Yañez and LaBrecque (1997, p. 7947) stated that the "simple [Vine-Matthews] mechanism cannot explain some characteristics of seafloor spreading magnetic anomalies nor the magnetization of dredged oceanic samples".

Though the Vine-Matthews hypothesis obviously contradicts many facts, it is still used as a cornerstone for reconstructions of assumed plate interactions and motions, and it has even been used as a basis for postulating plates that no longer exist (e.g., Kula). The comparative compositional uniformity of oceanic crust and the apparent simplicity of its tectonics (in all the four main aspects) are striking. But this uniformity should not be overestimated, as it is in some simple current geophysical models of oceanic-plate kinematics (Engebretson et al., 1985; DeMets et al., 1990).

Seismic profiles in oceanic areas show that the crystalline oceanic crust is heavily faulted, and these faults can have different sense and time of movement. Subparallel arrangement of some of these fault families may be a product of extensional tectonic regimes which may be related to the activity at spreading centers and to plate motions. Also evident are various types of propagating faults, compressional faults, as well as plicative features of crustal deformation (e.g., in the northern and southern segments of the oceanic Juan de Fuca plate system off the western continental margin of North America; see review in Lyatsky, 1996). The perception of plates as essentially undeformed entities has been proved untrue. Bending around vertical axis (e.g., in the Gorda plate west of California) and large-scale upward and downward warping (e.g., in the Indian plate) are the best-known manifestations of internal deformation of oceanic-crust plates.

The tendency to oversimplify the complex geologic phenomena in speculative hypotheses and models is now common, and it is dangerous because from these oversimplifications arise incorrect speculations (Oreskes et al., 1994). Simplified models based on arbitrary assumptions are limited data sets fail to assist the multi-aspect tectonic analysis, which requires meticulous analysis of real data and accurately described facts. Physical models must not drive the reasoning in practical tectonic analysis.

Picture of global tectonics from the achievements of practical tectonic studies

Tectonics of the continents has been grasped predominantly from rock-based facts obtained by direct geologic mapping of various regions. On continents like nowhere else, the huge variety of rock types and structures permits to comprehend the complexities of tectonic processes. Modern seismic images of sublithospheric mantle under the continents, coupled with geochemical and other considerations, have revealed roots exceeding 400 km in depth (Jordan, 1975; Anderson and Dziewonski, 1984; Gossler and Kind, 1996). Zones of reduced seismic S-wave velocity, correlated with belts of higher heat production, are traced under mid-ocean ridges to depths in excess of 300-400, reaching 600 km (Su and Dziewonski, 1992; Su et al., 1992). These findings point to great lateral dissimilarity between the continental-crust masses denoted conventionally as continents, and the oceanic-crust masses underlying most (not all) of the geographically defined oceans. Relief at the base of these lithospheric compartments is huge, much greater than that in the global subaerial and submarine topography.

The deep and cool roots of continents and upwellings of hot mantle material at spreading centers are probably fairly stationary in time. Deep high-velocity anomalies surrounding the Pacific and the Mediterranean belt across Eurasia reach even into the lower mantle. To suppose that intact slabs of subducted oceanic lithosphere can maintain their integrity to these depths is, however, unreasonable. More plausible is the idea that lateral inhomogeneity of the entire mantle (as reported by Forte and Woodward, 1997) reflects lower-mantle dynamics distinct from the dynamics of the sublithospheric upper mantle, which in turn differs from the dynamics of the lithosphere in general and the continental crust in particular. All of them are self-developing spatio-temporal continuums which interact with one another, as the whole planet Earth interacts with the Solar System. These interactions may be very complex, with some phenomena in the tectosphere induced (perhaps temporarily and accidentally) by the dynamics of the self-developing deep-mantle system. But mainly, tectonics of the lithosphere is a direct result of its own self-developing dynamic systems: (a) subcrustal upper mantle, which contains rocks of certain compositions, altered by great temperatures and pressures, but still recognizable as rocks; (b) lower crustal, beneath the brittle ductile transition but above the base of the crust, where crustal rocks experience specific types of kinematics - creep, flowage; and (c) upper crustal, least independent from the underlying spatio-temporal rock continuums, but most familiar through surface geologic mapping.

In the first half of the 20th century, tectonists believed that only the granite-rich continental crust, which contains a sufficient concentration of radioactive elements, is able to induce tectonism (cp. Rich, 1951). The screening effect of thick continental crust, blocking the heat from the Earth's deep interior, was acknowledged generally at that time. Subsequent measurements at the ocean bottom revealed a similarity of surface heat flow in both the continental and oceanic crust (Revelle and Maxwell, 1952; Bullard et al., 1956). This used to puzzle investigators, but the rise of new tectonic ideas in the 1960s distracted their attention: it became common to assume that the world's tectonism is concentrated primarily under the oceans and that the continents are influenced from there. But it was clear then and now that the greatest variety of rock types and tectonic manifestations known on Earth is found on continents, that radioactive elements are concentrated in granite-rich continental crust, and that heat from them compensates for the continental crust's ability to screen the heat from the mantle.

Processes related to activity of lithospheric plates (as specific systems) overlap with the processes related to cratons and mobile megabelts (as specific systems of lower hierarchical rank). Subduction-related processes along the boundaries of some plates and mobile-megabelt zones are now well recognized, in their Pacific-type and Mediterranean-type expressions. The largest epeirogenic warping and other, related tectonic manifestations (such as voluminous non-orogenic magmatism) are more likely to be related to deep mantle convection. Forte et al. (1993a) stressed that the effectsof such deep convection may be non-subduction subsidence of whole continents, if they happen to lie over large downgoing limbs of mantle convection cells.

A low-velocity seismic anomaly related to a hot zone in the mantle under the East Pacific Rise continues to a depth of 1,300 km, and an even more pronounced such anomaly is found at latitude ~10°S, crossing the Pacific Ocean (Su et al., 1994). Conversely, the relatively high-velocity and cold mantle lies under much of Canada and some other areas (e.g., the Caribbean). From teleseismic data, mantle velocities under the western Canadian Shield and northern Interior Platform are relatively high, and the deep seismic velocity domains seem unrelated to the tectonic disparities seen at the surface. In particular, no fundamental velocity changes are recorded across the boundaries of the Archean Slave craton, the Early Proterozoic Taltson magmatic zone and the Wopmay orogenic belt (Bostock and Cassidy, 1996). This implies either that the distribution of hot and cold mantle in this region during the Early Proterozoic was very different than later. It seems, therefore, that at least one drastic rearrangement of hot and cold mantle domains under the continental crust did take place in these Archean and Early Proterozoic regions of the western Canadian Shield. The fact that the modern distribution of hot and cold mantle zones has no relation to the configuration of ancient orogens and cratonic terrains supports the idea that at differenttimes the dynamic systems in the deep mantle, shallow mantle and crust varied.

Theory and practice in rock-based tectonics

The cratonic tectonic stages recognized from preserved platformal étages correspond to particular spatio-temporal geodynamical systems. Some tectonic stages were relatively local, others common to most of the North American craton, still others were manifested worldwide. The latter included stages marked by sharp changes in isotopic systems and mineral composition of rocks from the Archean to the Proterozoic, or appearance of volcano-sedimentary basins during the Proterozoic, or formation of continuous sedimentary cover in the Neoproterozoic; all of them correspond to spatio-temporal systems of a high order.

Structural-formational étages, whether in shield or on platforms, in their compositional and structural characteristics retain a memory of all tectonic stages, since the origin of the rock-bearing perisphere in the Eoarchean around 4 Ga. The challenge for geologists and tectonists is to decipher this tangled record and read the information from the rocks.

Tectonics is only one of many geological disciplines. All of them - mineralogy, petrology, hydrogeology, sedimentology, stratigraphy, geologically applied geophysics and geochemistry, etc. - focus on specific objects in the rock-made perisphere or on methods of investigation. Physics and chemistry of the Earth are planetary sciences. Only some of their aspects matter directly to geology, which concerns itself with rock-made bodies and their characteristics. Mostly, planetary physics (so-called "big geophysics") and planetary chemistry have no direct relationship to rock-oriented geology and tectonics. The so-called "plate tectonics" deals with kinematics and geodynamics of loosely specified geospheres, including sublithospheric mantle.

Geology, into which tectonics is included, deals with the rock-made outer shell, or perisphere, of the Earth (often, it is designated as lithosphere). It is generally strong and more elastic than the asthenosphere, although its base is defined differently by different authors (Barrell, 1914; Morgan, 1968; Jordan, 1975; McNutt and Menard, 1982; Watts, 1982). But wherever its base lies, tectonics of the perisphere (lithosphere) and its plates includes repeated deep changes of its structural-formational systems, and not just mechanical interactions of internally inert plates. Kinematic reconstructions of plates based on the assumed record of their movements in the magnetic stripes is not yet tectonics, although some results of this modeling, if verified by geological observations, may be of interest.

Bercovici (1993, p. 635) observed that "one of the major quandaries in geodynamics [still] concerns the relation between plate tectonics and mantle convection...". But this is a cornerstone of the entire theory of plate movements. Ideas about displacements of continental masses over the globe began as an attempt to explain the visible similarity in continent outlines across the Atlantic. Early in the 20th century, Wegener (1915) discussed only the similarity of coastlines. He thought continents drifted like rafts through the crust of oceanic regions. In the 1960s, a more plausible idea took hold, that continents are frozen into the lithosphere and move over the asthenosphere together with their oceanic-crust counterparts. In search for edges of continental-crust masses, which continue far beyond the coastlines, investigators shifted their attention to the continental slopes. Bullard et al. (1956) quantitatively compared the outlines of continental slopes on opposite sides of the Atlantic. Despite the overall similarity, this comparison demonstrated the existence of gaps and overlaps, which in rigid-plate tectonics are unacceptable. To this day, no conclusive explanation for them exists.

Long-distance drift of deep-rooted continents, and sharp reorganizations of plates across deep-rooted mid-ocean ridges, are as yet unexplained. The existing simplified physical and numerical models help little: subduction-slab downwelling does not seem to be related to simple convection cells, and upwelling is not sheet-like (Bercovici et al., 1989; Bercovici, 1993; Zhang and Christensen, 1993). Global geoid anomalies match the plate boundaries only in places (Bowin, 1983). Simple models of plate motions, relying on simple interpretation of magnetic stripes and presumed hot-spot tracks, still leave impermissible gaps and overlaps (e.g., in the Pacific; Stock and Molnar, 1988).

Despite all these uncertainties, an enormous volume of oceanic lithosphere - on the order of 13,000 line-km - was hypothesized to have been subducted under the western North American continent during just 180 m.y., since the Jurassic (Monger, 1993; Cook, 1995a). But no evidence exists to indicate that the supposed terrane accretion in the Cordillera indeed took place, and the amount of the exotic material there is much smaller than often supposed (cp. Ernst, 1988; Woodsworth et al., 1991). Field information from the Cordillera does not suggest that a huge volume of mafic material has been supplied to the Cordilleran crust-mantle system. Basaltic terrains in Oregon and Washington were previously treated as accreted oceanic crust (Duncan, 1982), and the appearance of the Rocky Mountain fold-and-thrust belt was also ascribed to plate convergence at the Oregon-Washington-British Columbia continental margin (Atwater, 1989). Later, however, it was shown that the basalts along the western Oregon and Washington continental margin erupted in a continental-rift setting (Babcock et al., 1992, 1994), and no oceanic crust have been defined there seismically (Miller et al., 1997). No subduction is taking place north of the Strait of Juan de Fuca at the British Columbia continental margin (Lyatsky, 1996), and the location of the great Rocky Mountain fold-and-thrust belt very far to the east negates its direct connection with subduction at the continental margin.

Welding science and practicality in rock-based regional tectonics: a historical look

In Antiquity, all knowledge of nature was in embryonic state. All of it was incorporated into philosophy, which was the only syncretic science of that time. Later, specialized sciences branched out into a number of concrete disciplines, but philosophy as a specific science (in modern sense) survives. Even fundamental and radical shifts in theoretical underpinnings do not eliminate a science if its principles are true to reality. Astronomy was for many centuries based on the incorrect geocentric model of Ptolemy, but in a modernized form it still survives. Einstein's physics was a major departure from the traditional understanding of nature, which revolutionized physics - but the updated Newtonian physics continues to exist and be useful. Geology has now branched out into a great number of specialized disciplines, but, as before, it remains part of the four fundamental sciences that still survive and flourish: the others are physics, chemistry and biology.

Tectonics is one of the geological sciences. It has its own objects (rock-made bodies), subjects (kinematics or geodynamics of these bodies), methods of separation and integration of relevant data from differentgeological fields (structural geology, geochemistry, geodesy, geophysics, etc.), its own principles, and so on. In the modern view, tectonics has a very practical side: it deals with tangible geologic bodies, their observable properties, etc. It has a theoretical sides as well, as it considers abstract systems and imaginary continuums. One of the very theoretical abstractions is the so-called *tectosphere* (a word used by Morgan, 1968 and many others). It is not an actual geosphere but an imaginary geodynamical system which produced the phenomena mappable at the ground surface. The geodynamical "tectosphere" includes parts of different geospheres, and it may also be influenced by extraterrestrial factors.

The globalization of tectonic thinking permits to see the lithosphere in its totality and understand the tectonic nature of continents better than ever before. Delineated by continental slopes and deep oceanic trenches, continents are capable of self-development, but today, they must also be considered in a broad context of plate and craton-mobile megabelt kinematics as well as their dynamics. Results of global kinematics studies must be consistent with the information from studies of rocks and rock-made bodies, which are represented in their full, rich variety only on continents. There, fortunately, they are accessible for direct geological observation.

Cratons, in their present extent, embrace most continental areas, and in the past they also included many areas that have since been orogenized in mobile megabelts. Cratonic lithosphere, chiefly crust, provides tectonists with the most reliable, compelling and uncorrupted geological data and facts, which compose the framework for studies of the more complicated continental mobile megabelts and the little-accessible ocean floor.

The cratonic stages recognized from preserved structural-formational étages correspond to particular spatio-temporal geodynamical systems. Some tectonic stages were relatively local, others common to most of a craton, still others (megastages) were manifested in cratonized continental-crust areas worldwide. The latter included megastages marked by sharp changes in isotopic systems and mineral composition of rocks from the Archean to the Proterozoic, or stages representing the global ~200-m.y. cycle and events like the appearance of volcano-sedimentary basins in the Early Proterozoic and of the continuous sedimentary cover in the Neoproterozoic-Early Cambrian. Structural-formational étages, whether in shield or platform areas, in their composition and structural characteristics retain a memory of all tectonic stages, since the origin of the rock-bearing perisphere in the Eoarchean, around 4,000 Ma. The challenge for geologists, and for tectonists, is to decipher this tangled record and read the information from the rocks properly.

Derived substantially from geologic field mapping while addressing the social priorities, practical tectonics deals with rocks and rock-made bodies as the foundation of knowledge about any particular region and the Earth's lithosphere as a whole. From rock-based studies, generalizations follow, subject to constraints from mappable facts. Such practical tectonics may look less glamorous than some current model-driven speculations, and it is certainly more labor-intensive, but it is definitely more useful.

There are countless material properties of rocks - chemical, physical and many others - that can now be analyzed in the laboratory to extract information about rock age, paleomagnetic history, thermal transformations, and so on. These analyses supplement, but do not replace, the information obtainable by observing rocks in outcrop during field mapping.

The power of modern computers has allowed to create a great variety of models. But only few of them are based on realistic assumptions drawn from observed facts and make use of strong empirical principles. Quantitative models offer important constraints on speculative reasoning. But the use of computers and numerical methods in geology can easily lead to their abuse, unless it is guided by proven relationships between the studies characteristics and phenomena. Behind the mathematical sophistication allowed by modern computers, the unrealistic nature of incorrect models sometimes escapes notice where there is no adequate control from practice. Even simple analog physical models are often difficult to produce in the laboratory in a way that realistically represents nature (Eisenstadt and Withjack, 1995).

Rock-based tectonic analysis, as illustrated in this book, permits to restore stage by stage the history of tectonic reorganizations of large regions, in particular, cratonic platforms. Integrated tectonic analysis yields improved understanding of tectonic history, from the combined viewpoint of tectono-sedimentary, -magmatic, -metamorphic and -deformational considerations. This analysis takes into account all manifestations of crustal deformation, faulting, folding, warping and block movements in mobile megabelts, and block displacements and warping in cratons. The timing of these events is specific for each tectonic region, discriminated from others by these sets of events. Episodic tectonic reworking of the crust in particular parts of orogenic belts and cratons leads to restructuring which separates evolutionary stages of the defined tectonic region, although the lateral extent of such a region may shift with time.

Practical tectonics employs field observations, geophysical imaging, chemical analysis and modeling results without prejudices, as long as there is no contradiction with the proven geologic facts.

REFERENCES

Adams, J. and Bell, J.S., 1991. Crustal stresses in Canada; *in* D.B. Slemmons, E.R. Engdahl, M.D. Zoback, and D.D. Blackwell (eds.), Neotectonics of North America; Geological Society of America, The Geology of North America, Decade Map Volume I, p. 367-386.

Ager, D.V., 1984. The stratigraphic code and what it implies; *in* W.A. Berggren and J.A. van Couvering (eds.), Catastrophes and Earth History; Princeton University Press, p. 91-100.

Ahern, J.L. and Ditmars, R.C., 1985. Rejuvenation of continental lithosphere beneath an intracratonic basin; Tectonophysics, v. 120, p. 21-35.

Ainemer, A.I. and Lyatsky, V.B., 1972. Identification of stationarity in the development of a shelf zone from map data by means of defining uniform clusters (abs.); *in* Mathematization and Automation in Geologic Studies; VSEGEI, Leningrad (St. Petersburg), p. 56-57 (in Russian).

Aitken, J.D., 1981. Stratigraphy and sedimentology of the Upper Proterozoic Little Dal Group, Mackenzie Mountains, Northwest Territories; *in* F.H.A. Campbell (ed.), Proterozoic Basins of Canada; Geological Survey of Canada, Paper 81-10, p. 47-71.

Aitken, J.D., 1989. Birth, growth and death of the middle Cambrian Cathedral carbonate lithosome, Southern Rocky Mountains; Bulletin of Canadian Petroleum Geology, v. 37, p. 316-333.

Aitken, J.D., 1993a. Tectonic framework; *in* D.F. Stott and J.D. Aitken (eds.), Sedimentary Cover of the Craton in Canada; Geological Society of America, The Geology of North America, v. D-1, p. 45-54.

Aitken, J.D., 1993b. Proterozoic sedimentary rocks; *in* D.F. Stott and J.D. Aitken (eds.), Sedimentary Cover of the Craton in Canada; Geological Society of America, The Geology of North America, v. D-1, p. 81-95.

Aitken, J.D., 1993c. Tectonic evolution and basin history; *in* D.F. Stott and J.D. Aitken (eds.), Sedimentary Cover of the Craton in Canada; Geological Society of America, The Geology of North America, v. D-1, p. 483-502.

Aitken, J.D. and McIlreath, I.A., 1982. Depositional environments of the Cathedral Escarpment, near Field, British Columbia; *in* M.E. Taylor (ed.), The Cambrian System in the Southern Canadian Rocky Mountains, Alberta and British Columbia; Second International Symposium on the Cambrian System, Field Trip Guidebook 2, p. 35-44.

Allmendinger, R.W., 1992. Fold and thrust tectonics of the western United States exclusive of the accreted terranes; *in* B.C. Burchfiel, P.W. Lipman, and M.L. Zoback (eds.), The Cordilleran Orogen: Conterminous U.S.; Geological Society of America, The Geology of North America, v. G-3, p. 583-607.

Alsharhan, A.S. and Nairn, A.E.M., 1997. Sedimentary Basins and Petroleum Geology of the Middle East; Elsevier, 978 p.

Ameen, M.S., 1992. Effect of basement tectonics on hydrocarbon generation, migration, and accumulation in northern Iraq; Bulletin of the American Association of Petroleum Geologists, v. 76, p. 356-370.

Anderson, D.L., 1989. Theory of the Earth; Blackwell, 366 p.

Anderson, D.L., 1995. Lithosphere, asthenosphere, and perisphere; Reviews of Geophysics, v. 33, p. 125-149.

Anderson, D.L. and Dziewonski, A.M., 1984. Seismic tomography; Scientific American, v. 251/4, p. 60-68.

Anderson, J.L., Wooden, J.L., and Bender, E.E., 1993. Mojave province of southern California and vicinity; in J.C. Reed, Jr., M.E. Bickford, R.S. Houston, P.K. Link, D.W. Rankin, P.K. Sims, and W.R. Van Schmus (eds.), Precambrian: Conterminous U.S.; Geological Society of America, The Geology of North America, v. C-2, p. 176-188.

Anderson, N.L. and Brown, J.B., 1991. Reconstruction of the Wabamun Group salt, southern Alberta, Canada; in T.D. Cavanaugh (ed.), Integrated Exploration Case Studies, North America; Geophysical Society of Tulsa, 1991 Spring Symposium, p. 145-177.

Anderson, N.L., Brown, R.J., and Hinds, R.C., 1988. Geophysical aspects of Wabamun salt distribution in southern Alberta; Canadian Journal of Exploration Geophysics, v. 24, p. 166-178.

Andrews, G.D., 1987. Devonian Leduc outcrop reef-edge models and their potential seismic expression; in N.M. McMillan, A.F. Embry, and D.J. Glass (eds.), Devonian of the World, Proceedings of the 2nd International Symposium on the Devonian System, Calgary; Canadian Society of Petroleum Geologists, Memoir 14 (II), p. 427-450.

Anfiloff, V., 1988. Polycyclic rifting - an interpretation of gravity and magnetics in the North West Shelf; in P.G. Purcell and R.R. Purcell (eds.), The North West Shelf, Australia; Proceedings of the North West Shelf Symposium, Perth, p. 443-455.

Ansdell, K.M., Lucas, S.B., Connors, K., and Stern, R.A., 1995. Kisseynew metasedimentary gneiss belt, Trans-Hudson orogen (Canada): back-arc origin and collisional inversion; Geology, v. 23, p. 1039-1043.

Ansdell, K.M., Lucas, S.B., Connors, K., and Stern, R.A., 1997. Kisseynew metasedimentary gneiss belt, Trans-Hudson orogen (Canada): back-arc origin and collisional inversion: reply; Geology, v. 25, p. 91-92.

Ansdell, K.M. and Norman, A.R., 1995. U-Pb geochronology and tectonic development of the southern flank of the Kisseynew Domain, Trans-Hudson Orogen, Canada; Precambrian Research, v. 72, p. 147-167.

Armstrong, R.L., 1988. Mesozoic and early Cenozoic magmatic evolution of the Canadian Cordillera; in S.P. Clark, B.C. Burchfiel, and J. Suppe (eds.), Processes in Continental Lithospheric Deformation; Geological Society of America, Special Paper 218, p. 55-91.

Armstrong, R.L., 1991. The persistent myth of crustal growth; Australian Journal of Earth Sciences, v. 38, p. 613-630.

Armstrong, R.L., Parrish, R.R., van der Heyden, P., Scott, K., and Runkle, D., 1991. Early Proterozoic basement exposures in the southern Canadian Cordillera: core gneiss of Frenchman Cap, Unit I of the Grand Fork Gneiss, and the Vaseaux formation; Canadian Journal of Earth Sciences, v. 28, p. 1169-1201.

Armstrong, R.L. and Ramaekers, P., 1985. Sr isotopic study of Helikian sediments and diabase dykes in the Athabasca Basin, northern Saskatchewan; Canadian Journal of Earth Sciences, v. 22, p. 399-407.

Armstrong, R.L. and Ward, P., 1991. Evolving geographic patterns of Cenozoic magmatism in the North American Cordillera: temporal and spatial association of magmatism and metamorphic core complexes; Journal of Geophysical Research, v. 96, p. 13,201-13,224.

Arnott, R.W.C., Hein, F.J., and Pemberton, S.G., 1995. Influence of the ancestral Sweetgrass Arch on sedimentation of the Lower Cretaceous Bootlegger Member, north-central Montana; Journal of Sedimentary Research, v. B65, p. 222-234.

Artyushkov, E.V. and Baer, M.A., 1989. Mechanism of formation of deep basins on continental crust; in R.A. Price (ed.), Origin and Evolution of Sedimentary Basins and Their Energy and Mineral Resources; American Geophysical Union, Monograph Series 48, p. 175-185.

Aspler, L.B. and Chiarenzelli, J.R., 1996. Stratigraphy, sedimentology, and physical volcanology of the Henik Group, central Ennadai-Rankin greenstone belt, Northwest Territories, Canada: Late Archean paleogeography of the Hearne Province and tectonic implications; Precambrian Research, v. 77, p. 59-89.

Ashton, K.E., Reilly, B.A., Slimmon, W.L., Thomas, D.J., and Heaman, L.M., 1996. Structural and metamorphic history of the southeastern Reindeer Zone, Saskatchewan (abs.); Summary of Investigations 1996, Saskatchewan Geological Survey, Miscellaneous Report 96-4, p. 171-172.

Atwater, T., 1970. Implications of plate tectonics for the Cenozoic tectonic evolution of western North America; Bulletin of the Geological Society of America, v. 81, p. 3518-3536.

Atwater, T., 1989. Plate tectonic history of the northeast Pacific and western North America; in E.L. Winterer, D.M. Hussong, and R.W. Decker (eds.), The Eastern Pacific Ocean and Hawaii; Geological Society of America, The Geology of North America, v. N, p. 21-72.

Aubouin, J., 1965. Geosynclines; Developments in Geotectonics - I, Elsevier, 335 p.

Aubry, M.-P., 1995. From chronology to stratigraphy: interpreting the Lower and Middle Eocene stratigraphic record in the Atlantic Ocean; in W.A. Berggren, D.V. Kent, M.-P. Aubry, and J. Hardenbol (eds.), Geochronology, Time Scales and Global Stratigraphic Correlation; SEPM (Society for Sedimentary Geology), Special Publication 54, p. 213-274.

Baars, D.L., 1988. Triassic and older stratigraphy; Southern Rocky Mountain and Colorado Plateau; *in* L.L. Sloss (ed.), Sedimentary Cover - North American Craton; U.S.; Geological Society of America, The Geology of North America, v. D-2, p. 53-64.

Babcock, E.A., 1973. Regional jointing in southern Alberta; Canadian Journal of Earth Sciences, v. 10, p. 1769-1781.

Babcock, E.A., 1974. Jointing in central Alberta; Canadian Journal of Earth Sciences, v. 11, p. 1181-1186.

Babcock, E.A. and Sheldon, L.G., 1979. Relationship between photo lineaments and geologic structures, Athabasca oil sands area, Northeast Alberta; *in* Proceedings of the Second International Conference on Basement Tectonics Denver, v. 2, p. 177-188.

Babcock, R.S., Burmester, R.F., Engebretson, D.C., Warnock, A.C., and Clark, K.P., 1992. A rifted margin origin for the Crescent basalts and related rocks in the northern Coast Range volcanic province, Washington and British Columbia; Journal of Geophysical Research, v. 97, p. 6799-6821.

Babcock, R.S., Suczek, C.A., and Engebretson, D.C., 1994. The Crescent "terrane", Olympic Peninsula and southern Vancouver Island; *in* R. Lasmanis and E.S. Cheney (eds.), Regional Geology of Washington State; Washington Division of Geology and Earth Resources, Bulletin 80, p. 141-157.

BABEL Working Group, 1991. Reflectivity in a Proterozoic shield: examples from BABEL seismic profiles across Fennoscandia; Continental Lithosphere: Deep Seismic Reflections; Geodynamics Series, v. 22; American Geophysical Union, p. 77-86.

Bailey, R.C., 1990. Trapping of aqueous fluids in the deep crust; Geophysical Research Letters, v. 17, p. 1129-1132.

Baird, D.J., Nelson, K.D., Knapp, J.H., Walters, J.J., and Brown, L.D., 1996. Crustal structure and evolution of the Trans-Hudson orogen: results from seismic reflection profiling; Tectonics, v. 15, p. 416-426.

Bally, A.W., 1975. A geodynamic scenario for hydrocarbon occurrences; 9th World Petroleum Congress, Tokyo, Proceedings, v. 2, p. 33-44.

Bally, A.W., 1981. Thoughts on the tectonics of folded belts; *in* K.R. McClay and N.J. Price (eds.), Thrust and Nappe Tectonics; Geological Society (London), Special Publication 9, p. 13-32.

Bally, A.W., 1989. Phanerozoic basins of North America; *in* A.W. Bally and A.R. Palmer (eds.), The Geology of North America - An Overview; Geological Society of America, The Geology of North America, v. A, p. 397-446.

Bally, A.W., Gordy, P.L., and Stewart, G.A., 1966. Structure, seismic data, and orogenic evolution of southern Canadian Rocky Mountains; Bulletin of Canadian Petroleum Geology, v. 14, p. 337-381.

Bally, A.W., Scotese, C.R., and Ross, M.L., 1989. North America; plate-tectonic setting and tectonic elements; *in* A.W. Bally and A.R. Palmer (eds.), The Geology of North America - An Overview; Geological Society of America, The Geology of North America, v. A, p. 1-16.

Bally, A.W. and Snelson, S., 1980. Realms of subsidence; *in* A.D. Miall (ed.), Facts and Principles of World Petroleum Occurrence; Canadian Society of Petroleum Geologists, Memoir 6, p. 9-94.

Bamber, E.W., Henderson, C.M., Richards, B.C., and McGugan, A., 1991. Carboniferous and Permian stratigraphy of the Foreland Belt; *in* H. Gabrielse and C.J. Yorath (eds.), Geology of the Cordilleran Orogen in Canada; Geological Society of America, The Geology of North America, v. G-2, p. 242-265.

Barclay, J.E., Krause, F.F., Campbell, R.I., and Utting, J., 1990. Dynamic casting and growth faults: Dawson Creek Graben Complex, Carboniferous-Permian Peace River Embayment, Western Canada; *in* S.C. O'Connell and J.S. Bell (eds.), Geology of the Peace River Arch; Bulletin of Canadian Petroleum Geology, v. 38A, p. 115-145.

Barrell, J., 1914. The strength of the earth's crust; Geology, v. 22, p. 28-48.

Barton, R.H., Christiansen, E.A., Kupsch, W.O., Mathews, W.H., Gravenor, C.P., and Bayrock, L.A., 1964. Quaternary; *in* R.G. McCrossan and R.P. Glaister (eds.), Geological History of Western Canada; Alberta Society of Petroleum Geologists, p. 195-200.

Basu, A., Young, S.W., Suttner, L.J., James, W.C., and Mack, G.H., 1975. Reevaluation of the use of undulatory extinction and polycrystallinity in detrital quartz for provenance interpretation; Journal of Sedimentary Petrology, v. 45, p. 873-882.

Bates, R.L. and Jackson, J.A. (eds.), 1987. Glossary of Geology, third edition; American Geological Institute, 788 p.

Beaumont, C., 1978. Evolution of sedimentary basins on a viscoelastic lithosphere: theory and examples; Geophysical Journal of the Royal Astronomical Society, v. 55, p. 471-497.

Beck, M.E., Burmester, R.F., and Furlong, P.O., 1997. Paleomagnetism of Miocene volcanic rocks near Mount Rainier and the paleomagnetic record of Cenozoic tectonism in the Washington Cascades; Washington Geology, v. 25, no. 2, p. 8-14.

Bell, J.S., 1996. In situ stresses in sedimentary rocks (Part 1): measurement techniques; Geoscience Canada, v. 23, p. 85-100.

Bell, S. and Babcock, E.A., 1986. The stress regime of the western Canada basin and implications for hydrocarbon production; Bulletin of Canadian Petroleum Geology, v. 34, p. 364-378.

Bell, J.S. and McLellan, P.J., 1995. Exploration and production implications of subsurface rock stresses in western Canada; *in* J.S. Bell and T.D. Bird (eds.), Proceedings of the Oil and Gas Forum '95 - Energy from Sediments; Geological Survey of Canada, Open File 3058, p. 1-5.

Bell, J.S., Price, P.R., and McLellan, P.J., 1994. In-situ stress in the Western Canada Sedimentary Basin; *in* G.D. Mossop and I. Shetsen (comps.), Geological Atlas of the Western Canada Sedimentary Basin; Canadian Society of Petroleum Geologists & Alberta Research Council, p. 439-446.

Beloussov, V.V., 1962. Basic Problems in Geotectonics; McGraw-Hill, 809 p.

Belyea, H.R., 1955. Cross Section Through the Devonian System of the Alberta Plains; Geological Survey of Canada, Paper 55-3.

Beratan, K.K. (ed.), 1996. Reconstructing the History of Basin and Range Extension Using Sedimentology and Stratigraphy; Geological Society of America, Special Paper 303, 212 p.

Bercovici, D., 1993. A simple model of plate generation from mantle flow; Geophysical Journal International, v. 114, p. 635-650.

Bercovici, D., Schubert, G., and Glatzmaier, G.S., 1989. Three-dimensional spherical models of convection in the Earth's mantle; Science, v. 244, p. 950-955.

Berggren, W.A., Kent, D.V., Aubry, M.-P., and Hardenbol, J. (eds.), 1995. Geochronology, Time Scales and Global Stratigraphic Correlation; SEPM (Society for Sedimentary Geology), Special Publication 54, 386 p.

Bergman, K.M. and Walker, R.G., 1995. Influence of basement trends on Cretaceous linear sandbodies, Western Canada Sedimentary Basin: some preliminary observations; in G.M. Ross (ed.), Alberta Basement Transects Workshop, Lithoprobe Report 47, p. 228-249.

Bickford, M.E., 1988. The formation of continental crust: Part 1. A review of some principles; Part 2. An application to the Proterozoic evolution of southern North America; Bulletin of the Geological Society of America, v. 100, p. 1375-1391.

Bickford, M.E., Collerson, K.D., and Van Schmus, W.R., 1990. Proterozoic collisional tectonism in the Trans-Hudson Orogen, Saskatchewan; Geology, v. 18, p. 14-18.

Bickford, M.E., Van Schmus, W.R., Macdonald, R., Lewry, J.F., and Pearson, J.G., 1986. U-Pb zircon geochronology project for the Trans-Hudson Orogen: current sampling and recent results; in Summary of Investigations 1986, Saskatchewan Geological Survey, Miscellaneous Report 86-4, p. 101-107.

Bilibin, Y.A., 1955 (translated in 1967). Metallogenic provinces and metallogenic epochs; Geological Bulletin, Dept. of Geology, Queen's College Press, Flushing, NY, 35 p.

Binda, P.L., Sparks, D.E., Beaudoin, N.C., Stasiuk, L.D., Bend, S.L., and Buchanan, A.A., 1996. Preliminary observations on the acid-resistant microfossils from the lower Paleozoic of Southern Saskatchewan; in Summary of Investigations 1996; Saskatchewan Geological Survey, Miscellaneous Report 96-4, p. 157-165.

Blakely, R.J. and Simpson, R.W., 1986. Locating edges of source bodies from magnetic and gravity anomalies; Geophysics, v. 51, p. 1494-1498.

Bleeker, W., 1990. New structural-metamorphic constraints on Early Proterozoic oblique collision along the Thompson nickel belt, Manitoba, Canada; in J.F. Lewry and M.R. Stauffer (eds.), The Early Proterozoic Trans-Hudson Orogen of North America; Geological Association of Canada, Special Paper 37, p. 57-73.

Bleeker, W. and Stern, R., 1997. The Acasta gneisses: an imperfect sample of Earth's oldest crust; in F. Cook and P. Erdmer (comps.), Lithoprobe Slave-Northern

Cordillera Lithospheric Evolution (SNORCLE) and Cordilleran Tectonics Workshop; Lithoprobe Report 56, p. 32-34.

Blockley, J.G., Trendall, A.F., and Thorne, A.M., 1989. Early Precambrian crustal evolution and mineral deposits, Pilbara Craton and adjacent Ashburton Trough; *in* R.A. Price (ed.), Origin and Evolution of Sedimentary Basins and Their Energy and Mineral Resources; American Geophysical Union, Geophysical Monograph Series, v. 48 & International Union of Geodesy and Geophysics, v. 3, p. 159-167.

Boerner, D., Kurtz, R., Craven, J., and Jones, F.W., 1996. Electromagnetic results from the Alberta basement Lithoprobe transect; *in* G.M. Ross (comp.), Alberta Basement Transects Workshop, Lithoprobe Report 51, p. 61-70.

Boerner, D.E., Kurtz, R.D., Craven, J.A., Rondenay, S., and Qian, W., 1995. A buried Proterozoic foredeep under the Western Canada Sedimentary Basin?; Geology, v. 23, p. 297-300.

Bonatti, E., 1987. The rifting of continents; Scientific American, v. 256/3, p. 97-103.

Bond, G.C. and Kominz, M.A., 1984. Construction of tectonic subsidence curves for the early Paleozoic miogeocline, southern Canadian Rocky Mountains: implications for subsidence mechanisms, age of breakup, and crustal thinning; Bulletin of the Geological Society of America, v. 95, p. 155-173.

Bond, G.C., Nickeson, P.A., and Kominz, M.A., 1984. Breakup of a supercontinent between 625 Ma and 555 Ma: new evidence and implications for continental histories; Earth and Planetary Science Letters, v. 70, p. 325-345.

Borm, G.W, Engeser, B., Hoffers, B., Kutter, H.K., and Lempp, Ch., 1997. Borehole instabilities in the KTB main borehole; Journal of Geophysical Research, v. 102, p. 18,507-18,518.

Bostock, M.G. and Cassidy, J.F., 1997. Upper mantle stratigraphy beneath the southern Slave craton; Canadian Journal of Earth Sciences, v. 34, p. 577-587.

Bostock, H.H. and van Breemen, O., 1994. Ages of detrital and metamorphic zircons and monazites from a pre-Taltson magmatic zone basin at the western margin of Rae Province; Canadian Journal of Earth Sciences, v. 31, p. 1353-1364.

Bostock, H.H., van Breemen, O., and Loveridge, W.D., 1991. Further geochronology of plutonic rocks in northern Taltson magmatic zone, District of Mackenzie, N.W.T.; *in* Radiogenic Age and Isotopic Studies, Report 4; Geological Survey of Canada, Paper 90-2, p. 67-78.

Bostock, H.S., 1970. Physiographic subdivisions of Canada; *in* R.J.W. Douglas (ed.), Geology and Economic Minerals of Canada; Geological Survey of Canada, Economic Geology Report 1, p. 10-30.

Bourcart, J., 1955. Les sables profonds de la Méditerranée; Archives Science (Geneva), v. 8, p. 5-13.

Bowin, C., 1983. Depth of principal mass anomalies contributing to the Earth's geoidal undulations and gravity anomalies; Marine Geodesy, v. 7, p. 61-100.

Bowring, S.A., Williams, A.S., and Compton, W., 1989. 3.96 Ga gneisses from the Slave province, Northwest Territories, Canada; Geology, v. 17, p. 971-975.

Brown, R.L., Carr, S.D., Johnson, B.J., Coleman, V.J., Cook, F.A., and Varsek, J.L., 1992. The Monashee decollement of the southern Canadian Cordillera: a crustal-scale shear zone linking the Rocky Mountain foreland belt to lower crust beneath accreted terranes; in K. McClay (ed.), Thrust Tectonics; Chapman & Hall, London, p. 357-364.

Brown, R.W., 1962. Paleocene Flora of the Rocky Mountains and Great Plains; U.S. Geological Survey, Professional Paper 375, 119 p.

Buchanan, J.G. and Buchanan, P.G. (eds.), 1995. Basin Inversion; Geological Society (London), Special Publication 88, 596 p.

Bullard, E.C., Maxwell, A.E., and Revelle, R., 1956. Heat flow through the deep sea floor; Advances in Geophysics, v. 3, p. 153-181.

Burchfiel, B.C., Cowan, D.S., and Davis, G.A., 1992. Tectonic overview of the Cordilleran orogen in the western United States; in B.C. Burchfiel, P.W. Lipman, and M.L. Zoback (eds.), The Cordilleran Orogen: Conterminous U.S.; Geological Society of America, The Geology of North America, v. G-3, p. 407-479.

Burchfiel, B.C. and Davis, G.A., 1975. Nature and controls of Cordilleran orogenesis, western United States: extensions of an earlier synthesis; American Journal of Science, v. 275-A, p. 363-396.

Burianyk, M.J.A. and Kanasewich, E.R., 1995. Crustal velocity structure of the Omineca and Intermontane Belts, southeastern Canadian Cordillera; Journal of Geophysical Research, v. 100, p. 15,303-15,316.

Burwash, R.A., Baadsgaard, H., Peterman, Z.E., and Hunt, G.H., 1964. Precambrian; in R.G. McCrossan and R.P. Glaister (eds.), Geological History of Western Canada; Alberta Society of Petroleum Geologists, p. 14-19.

Burwash, R.A., Chacko, T., and Muehlenbachs, K., 1995. Tectonic interpretation of Kimiwan anomaly, northwestern Alberta; in G.M. Ross (ed.), Alberta Basement Transects Workshop, Lithoprobe Report 47, p. 340-349.

Burwash, R.A., Chacko, T., and Muehlenbachs, K., 1996. Paleozoic reactivation of mid-Proterozoic Kimiwan Structure, Peace River Arch, northwestern Alberta; in G.M. Ross (comp.), Alberta Basement Transects Workshop, Lithoprobe Report 51, p. 204-212.

Burwash, R.A. and Culbert, R.R., 1976. Multivariate geochemical and mineral patterns in the Precambrian basement of western Canada; Canadian Journal of Earth Sciences, v. 13, p. 1-18.

Burwash, R.A., Green, A.G., Jessop, A.M., and Kanasewich, E.R., 1993. Geophysical and petrophysical characteristics of the basement rocks of the western Canada sedimentary basin; in D.F. Stott and J.D. Aitken (eds.), Sedimentary Cover of the Craton in Canada; Geological Society of America, The Geology of North America, v. D-1, p. 55-77.

Burwash, R.A., McGregor, C.R., and Wilson, J.A., 1994. Precambrian basement beneath the Western Canada Sedimentary Basin; *in* G.D. Mossop and I. Shetsen (comps.), Geological Atlas of the Western Canada Sedimentary Basin; Canadian Society of Petroleum Geologists & Alberta Research Council, p. 49-56.

Burwash, R.A. and Muehlenbachs, K., 1997. Tectonic setting of eastern Alberta basement granites inferred from Pearce trace element discrimination diagrams; *in* G.M. Ross (comp.), Alberta Basement Transects Workshop, Lithoprobe Report 59, p. 35-49.

Burwash, R.A. and Power, M.A., 1990. Trout Mountain anomaly, northern Alberta: its role in the northwest foreland of the Trans-Hudson Orogen; *in* J.F. Lewry and M.R. Stauffer(eds.), The Early Proterozoic Trans-Hudson Orogen of North America; Geological Association of Canada, Special Paper 37, p. 301-311.

Bustin, R.M., 1991. Organic maturity in the western Canada sedimentary basin; International Journal of Coal Geology, v. 19, p. 319-358.

Butler, R.F., Gehrels, G.E., McClelland, W.C., May, S.R., and Klepacki, D., 1989. Discordant paleomagnetic poles from the Canadian Coast Plutonic Complex: regional tilt rather than large-scale displacement?; Geology, v. 17, p. 691-694.

Caldwell, W.G.E. and Kauffman, E.G. (eds.), 1993. Evolution of the Western Interior Basin; Geological Association of Canada, Special Paper 39, 680 p.

Calvert, A.J., Sawyer, E.W., Davis, W.J., and Ludden, J.N., 1995. Archean subduction inferred from seismic images of a mantle suture in the Superior Province; Nature, v. 375, p. 670-674.

Camfield, P.A., Gupta, J.C., Jones, A.G., Kurtz, R.D., Krentz, D.H., Ostrowski, J.A., and Craven, J.A., 1989. Electromagnetic sounding and crustal electrical conductivity in the region of the Wopmay Orogen, Northwest Territories, Canada; Canadian Journal of Earth Sciences, v. 26, p. 2385-2395.

Campbell, F.H.A. (ed.), 1981. Proterozoic Basins of Canada; Geological Survey of Canada, Paper 81-10.

Cande, S.C., 1976. A paleomageticpole from Late Cretaceous marine magnetic anomalies in the Pacific; Geophysical Journal of the Royal Astronomical Society, v. 44, p. 547-566.

Cande, S.C., LaBrecque, R.L., Larson, R.L., Pittman III, W.C., Golochenko, X., and Haxby, W., 1989. Magnetic Lineations of the World's Ocean Basins; American Association of Petroleum Geologists, Special Map.

Cant, D.J., 1988. Regional structure and development of the Peace River Arch, Alberta: a Paleozoic failed-riftsystem?; Bulletin of Canadian Petroleum Geology, v. 36, p. 284-295.

Cant, D.J., 1989. Lower Zuni Sequence: Middle Jurassic to middle Cretaceous; *in* B.D. Ricketts (ed.), Western Canada Sedimentary Basin, A Case History; Canadian Society of Petroleum Geologists, Special Publication 30, p. 251-269.

Cant, D.J. and Abrahamson, B., 1996. Regional distribution and internal stratigraphy of the Lower Mannville; Bulletin of Canadian Petroleum Geology, v. 44, p. 508-529.

Carl, C., von Pechmann, E., Höhndorf, A., and Ruhrman, G., 1992. Mineralogy and U/Pb, Pb/Pb, and Sm/Nd geochronology of the Key Lake uranium deposit, Athabasca Basin, Saskatchewan, Canada; Canadian Journal of Earth Sciences, v. 29, p. 879-895.

Carr, S.D., 1992. Tectonic setting and U-Pb geochronology of the early Tertiary Ladybird leucogranite suite, southern Omineca Belt, British Columbia; Tectonics, v. 11, p. 258-278.

Carr, S.D., 1995. The southern Omineca Belt, British Columbia: new perspectives from the Lithoprobe geoscience program; Canadian Journal of Earth Sciences, v. 32, p. 1720-1739.

Cecile, M.P., Morrow, D.W., and Williams, G.K., 1997. Early Paleozoic (Cambrian to Early Devonian) tectonic framework, Canadian Cordillera; Bulletin of Canadian Petroleum Geology, v. 45, p. 54-74.

Cecile, M.P. and Norford, B.S., 1993. Ordovician and Silurian; in D.F. Stott and J.D. Aitken (eds.), Sedimentary Cover of the North American Craton in Canada; Geological Society of America, The Geology of North America, v. D-1, p. 125-149.

Chacko, T., King, R.W., Muehlenbachs, K., and Burwash, R.A., 1995. The Kimiwan isotope anomaly, a low ^{18}O zone in the Precambrian basement of Alberta: constraints on the timing of ^{18}O depletion from K-Ar and Rb-Sr data; in G.M. Ross (ed.), Alberta Basement Transects Workshop; Lithoprobe Report 47, p. 336-339.

Chang, K.H., 1975. Unconformity-bounded stratigraphic units; Bulletin of the Geological Society of America, v. 86, p. 1544-1552.

Chiarenzelli, J., Aspler, L., Villeneuve, M., and Lewry, J., 1998. Early Proterozoic evolution of the Saskatchewan Craton and its allochthonous cover, Trans-Hudson Orogen; Journal of Geology, v. 106, p. 247-267.

Christiansen, R.L. and Yeats, R.S., 1992. Post-Laramide geology of the U.S. Cordilleran region; in B.C. Burchfiel, P.W. Lipman, and M.L. Zoback (eds.), Geological Society of America, The Geology of North America, v. G-3, p. 261-406.

Christensen, N.I. and Mooney, W.D., 1995. Seismic velocity structure and composition of the continental crust: a global view; Journal of Geophysical Research, v. 100, p. 9761-9788.

Clauser, C. and Huenges, E., 1993. KTB thermal regime and heat transport mechanism - current knowledge; Scientific Drilling, v. 3, p. 271-281.

Cloos, H., 1948. The ancient European basement blocks - preliminary note; Transactions of the American Geophysical Union, v. 29, p. 99-103.

Clowes, R.M., 1996. Lithoprobe Phase IV: multidisciplinary studies of the evolution of a continent - a progress report; Geoscience Canada, v. 23, p. 109-123.

Clowes, R.M., Baird, D.J., and Dehler, S.A., 1997. Crustal structure of the Cascadia subduction zone, southwestern British Columbia, from potential field and seismic studies; Canadian Journal of Earth Sciences, v. 34, p. 317-335.

Clowes, R.M., Brandon, M.T., Green, A.G., Yorath, C.J., Sutherland Brown, A., Kanasewich, E.R., and Spencer, C., 1987. LITHOPROBE - southern Vancouver Island: Cenozoic subduction complex imaged by deep seismic reflections: Canadian Journal of Earth Sciences, v. 24, p. 31-51.

Clowes, R.M., Zelt, C.A., Amor, J.R., and Ellis, R.M., 1995. Lithospheric structure in the southern Canadian Cordillera from a network of seismic refraction lines; Canadian Journal of Earth Sciences, v. 32, p. 1485-1513.

Cogley, J.G., 1984. Continental margins and the extent and number of continents; Reviews of Geophysics, v. 22, p. 101-122.

Cohen, Z., Kaptzan, V., and Flexer, A., 1990. The tectonic mosaic of the southern Levant: implications for hydrocarbon prospects; Journal of Petroleum Geology, v. 13, p. 437-462.

Collerson, K.D., Lewry, J.F., Bickford, M.E., and Van Schmus, W.R., 1990. Crustal evolution of the buried Precambrian of southern Saskatchewan: implications for diamond exploration; in L.S. Beck and C.T. Harper (eds.), Modern Exploration Techniques; Saskatchewan Geological Society, Special Publication 10, p. 150-165.

Collerson, K.D., Van Schmus, R.W., Lewry, J.F., and Bickford, M.E., 1988. Buried Precambrian basement in south-central Saskatchewan: provisional results from Sm-Nd model ages and U-Pb zircon geochronology; in Saskatchewan Geological Survey, Miscellaneous Report 88-4, p. 142-150.

Colpron, M., Price, R.A., Archibald, D.A., and Carmichael, D.M., 1996. Middle Jurassic exhumation along the western flank of the Selkirk fan structure: Thermobarometric and thermochronometric constraints from the Illecillewaet synclinorium, southeastern British Columbia; Bulletin of the Geological Society of America, v. 108, p. 1372-1392.

Committee on Stratigraphic Nomenclature, 1933. Classification and Nomenclature of rock units; Bulletin of the Geological Society of America, v. 44, p. 423-459 and Bulletin of the American Association of Petroleum Geologists, v. 17, p. 843-868.

Coney, P.J., Jones, D.L., and Monger, J.W.H., 1980. Cordilleran suspect terranes; Nature, v. 288, p. 329-333.

Constenius, K.N., 1996. Late Paleogene extensional collapse of the Cordilleran foreland fold and thrust belt; Bulletin of the Geological Society of America, v. 108, p. 20-39.

Cook, D.G. and Mayers, I.R., 1990. Precambrian structure and stratigraphy based on seismic interpretation, Colville Hills region, Northwest Territories; in Current Research, Part C; Geological Survey of Canada, Paper 90-1C, p. 339-348.

Cook, D.G. and Mayers, I.R., 1991. Precambrian structure and stratigraphy based on seismic interpretation, Colville Hills region, Northwest Territories: reply; in Current Research, Part E; Geological Survey of Canada, Paper 91-1E, p. 239-241.

Cook, F.A., 1995a. The Southern Canadian Cordillera Transect of Lithoprobe: Introduction; Canadian Journal of Earth Sciences, v. 32, p. 1483-1484.

Cook, F.A., 1995b. The reflection Moho beneath the southern Canadian Cordillera; Canadian Journal of Earth Sciences, v. 32, p. 1520-1530.

Cook, F.A., 1995c. Lithospheric processes and products in the southern Canadian Cordillera: a Lithoprobe perspective; Canadian Journal of Earth Sciences, v. 32, p. 1803-1824.

Cook, F.A., van der Velden, A.J., Hall, K.W., and Roberts, B.R., 1997. Lithoprobe SNORCLE-ing beneath the northwestern Canadian Shield: deep lithospheric reflection profiles from the Cordillera to the Archean Slave Province; Lithoprobe Seismic Processing Facility Newsletter, University of Calgary, v. 10, no. 1, p. 59-64.

Cope, J.C.W., 1996. The role of Secondary Standard in stratigraphy; Geological Magazine, v. 133, p. 107-110.

Cordell, L. and Grauch, V.J.S., 1985. Mapping basement magnetization zones from aeromagnetic data in the San Juan basin, New Mexico; *in* W.J. Hinze (ed.), The Utility of Regional Gravity and Magnetic Anomaly Maps; Society of Exploration Geophysicists, p. 181-197.

Cowan, D.S. and Bruhn, R.L., 1992. Late Jurassic to late Early Cretaceous geology of the U.S. Cordillera; *in* B.C. Burchfiel, P.W. Lipman, and M.L. Zoback (eds.), The Cordilleran Orogen: Conterminous U.S.; Geological Society of America, The Geology of North America, v. G-3, p. 169-203.

Coward, M.P., 1991. Inversion tectonics in NW Europe (abs.); *in* 6th Meeting of the European Union of Geosciences; Terra Abstracts, v. 3, p. 229.

Coward, M.P., Dietrich, D., and Park, R.G. (eds.), 1989. Alpine Tectonics; Geological Society (London), Special Publication 45.

Cross, T.A. and Homewood, P.W., 1997. Amanz Gressly's role in founding modern stratigraphy; Bulletin of the Geological Society of America, v. 109, p. 1617-1630.

Crough, S.T., 1981. Mesozoic hotspot epeirogeny in eastern North America; Geology, v. 9, p. 2-6.

Cumming, G.L. and Krstic, D., 1991. Geochronology at the Namew Lake Ni-Cu orebody, Flin Flon area, Manitoba, Canada: thermal history of a metamorphic terrane; Canadian Journal of Earth Sciences, v. 28, p. 309-325.

Cumming, G.L., Krstic, D., and Wilson, J.A., 1987. Age of the Athabasca Group, northern Alberta (abs.); Geological Association of Canada & Mineralogical Association of Canada, Program with Abstracts, v. 12, p. 35.

Dahlstrom, C.D.A., 1970. Structural geology in the eastern margin of the Canadian Rocky Mountains; Bulletin of Canadian Petroleum Geology, v. 18, p. 332-406.

Dalziel, I., 1995. La Terre avant la Pangee (The Earth before Pangea); Pour la Science, v. 209, p. 68-73.

Dana, J.D., 1873. On some results of the Earth's contraction from cooling, including a discussion of the origin of mountains and the nature of the Earth's interior; American Journal of Science, 3rd series, v. 5, p. 423-443; v. 6, p. 6-14, 104-115, 161-172.

Davidson, A., 1986. A new look at the Grenville Front in Ontario; Geological Association of Canada, Field Trip Guidebook 15, 31 p.

Davies, G.R. (ed.), 1975a. Devonian Reef Complexes of Canada I, Rainbow, Swan Hills; Canadian Society of Petroleum Geologists, Reprint Series 1, 229 p.

Davies, G.R. (ed.), 1975b. Devonian Reef Complexes of Canada II, Leduc-Cairn, Mercy Bay; Canadian Society of Petroleum Geologists, Reprint Series 1, 246 p.

Davis, E.E., 1982. Evidence for extensive basalt flows on the sea floor; Bulletin of the Geological Society of America, v. 93, p. 1023-1029.

De Sitter, L.U., 1956. Structural Geology; McGraw-Hill, 552 p.

Dehler, S.A. and Clowes, R.M., 1992. Integrated geophysical modelling of terranes and other structural features along the western Canadian margin; Canadian Journal of Earth Sciences, v. 29, p. 1492-1508.

Delaney, G., Tisdale, D., and Davies, H., 1996. Stratigraphic relationships and base metal mineralization in the Lower Proterozoic supracrustal assemblage along the Archean Johnson River inlier, Wollaston domain, Saskatchewan; in Summary of Investigations 1996, Saskatchewan Geological Survey, Miscellaneous Report 96-4, p. 3-11.

DeMets, C., Gordon, R.G., Argus, D.F., and Stein, S., 1990. Current plate motions; Geophysical Journal International, v. 101, p. 425-478.

Deroo, G., Powell, T.G., Tissot, B., and McCrossan, R.G., 1977. The Origin and Migration of Petroleum in the Western Canada Sedimentary Basin, Alberta - a Geochemical and Thermal Maturation Study; Geological Survey of Canada, Bulletin 262, 136 p.

Devlin, W.J. and Bond, G.C., 1988. The initiation of the early Paleozoic Cordilleran miogeocline: evidence from the uppermost Proterozoic - Lower Cambrian Hamill Group of southeastern British Columbia; Canadian Journal of earth Sciences, v. 25, p. 371-412.

Dewey, J.F. and Burke, K.C.A., 1973. Tibetan, Variscan and Precambrian basement reactivation: products of continental collision; Journal of Geology, v. 81, p. 683-692.

Dickinson, W.R., 1976. Sedimentary basins developed during evolution of Mesozoic-Cenozoic arc-trench system in western North America; Canadian Journal of Earth Sciences, v. 13, p. 1268-1287.

Dietrich, J.R. and Palmer, B., 1996. The Lower Paleozoic of central Alberta: insights from Lithoprobe seismic data; in G.M. Ross (comp.), Alberta Basement Transect Workshop; Lithoprobe Report 51, p. 123-132.

Dietz, R.S., 1966. Passive continents, spreading sea floors, and collapsing continental rises; American Journal of Science, v. 264, p. 177-193.

Dietz, R.S., 1972. Geosynclines, mountains and continent-building; Scientific American, v. 226, p. 30-38.

Dietz, R.S. and Holden, J.C., 1974. Collapsing continental rises: actualistic concept of geosynclines - a review; in R.H. Dott, Jr. and R.H. Shaver (eds.), Modern and Ancient Geosynclinal Sedimentation, SEPM (Society of Economic Paleontologists and Mineralogists), Special Publication 19, p. 14-25.

Digel, S., Ashton, K.E., and Wilcox, K.H., 1991. Metamorphic P-T investigations in the southern part of Kisseynew and western part of Flin Flon domains; *in* Summary of Investigations 1991, Saskatchewan Geological Survey, Miscellaneous Report 91-4, p. 41-46.

Digel, S.G. and Gordon, T.M., 1993. Quantitative estimates of pressure-temperature-fluid composition in metabasites from prehnite-pumpellyite to amphibolite facies near Flin Flon, Manitoba; *in* Z. Hajnal and J. Lewry (eds.), Proceedings, Trans-Hudson Orogen Transect Meeting; Lithoprobe Report 34, p. 139-156.

Dix, G.R., 1990. Stages of platform development in the Upper Devonian (Frasnian) Leduc Formation, Peace River Arch, Alberta; *in* S.C. O'Connell and J.S. Bell (eds.), Geology of the Peace River Arch; Bulletin of Canadian Petroleum Geology, v. 38A, p. 66-92.

Dods, S.D., Teskey, D.J., and Hood, P.J., 1989. Magnetic Anomaly Map of Canada, scale 1:10,000,000; Geological Survey of Canada, Canadian Geophysical Atlas, Map 11.

Doig, R., 1991. U-Pb zircon dates of Morin anorthosite suite rocks, Grenville Province, Quebec; Journal of Geology, v. 99, p. 729-738.

Dott, Jr., R.H. (ed.), 1992. Eustasy: the Historical Ups and Downs of a Major Geological Concept; Geological Society of America, Memoir 180, 111 p.

Douglas, R.J.W. (ed.), 1970. Geology and Economic Minerals of Canada; Geological Survey of Canada, Economic Geology Report No. 1, 838 p.

Douglas, R.J.W., Gabrielse, H., Wheeler, J.O., Stott, D.F., and Belyea, H.R., 1970. Geology of Western Canada; *in* R.J.W. Douglas (ed.), Geology and Economic Minerals of Canada; Geological Survey of Canada, Economic Geology Report No. 1, p. 367-488.

Dover, J.H., 1994. Geology of part of east-central Alaska; *in* G. Plafker and H.C. Berg (eds.), Geology of Alaska; Geological Society of America, The Geology of North America, v. G-1, p. 153-204.

Downes, H., 1993. The nature of the lower continental crust in Europe: petrological and geochemical evidence from xenoliths; Physics of the Earth and Planetary Interiors, v. 79, p. 195-218.

Drury, M.J., 1988. Tectonothermics of the North American Great Plains basement; Tectonophysics, v. 148, p. 299-307.

Duke, E.F., Shearer, C.K., Redden, J.A., and Papike, J.J., 1990. Proterozoic granite-pegmatite magmatism, Black Hills, South Dakota: structure and geochemical zonation; *in* J.F. Lewry and M.R. Stauffer (eds.), The Early Proterozoic Trans-Hudson Orogen in Canada; Geological Association of Canada, Special Paper 37, p. 253-269.

Duncan, R.A., 1982. A captured island chain in the Coast Range of Oregon and Washington; Journal of Geophysical Research, v. 87, p. 10,827-10,837.

Dunlop, D.J., 1995. Magnetism in rocks; Journal of Geophysical Research, v. 100, p. 2161-2174.

Dutch, S.I. and Nielsen, P.A., 1990. The Archean Wyoming Province and its relations with adjacent Proterozoic provinces; *in* J.F. Lewry and M.R. Stauffer(eds.), The Early Proterozoic Trans-Hudson Orogen in Canada; Geological Association of Canada, Special Paper 37, p. 287-300.

Eardley, A.J., 1962. Structural Geology of North America, 2nd edition; Harper & Row.

Eaton, D.W., Milkereit, B., Ross, G.M., Kanasewich, E.R., Geis, W., Edwards, D.J., Kelsch, L., and Varsek, J., 1995. Lithoprobe basin-scale seismic profiling in central Alberta: influence of basement on the sedimentary cover; Bulletin of Canadian Petroleum Geology, v. 43, p. 65-77.

Ebner, E., Peirce, J., and Marchand, N., 1995. Interpretation of aeromagnetic data; Recorder (Canadian Society of Exploration Geophysicists), v. 20/7, p. 8-11.

Edwards, D.E., Barclay, J.E., Gibson, D.W., Kvill, G.E., and Halton, E., 1994. Triassic strata of the Western Canada Sedimentary Basin; *in* G.D. Mossop and I. Shetsen (comps.), Geological Atlas of the Western Canada Sedimentary Basin; Canadian Society of Petroleum Geologists & Alberta Research Council, p. 259-275.

Edwards, D.J., Lyatsky, H.V., and Brown, R.J., 1995. Basement fault control on Phanerozoic stratigraphy in the Western Canada Sedimentary Province: integration of potential-field and lithostratigraphic data; *in* G.M. Ross (ed.), Alberta Basement Transects Workshop, Lithoprobe Report 47, p. 181-224.

Edwards, D.J., Lyatsky, H.V., and Brown, R.J., 1998. Regional interpretation steep faults in the Alberta Basin from public-domain gravity and magnetic data: an update; Recorder (Canadian Society of Exploration Geophysicists), v. XXIII, no. 1, p. 15-24.

Eisbacher, G.H., 1977. Mesozoic-Tertiary basin models for the Canadian Cordillera and their geological constraints; Canadian Journal of Earth Sciences, v. 14, p. 2414-2421.

Eisbacher, G.H., 1983. Devonian-Mississippian sinistral transcurrent faulting along the cratonic margin of western North America: a hypothesis; Geology, v. 11, p. 7-10.

Eisenstadt, G. and Withjack, M.O., 1995. Estimating inversion: results from clay models; *in* J.G. Buchanan and P.G. Buchanan (eds.), Basin Inversion; Geological Society (London), Special Publication 88, p. 119-136.

Elliott, C.G., 1996. Phanerozoic deformation in the "stable" craton, Manitoba, Canada; Geology, v. 24, p. 909-912.

Elliott, C.G., 1997. The Tabbernor fault in four dimensions; *in* Z. Hajnal and J. Lewry (eds.), Trans-Hudson Orogen Transect, Report of Sixth Transect Meeting; Lithoprobe Report 55, p. 5-9.

Elston, D.P. and Link, P.K., 1993. Middle and Late Proterozoic depositional and tectonic synthesis; *in* J.C. Reed, Jr., M.E. Bickford, R.S. Houston, P.K. Link, D.W. Rankin, P.K. Sims, and W.R. Van Schmus (eds.), Precambrian: Conterminous U.S.; Geological Society of America, The Geology of North America, v. C-2, p. 569-575.

Emmermann, R. and Lauterjung, J., 1997. The German Continental Deep Drilling Program KTB: overview and major results; Journal of Geophysical Research, v. 102, p. 18,179-18,201.

Engebretson, D.C., Cox, A., and Gordon, R.G., 1985. Relative Motions Between Oceanic and Continental Plates in the Pacific Basin; Geological Society of America, Special Paper 206, 59 p.

England, T.D. and Bustin, R.M., 1986. Thermal maturation of the western Canadian sedimentary basin south of the Red Deer River: 1, Alberta Plains; Bulletin of Canadian Petroleum Geology, v. 34, p. 71-90.

Ernst, R.E., Buchan, K.L., West, T.D., and Palmer, H.C., 1996. Diabase (Dolerite) Dyke Swarms of the World: First Edition; Geological Survey of Canada, Open File 3241, 1:35,000,000 map and 104-page report.

Ernst, W.G., 1988. Metamorphic terranes, isotopic provinces, and implications for crustal growth of the western United States; Journal of Geophysical Research, v. 93, p. 7634-7642.

Fahrig, W.F. and Loveridge, W.D., 1981. Rb-Sr isochron age of weathered pre-Athabasca Formation basement gneiss, northern Saskatchewan; in Current Research, Part C; Geological Survey of Canada, Paper 81-1C, p. 127-129.

Fedorowich, J.S., Kerrich, R., and Stauffer, M.R., 1995. Geodynamic evolution and thermal history of the central Trans-Hudson Orogen: constraints from structural development, $^{40}Ar/^{39}Ar$, and stable isotope geochemistry; Tectonics, v. 14, p. 472-503.

Fenton, M.M., Pawlowicz, J.G., and Dufresne, M., 1996. Till mineralogy and geochemistry in northern Alberta (abs.); 5th Annual Calgary Mining Forum, Abstracts, p. 25.

Fenton, M.M., Schreiner, B.T., Nielsen, E., and Pawlowicz, J.G., 1994. Quaternary geology of the Western Plains; in G.D. Mossop and I. Shetsen (comps.), Geological Atlas of the Western Canada Sedimentary Basin; Canadian Society of Petroleum Geologists & Alberta Research Council, p. 413-420.

Finch, J.C. and Baldwin, D.O., 1984. Stratigraphy of the Prichard Formation, Belt Supergroup (abs.); in S.W. Hobbs (ed.), The Belt; Montana Bureau of Mines and Geology, Special Publication 90, p. 5-7.

Finlayson, D.M., Wright, C., Leven J.H., Collins, C.N.D., Wake-Dyster, K.D., and Johnstone, D.W., 1989. Basement features under four intra-continental basins in central and eastern Australia; in R.A. Price (ed.), Origin and Evolution of Sedimentary Basins and Their Energy and Mineral Resources; American Geophysical Union, Geophysical Monograph 48 & International Union of Geodesy and Geophysics, v. 3, p. 43-55.

Fipke, C.E., Gurney, J.J., and Moore, R.O., 1995. Diamond Exploration Techniques Emphasising Indicator Mineral Geochemistry and Canadian Examples; Geological Survey of Canada, Bulletin 423, 86 p.

Flach, P.D., 1984. Oil Sands Geology - Athabasca Deposit North; Alberta Geological Survey, Bulletin 46, 31 p.

Flint, R.F. and Skinner, B.J., 1977. Physical Geology, second edition; John Wiley & Sons, 571 p.

Forte, A.M., Peltier, W.R., Dziewonski, A.M., and Woodward, R.L., 1993a. Dynamic surface topography: a new interpretation based upon mantle flow models derived from seismic tomography; Geophysical Research Letters, v. 20, p. 235-238.

Forte, A.M., Peltier, W.R., Dziewonski, A.M., and Woodward, R.L., 1993b. Reply to comment by M. Gurnis on "Dynamic surface topography: a new interpretation based upon mantle flow models derived from seismic tomography"; Geophysical Research Letters, v. 20, p. 1665-1666.

Forte, A.M. and Woodward, R.L., 1997. Seismic-geodynamic constraints on three-dimensional structure, vertical flow, and heat transfer in the mantle; Journal of Geophysical Research, v. 102, p. 17,981-17,994.

Fountain, D.M., Boundy, T.M., Austrheim, H., and Rey, P., 1994. Eclogite-facies shear zones: deep crustal reflectors?; Tectonophysics, v. 232, p. 411-424.

Fountain, D.M. and Christensen, N.I., 1989. Composition of the continental crust and upper mantle: a review; in L.C. Pakiser and W.D. Mooney (eds.), Geophysical Framework of the Continental United States; Geological Society of America, Memoir 172, p. 711-742.

Friedman, G.M., 1955. Petrology of the Memesagamesing Lake norite mass, Ontario, Canada; American Journal of Science, v. 253, p. 590-608.

Friedman, G.M., 1957. Structure and petrology of the Caribou Lake intrusive body, Ontario, Canada; Bulletin of the Geological Society of America, v. 68, p. 1531-1568.

Friedman, G.M., 1987a. Deep-burial diagenesis: its implications for vertical movements of the crust, uplift of the lithosphere and isostatic unroofing: a review; Sedimentary Geology, v. 50, p. 67-94.

Friedman, G.M., 1987b. Addendum: deep-burial diagenesis: its implications for vertical movements of the crust, uplift of the lithosphere and isostatic unroofing: a review; Sedimentary Geology, v. 54, p. 165-167.

Friedman, G.M., 1987c. Vertical movements of the crust: case histories from the northern Appalachian Basin; Geology, v. 15, p. 1130-1133.

Friedman, G.M., 1988a. The Catskill tectonic fan-delta complex: northern Appalachian Basin; Northeastern Geology, v. 10. p. 254-257.

Friedman, G.M., 1988b. Comments on "Orogeny and epeirogeny in the study of Phanerozoic and Archean rocks"; Geoscience Canada, v. 15, p. 230-231.

Friedman, G.M. and Sanders, J.E., 1982. Time-temperature-burial significance of Devonian anthracite implies former great (6.5 km) depth of burial of Catskill Mountains, New York; Geology, v. 10, p. 93-96.

Friedman, G.M. and Sanders, J.E., 1978. Principles of Sedimentology; John Wiley & Sons, New York, 792 p.

Friedman, G.M., Sanders, J.E., and Kopaska-Merkel, D.C., 1992. Principles of Sedimentary Deposits; Macmillan, 717 p.

Fritz, P. and Frape, S.K. (eds.), 1987. Saline Water and Gases in Crystalline Rocks; Geological Association of Canada, Special Paper 33, 259 p.

Frodeman, R., 1995. Geological reasoning: geology as an interpretive and historical science; Bulletin of the Geological Society of America, v. 107, p. 960-968.

Frost, C.D. and Burwash, R.A., 1986. Nd evidence for extensive Archean basement in the western Churchill Province, Canada; Canadian Journal of Earth Sciences, v. 23, p. 1433-1437.

Fueten, F. and Redmond, D.J., 1997. Documentation of a 1450 Ma contractional orogeny preserved between the 1850 Ma Sudbury structure and the 1 Ga Grenville orogenic front, Ontario; Bulletin of the Geological Society of America, v. 109, p. 268-179.

Fueten, F. and Robin, P.F., 1989. Structural petrology along a transect across the Thompson belt, Manitoba: dip slip at the western Churchill-Superior boundary; Canadian Journal of Earth Sciences, v. 26, p. 1976-1989.

Gabrielse, H., 1991. Fault-controlled basins; in H. Gabrielse and C.J. Yorath (eds.), Geology of the Cordilleran Orogen in Canada; Geological Society of America, The Geology of North America, v. G-2, p. 360-365.

Gabrielse, H. and Yorath, C.J. (eds.), 1991. Geology of the Cordilleran Orogen in Canada; Geological Society of America, The Geology of North America, v. G-2, 823 p.

Galloway, W.E., 1974. Deposition and diagenetic alteration of sandstone in northeast Pacific arc-related basin: implications for graywacke genesis; Bulletin of the Geological Society of America, v. 85, p. 379-390.

Garland, G.D. and Bower, M.E., 1959. Interpretation of aeromagnetic anomalies in Northeastern Alberta; World Petroleum Congress, Proceedings, v. 10, no. 18, p. 787-800.

Gary, M., McAfee, Jr., R., and Wolf, C.L. (eds.), 1972. Glossary of Geology; second edition; American Geological Institute, 805 p.

Gay, S.P., 1973. Pervasive Orthogonal Fracturing in Earth's Continental Crust; American Map Company, Salt Lake City, 123 p.

Geis, W.T., Cook, F.A., Green, A.G., Milkereit, B., Percival, J.A., and West, G.F., 1990. Thin thrust sheet formation of the Kapuskasing structural zone revealed by Lithoprobe seismic reflection data; Geology, v. 18, p. 513-516.

Geldsetzer, H.H.J., Mountjoy, E.W., Tebbutt, G.E., and Burrowes, O.G., 1982. Upper Devonian Stratigraphy and Sedimentology, Southern Alberta Rocky Mountains; 11th International Congress on Sedimentology, McMaster University, Hamilton, Ontario, Canada, Field Excursion Guidebook 28B, 69 p.

Geological Survey of Canada, 1990. Canadian Geophysical Atlas; 15 maps, scale 1:10,000,000.

Gerhard, L.C., Fischer, D.W., and Anderson, S.B., 1990. Petroleum geology of the Williston Basin; in M.W. Leighton, D.R. Kolata, D.F. Oltz, and J.J. Eidel (eds.), Interior Cratonic Basins; American Association of Petroleum Geologists, Memoir 51, p. 507-559.

Gibb, R.A. and Thomas, M.D., 1976. Gravity signature of fossil plate boundaries on the Canadian Shield; Nature, v. 262, p. 199-200.

Gibb, R.A., Thomas, M.D., Lapointe, P.L., and Mukhopadhyay, M., 1983. Geophysics of proposed Proterozoic sutures in Canada; Precambrian Research, v. 19, p. 349-384.

Gibb, R.A., Thomas, M.D., and Mukhopadhyay, M., 1980. Proterozoic sutures in Canada; Geoscience Canada, v. 7, p. 149-154.

Gibson, D.W., 1977. Upper Cretaceous and Tertiary Coal-Bearing Strata in the Drumheller-Ardley Region, Red Deer River Valley, Alberta; Geological Survey of Canada, Paper 76-35, 41 p.

Gibson, D.W. and Barclay, J.E., 1989. Middle Absaroka Sequence: the Triassic stable craton; in B.D. Ricketts (ed.), The Western Canada Sedimentary Basin - A Case History; Canadian Society of Petroleum Geologists, Special Publication 30, p. 219-232.

Gilbert, G.K., 1890. Lake Bonneville; U.S. Geological Survey, Monograph, v. 1, 438 p.

Gilluly, J., 1949. Distribution of mountain building in geologic time; Bulletin of the Geological Society of America, v. 60, p. 561-590.

Gilluly, J., 1950. Reply to discussion by H. Stille; Geologische Rundschau, v. 38, p. 103-107.

Gilluly, J., 1963. The tectonic evolution of the western United States; Quarterly Journal of the Geological Society of London, v. 119, p. 133-174.

Gilluly, J. and Gates, O., 1965. Tectonic and Igneous Geology of the Northern Shoshone Range, Nevada; U.S. Geological Survey, Professional Paper 465, 153 p.

Gol'braykh, I.G., Mirkin, G.R., and Zabalnyer, V.V., 1966. Tectonic analysis of megajointing; International Geology Review, v. 8, p. 1009-1016.

Goodacre, A.K., Grieve, R.A.F., Halpenny, J.F., and Sharpton, V.L., 1987a. Bouguer Gravity Anomaly Map of Canada, scale 1:10,000,000; Geological Survey of Canada, Canadian Geophysical Atlas, Map 3.

Goodacre, A.K., Grieve, R.A.F., Halpenny, J.F., and Sharpton, V.L., 1987b. Horizontal Gradient of the Bouguer Gravity Anomaly Map of Canada, scale 1:10,000,000; Geological Survey of Canada, Canadian Geophysical Atlas, Map 5.

Goodfellow, W.D., Cecile, M.P., and Leybourne, M.I., 1995. Geochemistry, petrogenesis, and tectonic setting of lower Paleozoic alkalic and potassic volcanic rocks, Northern Canadian Cordilleran Miogeosyncline; Canadian Journal of Earth Sciences, v. 32, p. 1236-1254.

Gordey, S.P., 1991. Devonian-Mississippian clastics of the Foreland and Omineca belts; in H. Gabrielse and C.J. Yorath (eds.),Geology of the Cordilleran Orogen in Canada; Geological Society of America, The Geology of North America, v. G-2, p. 230-242.

Gordey, S.P., Abbott, J.G., Tempelman-Kluit, D.J., and Gabrielse, H., 1987. "Antler" clastics in the Canadian Cordillera; Geology, v. 15, p. 103-107.

332

Gordon, T.M., Hunt, P.A., Bailes, A.H., and Syme, E.C., 1990. U-Pb ages from the Flin Flon and Kisseynew belts, Manitoba: chronology of crust formation at an Early Proterozoic accretionary margin; *in* J.F. Lewry and M.R. Stauffer (eds.), The Early Proterozoic Trans-Hudson Orogen of North America; Geological Association of Canada, Special Paper 37, p. 177-199.

Gordon, T.M., Aranovich, L.Ya., and Fed'kin, V.V., 1994. Exploratory data analysis in thermobarometry: an example from the Kisseynew sedimentary gneiss belt, Manitoba, Canada; American Mineralogist, v. 79, p. 973-982.

Gossler, J. and Kind, R., 1996. Seismic evidence for very deep roots of continents; Earth and Planetary Science Letters, v. 138, p. 1-13.

Grand, S.P., van der Hilst, R.D., and Widiyantoro, S., 1997. Global seismic tomography: a snapshot of convection in the Earth; GSA Today, v. 7/4, p. 1-7.

Grant, A.C., 1987. Inversion tectonics on the continental margin east of Newfoundland; Geology, v. 15, p. 845-848.

Gray, M.B. and Zeitler, P.K., 1997. Comparison of clastic wedge provenance in the Appalachian foreland using U/Pb ages of detrital zircons; Tectonics, v. 16, p. 151-160.

Grayston, L.D., Sherwin, D.F., and Allan, J.F., 1964. Middle Devonian; *in* R.G. McCrossan and R.P. Glaister (eds.), Geological History of Western Canada; Alberta Society of Petroleum Geologists, p. 49-59.

Green, A.G., Hajnal, Z., and Weber, W., 1985a. An evolutionary model of the western Churchill Province and western margins of the Superior Province in Canada and the north-central United States; Tectonophysics, v. 116, p. 281-322.

Green, A.G., Hajnal, Z., and Weber, W., 1986. An evolutionary model of the western Churchill Province and western margins of the Superior province in Canada and the north-central United States - reply; Tectonophysics, v. 131, p. 188-197.

Green, A.G., Weber, W., and Hajnal, Z., 1985b. Evolution of Proterozoic terrains beneath the Williston basin; Geology, v. 13, p. 624-628.

Greenwood, H.J., Woodsworth, G.J., Read, P.B., Ghent, E.D., and Evenchick, C.A., 1991. Metamorphism; *in* H. Gabrielse and C.J. Yorath (eds.), Geology of the Cordilleran Orogen in Canada; Geological Society of America, The Geology of North America, v. G-2, p. 533-570.

Greggs, R.G. and Greggs, D.H., 1989. Fault-block tectonism in the Devonian subsurface, Western Canada Basin; Journal of Petroleum Geology, v. 12, p. 377-404.

Gregor, V.A., 1997. Mannville linears in the Lloydminster heavy oil area and their relationship to fractures and fluid flow in the Western Canada Sedimentary Basin; *in* S.G. Pemberton and D.P. James (eds.), Petroleum Geology of the Cretaceous Mannville Group, Western Canada; Canadian Society of Petroleum Geologists, Memoir 18, p. 428-474.

Gressly, A., 1838. Observations géologiques sur le Jura soleurois: Nouveaux mémoires de la Sociéte Helvetique des Sciences natirelles, Neuchâtel, v. 2, 349 p.

Griffin, W.L. and O'Reilly, S.Y., 1987. The composition of the lower crust and the nature of the continental Moho; xenolith evidence; *in* P.H. Nixon (ed.), Mantle Xenoliths; John Wiley & Sons, p. 413-431.

Gupta, J.C., Kurtz, R.D., Camfield, P.A., and Niblett, E.R., 1985. A geomagnetic induction anomaly from IMS data near Hudson Bay, and its relation to crustal electrical conductivity in central North America; Geophysical Journal of the Royal Astronomical Society, v. 81, p. 33-46.

Gurnis, M.., 1993. Comment on "Dynamic surface topography: a new interpretation based upon mantle flow models derived from seismic tomography"; Geophysical Research Letters, v. 20, p. 1663-1664.

Haidl, F.M., 1991. Note on Ordovician-Silurian boundary in southeastern Saskatchewan; Saskatchewan Energy and Mines, Miscellaneous Report 91-4, p. 205-207.

Hajnal, Z., Lucas, S., White, D., Lewry, J., Bezdan, S., Stauffer, M.R., and Thomas, M.D., 1996. Seismic reflection images of high-angle faults and linked detachments in the Trans-Hudson Orogen; Tectonics, v. 15, p. 427-439.

Hajnal, Z., Nemeth, B., Clowes, R.M., Ellis, R.M., Spence, G.D., Burianyk, M.J.A., Asudeh, I., White, D.J., and Forsyth, D.A., 1997. Mantle involvement in lithospheric collision: seismic evidence from the Trans-Hudson Orogen, western Canada; Geophysical Research Letters, v. 24, p. 2079-2082.

Halbertsma, H.L., 1994. Devonian Wabamun Group of the western Canada sedimentary basin; *in* G.D. Mossop and I. Shetsen (comps.), Geological Atlas of the Western Canada Sedimentary Basin; Canadian Society of Petroleum Geologists and Alberta Research Council, Calgary, p. 203-220.

Halchuk, S.C. and Mereu, R.F., 1990. A seismic investigation of the crust and Moho underlying the Peace River Arch, Canada; Tectonophysics, v. 185, p. 1-19.

Halden, N.M., Clark, G.S., Corkery, M.T., and Schledewitz, D.C.P., 1990. Geochemical and Rb-Sr whole rock systematics of granite magmatism related to the origin of the Wathaman Batholith; *in* J.F. Lewry and M.R. Stauffer (eds.), The Early Proterozoic Trans-Hudson Orogen of North America; Geological Association of Canada, Special Paper 37, p.

Hall, J., 1859. Paleontology: vol. III, containing descriptions and figures of the organic remains of the Lower Helderberg Group and the Oriskany Sandstone; New York Geological Survey, Natural History of New York, Part 6, 532 p.

Halls, H.C., 1995. 1994-1995 annual report of the paleomagnetism - rock magnetism working group of the Canadian Geophysical Union; Elements, Newsletter of the Canadian Geophysical Union, v. 13, no. 2, p. 9-11.

Halls, H.C. and Fahrig, W.F. (eds.), 1987. Mafic Dyke Swarms; Geological Association of Canada, Special Paper 34.

Halls, H.C. and Palmer, H.C., 1990. The tectonic relationship of two Early Proterozoic dyke swarms to the Kapuskasing Structural Zone: a paleomagnetic and petrographic study; Canadian Journal of Earth Sciences, v. 27, p. 87-103.

Hamilton, W.B., 1993. Evolution of the Archean mantle and crust; *in* J.C. Reed, Jr., M.E. Bickford, R.S. Houston, P.K. Link, D.W. Rankin, P.K. Sims, and W.R. Van Schmus (eds.), Precambrian: Conterminous U.S.; Geological Society of America, The Geology of North America, v. C-2, p. 597-614.

Handa, S. and Camfield, P.A., 1984. Crustal electrical conductivity in north-central Saskatchewan: the North American Central Plains anomaly and its relation to a Proterozoic plate margin; Canadian Journal of Earth Sciences, v. 21, p. 533-543.

Hanmer, S.K., 1987. Granulite facies mylonites: a brief structural reconnaissance north of Stony Rapids, northern Saskatchewan; *in* Current Research, Part A; Geological Survey of Canada, Paper 87-1A, p. 563-572.

Hanmer, S.K., 1988. Great Slave Lake shear zone, Canadian Shield: reconstructed vertical profile of a crustal-scale fault zone; Tectonophysics, v. 149, p. 245-264.

Hanmer, S. and Connelly, J.N., 1986. Mechanical role of the syntectonic Laloche Batholith in the Great Slave Lake Shear Zone, District of Mackenzie, N.W.T.; *in* Current Research, Part B; Geological Survey of Canada, Paper 86-1B, p. 811-826.

Hanmer, S., Parrish, R., Williams, M., and Kopf, C., 1994. Striding-Athabasca mylonite zone: complex Archean deep-crustal deformation in the East Athabasca mylonite triangle, northern Saskatchewan; Canadian Journal of Earth Sciences, v. 31, p. 1287-1300.

Haq, B.U., Hardenbol, J., and Vail, P.R., 1987. Chronology of fluctuating sea levels since the Triassic; Science, v. 235, p. 1156-1167.

Harland, W.B., 1992. Stratigraphic regulation and guidance: a critique of current tendencies in stratigraphic codes and guides; Bulletin of the Geological Society of America, v. 104, p. 1231-1235.

Harland, W.B., Armstrong, R.L., Cox, A.V., Craig, L.E., Smith, A.G., and Smith, D.G., 1990. A Geological Time Scale 1989; Cambridge University Press, 263 p.

Harrison, J.E., 1972. Precambrian Belt basin of northwestern United States: its geometry, sedimentation, and copper occurrences; Bulletin of the Geological Society of America, v. 83, p. 1215-1240.

Harrison, J.E., 1972. Precambrian Belt basin of the northwestern United States: its geometry, sedimentation and copper occurrences; Bulletin of the Geological Society of America, v. 83, p. 1215-1240.

Harrison, J.E., Griggs, A.B., and Wells, J.D., 1974. Tectonic Features of the Precambrian Belt Basin and Their Influence on Post-Belt Structures; U.S. Geological Survey, Professional Paper 866, 15 p.

Hatcher, R.D., Jr., 1989. Appalachians introduction; *in* R.D. Hatcher, Jr., W.A. Thomas, and G.V. Viele (eds.), The Appalachian-Ouachita Orogen in the United States; Geological Society of America, The Geology of North America, v. F-2, p. 1-6.

Hayes, B.J.R., Christopher, J.E., Rosenthal, L., Los, G., McKercher, B., Minken, D., Tremblay, Y.M., and Fennel, J., 1994. Cretaceous Mannville Group of the Western Canada sedimentary basin; *in* G.D. Mossop and I. Shetsen (comps.), Geological

Atlas of the Western Canada Sedimentary Basin; Canadian Society of Petroleum Geologists & Alberta Research Council, p. 317-334.

Heaman, L.M., 1997. Global mafic magmatism at 2.45 Ga: remnants of an ancient large igneous province?; Geology, v. 25, p. 299-302.

Heaman, L.M. and Le Cheminant, A.N., 1988. U-Pb baddeleyite ages of the Muskox Intrusion and Mackenzie Dyke Swarm, N.W.T., Canada (abs.); Geological Association of Canada-Mineralogical Association of Canada, Program with Abstracts, v. 12, p. 53.

Hedberg, H.D. (ed.), 1976. International Stratigraphic Guide: a Guide to Stratigraphic Classification, Terminology, and Procedure; Wiley, 200 p.

Hein, F.J., 1987. Tidal/littoral offshore shelf deposits - Lower Cambrian Gog Group, Southern Rocky Mountains, Canada; Sedimentary Geology, v. 52, p. 155-182.

Hein, F.J., Nowlan, G.S., Norford, B.S., Fritz, W.H., and Wicander, R., 1995. Litho- and biostratigraphy of Cambrian and Ordovician strata in the subsurface of the Western Canada Basin in central Alberta; in Proceedings of the Oil and Gas Forum; Geological Survey of Canada, Open File 3058, p. 43-48.

Heirtzler, J.R., Dickson, G.O., Herron, E.M., Pitman, W.C. III, and Le Pichon, X., 1968. Marine magnetic anomalies, geomagnetic field reversals and motions of the ocean floor and continents; Journal of Geophysical Research, v. 73, p. 2119-2136.

Henderson, C.M., 1989. The Lower Absaroka Sequence: Upper Carboniferous and Permian; in B.D. Ricketts (ed.), Western Canada Sedimentary Basin, A Case Study; Canadian Society of Petroleum Geologists, Special Publication 30, p. 203-217.

Henderson, C.M., Bamber, E.W., Richards, B.C., Higgins, A.C., and McGugan, A., 1993. Permian; in D.F. Stott and J.D. Aitken (eds.), Sedimentary Cover of the Craton in Canada; Geological Society of America, The Geology of North America, v. D-1, p. 272-293.

Henderson, C.M., Richards, B.C., and Barclay, J.E., 1994. Permian Strata of the Western Canada Sedimentary Basin; in G.D. Mossop and I. Shetsen (comps.), Geological Atlas of the Western Canada Sedimentary Basin; Canadian Society of Petroleum Geologists & Alberta Research Council, p. 251-258.

Hermes, J.J. and Borradaile, G.J., 1985. On orogeny and epeirogeny in the study of Phanerozoic and Archean rocks; Geoscience Canada, v. 12, p. 148-151.

Hinze, W.J. (ed.), 1985. The Utility of Regional Gravity and Magnetic Anomaly Maps; Society of Exploration Geophysicists, 454 p.

Hinze, W.J. and Braile, L.W., 1988. Geophysical aspects of the craton: U.S.; in L.L. Sloss (ed.), Sedimentary Cover - North American Craton; U.S.; Geological Society of America, The Geology of North America, v. D-2, p. 5-51.

Hiscott, R.N., Pickering, K.T., Bouma, A.H., Hand, B.M., Kneller, B.C., Postma, G., and Soh, W., 1997. Basin-floor fans in the North Sea: sequence stratigraphic models vs. sedimentary facies: discussion; Bulletin of the American Association of Petroleum Geologists, v. 81, p. 662-665.

Hitchon, B., Bachu, S., and Underschultz, J.R., 1990. Regional subsurface hydrogeology, Peace River Arch area, Alberta and British Columbia; *in* S.C. O'Connell and J.S. Bell (eds.), Geology of the Peace River Arch; Bulletin of Canadian Petroleum Geology, v. 38A, p. 196-217.

Hobbs, S.W. (ed.), 1984. The Belt; Montana Bureau of Mines and Geology, Special Publication 90.

Hobbs, W.H., 1904. Lineaments of the Atlantic border regions; Bulletin of the Geological Society of America, v. 15, p. 483-506.

Hobbs, W.H., 1911. Repeating patterns in the relief and in the structure of the land; Bulletin of the Geological Society of America, v. 22, p. 123-176.

Hoffman, P.F., 1981. Autopsy of Athapuscow aulacogen; a failed arm affected by three collisions; *in* F.H.A. Campbell (ed.), Proterozoic Basins of Canada; Geological Survey of Canada, Paper 81-10, p. 543-603.

Hoffman, P.F., 1988. United plates of America - Early Proterozoic assembly and growth of Laurentia; Annual Reviews of Earth and Planetary Sciences, v. 16, p. 543-603.

Hoffman, P.F., 1989. Precambrian geology and tectonic history of North America; *in* A.W. Bally and A.R. Palmer (eds.), The Geology of North America - an Overview; Geological Society of America, The Geology of North America, v. A, p. 447-512.

Hoffman, P.F., 1990. Subdivision of the Churchill Province and extent of the Trans-Hudson orogen; *in* J.F. Lewry and M.R. Stauffer (eds.), The Early Proterozoic Trans-Hudson Orogen in North America; Geological Association of Canada, Special Publication 37, p. 15-39.

Hoffman, P.F., 1991. Did the breakout of Laurentia turn Gondwanaland inside-out?; Science, v. 252, p. 1409-1412.

Holland, S.S., 1964. Landforms of British Columbia - a Physiographic Outline; British Columbia Department of Mines and Petroleum Resources, Bulletin 48, 138 p.

Hood, P.J. and Teskey, D.J., 1989. Aeromagnetic gradiometer program of the Geological Survey of Canada; Geophysics, v. 54, p. 1012-1022.

Hopkins, J.C., 1981. Sedimentology of quartzose sandstones of Lower Mannville and associated units, Medicine River area, Central Alberta; Bulletin of Canadian Petroleum Geology, v. 29, p. 12-41.

Hopkins, J.C., 1987. Contemporaneous subsidence and fluvial channel sedimentation: Upper Mannville C pool, Berry field, Lower Cretaceous of Alberta; Bulletin of the American Association of Petroleum Geologists, v. 71, p. 334-345.

Hoppin, R.A., 1974. Lineaments; their role in tectonics of Central Rocky Mountains; Bulletin of the American Association of Petroleum Geologists, v. 58, p. 2260-2273.

Horváth, F., 1988. Neotectonic behavior of the Alpine-Mediterranean region; *in* L.H. Royden and F. Horváth (eds.), The Pannonian Basin: a Study in Basin Evolution; American Association of Petroleum Geologists, Memoir 45, p. 49-55.

Höy, T., 1982. The Purcell Supergroup in southeastern British Columbia: sedimentation, tectonics and stratiform lead-zinc deposits; *in* R.W. Hutchinson, C.D. Spence, and J.M. Franklin (eds.), Precambrian Sulphide Deposits; Geological Association of Canada, Special Paper 25, p. 11-37.

Höy, T., 1984. The Purcell Supergroup near the Rocky Mtn. Trench, southeastern British Columbia; *in* W.S. Hobbs (ed.), The Belt; Montana Bureau of Mines and Geology, Special Publication 90, p. 36-38.

Huang, T.K., 1980. An outline of the tectonic characteristics of China; *in* Continental Tectonics; National Academy of Sciences, Washington, DC, p. 184-197.

Hubbard, R.J., Pape, J., and Roberts, D.G., 1985a. Depositional sequence mapping as a technique to establish tectonic and stratigraphic framework and evaluate hydrocarbon potential on a passive continental margin; *in* O.R. Berg and D. Woolverton (eds.), Seismic Stratigraphy - II; American Association of Petroleum Geologists, Memoir 39, p. 79-91.

Hubbard, R.J., Pape, J., and Roberts, D.G., 1985b. Depositional sequence mapping to illustrate the evolution of a passive continental margin; *in* O.R. Berg and D. Woolverton (eds.), Seismic Stratigraphy - II; American Association of Petroleum Geologists, Memoir 39, p. 93-115.

Huenges, E., Lauterjung, J., Bücker, C., Lippmann, E., and Kern, H., 1997. Seismic velocity, density, thermal conductivity and heat production of cores from the KTB pilot hole; Geophysical Research Letters, v. 24, p. 345-348.

Hughes, J.D., 1984. Geology and Depositional Setting of the Late Cretaceous, Upper Bearpaw and Lower Horseshoe Canyon Formations in the Dodds-Round Hill Coalfield of Central Alberta - A Computer-Based Study of Closely-Spaced Exploration Data; Geological Survey of Canada, Bulletin 361, 81 p.

Hulbert, L., Williamson, B., and Theriault, R., 1993. Geology of Middle Proterozoic MacKenzie diabase suites from Saskatchewan: an overview and their potential to host Noril'sk-type Ni-Cu-PGE mineralization; *in* Summary of Investigations 1993; Saskatchewan Geological Survey, Miscellaneous Report 93-4, p. 112-126.

Hyndman, R.D., 1995. The Lithoprobe corridor across the Vancouver Island continental margin: the structural and tectonic consequences of subduction; Canadian Journal of Earth Sciences, v. 32, p. 1777-1802.

Hyndman, R.D., Yorath, C.J., Clowes, R.M., and Davis, E.E., 1990. The northern Cascadia subduction zone at Vancouver Island: seismic structure and tectonic history; Canadian Journal of Earth Sciences, v. 27, p. 313-329.

Igolkina, N.S. (ed.), 1981. Geologic Formations of the Sedimentary Cover of the Russian Platform; VSEGEI, Trudy, v. 296, Nedra Publishing (in Russian).

Illies, H., 1981. Mechanism of graben formation; Tectonophysics, v. 73, p. 249-266.

Imlay, R.W., 1984. Jurassic ammonite successions in North America and biogeographic implications; *in* G.E.G. Westermann (ed.), Jurassic-Cretaceous Biochronology and Paleogeography of North America; Geological Association of Canada, Special Paper 27, p. 1-12.

Isachsen, Y.W., 1992. The Adirondacks: still rising after all these years; Natural History, v. 5, p. 383-387.

Isachsen, Y.W., Geraghty, E.P., and Wiener, R.W., 1983. Fracture domains associated with a neotectonic basement-cored dome - the Adirondack Mountains, New York; 4th International Basement Tectonics Conference, Salt Lake City, p. 287-306.

Isacks, B., Oliver, J., and Sykes, L.R., 1968. Seismology and the New Global Tectonics; Journal of Geophysical Research, v. 18, p. 5855-5899.

Issler, D.R., Beaumont, C., Willett, S.D., Donelick, R.A., Mooers, J., and Grist, A., 1990. Preliminary evidence from apatite fission-track data concerning the thermal history of the Peace River Arch region, Western Canada Sedimentary Basin; in S.C. O'Connell and J.S. Bell (eds.), Geology of the Peace River Arch; Bulletin of Canadian Petroleum Geology, v. 38A, p. 250-269.

Jackson, P.C., 1981. Geological Highway Map of Alberta, second edition; Canadian Society of Petroleum Geologists, Calgary.

Jackson, P.C., 1984. Paleogeography of the Lower Cretaceous Mannville Group of western Canada; in J.A. Masters (ed.), Elmworth - Case Study of a Deep Basin Gas Field; American Association of Petroleum Geologists, Memoir 38, p. 49-78.

Janecke, S.U., 1994. Sedimentation and paleogeography of an Eocene to Oligocene rift zone, Idaho and Montana; Bulletin of the Geological Society of America, v. 106, p. 1083-1095.

Jardine, D., 1974. Cretaceous oil sands of western Canada; in L.V. Hills (ed.), Oil Sands - Fuel of the Future; Canadian Society of Petroleum Geologists, Memoir 3, p. 50-67.

Jeanloz, R. and Romanowicz, B., 1997. Geophysical dynamics at the center of the Earth; Physics Today, v. 50/8, p. 22-27.

Jeletzky, J.A., 1975. Jurassic and Lower Cretaceous Paleogeography and Depositional Tectonics of Porcupine Plateau, Adjacent Areas of Northern Yukon and Those of Mackenzie District, Northwest Territories; Geological Survey of Canada, Paper 74-16, 52 p.

Jerzykiewicz, T., 1992. Controls on the distribution of coal in the Campanian to Paleocene post-Wapiabi strata of the Rocky Mountain Foothills, Canada; in P.J. McCabe and J.T. Parrish (eds.), Controls on the Distribution and Quality of Cretaceous Coals; Geological Society of America, Special Paper 267, p. 139-150.

Jerzykiewicz, T. and Norris, D.K., 1992. Anatomy of the Laramide Foredeep and the structural style of the adjacent foreland thrust belt in Southern Alberta; Guidebook for Canadian Society of Petroleum Geologists Field Trip #3 for the American Association of Petroleum Geologists Annual Convention, Calgary, 88 p.

Jerzykiewicz, T. and Norris, D.K., 1994. Stratigraphy, structure and syntectonic sedimentation of the Campanian 'Belly River' clastic wedge in the southern Canadian Cordillera; Cretaceous Research, v. 15, p. 367-399.

Johnson, R.A., Karlstrom, K.E., Smithson, S.B., and Houston, R.S., 1984. Gravity profiles across the Cheyenne Belt, a Precambrian crustal suture in southern Wyoming; Journal of Geodynamics, v. 1, p. 445-472.

Jones, A.G., 1988. Discussion of "A magnetotelluric investigation under the Williston Basin of south-eastern Saskatchewan"; Canadian Journal of Earth Sciences, v. 25, p. 1132-1139.

Jones, A.G. and Craven, J.A., 1990. The North American Central Plains conductivity anomaly and its correlation with gravity, magnetic, seismic, and heat flow data in Saskatchewan, Canada; Physics of the Earth and Planetary Interiors, v. 60, p. 169-194.

Jones, A.G., Craven, J.A., McNeice, G.A., Ferguson, I.J., Boyce, T., Farquarson, C., and Ellis, R.G., 1993. The North American Central Plains conductivity anomaly within the Trans-Hudson orogen in northern Saskatchewan; Geology, v. 21, p. 1027-1030.

Jones, A.G. and Gough, I.D., 1995. Electromagnetic images of crustal structures in southern and central Canadian Cordillera; Canadian Journal of Earth Sciences, v. 32, p. 1541-1563.

Jones, A.G. and Savage, P.J., 1986. North American Central Plains conductivity anomaly goes east; Geophysical Research Letters, v. 13, p. 685-688.

Jones, D.L., Silberling, N.J., and Hillhouse, J., 1977. Wrangellia - a displaced terrane in northwestern North America; Canadian Journal of Earth Sciences, v. 14, p. 2566-2577.

Jones, R.M.P., 1980. Basinal isostatic adjustment faults and their petroleum significance; Bulletin of Canadian Petroleum Geology, v. 28, p. 211-251.

Jordan, T.H., 1975. The continental tectosphere; Geophysics and Space Physics, v. 13, p. 1-12.

Kalkreuth, W. and McMechan, M.E., 1984. Regional pattern of thermal maturation as determined from coal-rank studies, Rocky Mountain Foothills and Front Ranges north of Grande Cache, Alberta - implications for petroleum exploration; Bulletin of Canadian Petroleum Geology, v. 32, p. 249-271.

Kalkreuth, W. and McMechan, M., 1988. Burial history and thermal maturity, Rocky Mountain Front Ranges, Foothills and Foreland, east-central British Columbia and adjacent Alberta, Canada; Bulletin of the American Association of Petroleum Geologists, v. 72, p. 1395-1410.

Kanasewich, E.R., Burianyk, M.J.A., Dubuc, G.P., Lemieux, J.F., and Kalantzis, F., 1995. Three-dimensional seismic reflection studies of the Alberta basement; Canadian Journal of Exploration Geophysics, v. 31, p. 1-10.

Kanasewich, E.R., Burianyk, M.J.A., Ellis, R.M., Clowes, R.M., White, D.T., Côte, T., Forsyth, D.A., Luetgert, J.H., and Spence, G.D., 1994. Crustal velocity structure of the Omineca belt, southeastern Canadian Cordillera; Journal of Geophysical Research, v. 99, p. 2653-2670.

Kanasewich, E.R., Clowes, R.M., and McCloughan, C.H., 1969. A buried Precambrian rift in western Canada; Tectonophysics, v. 8, p. 513-527.

Karlstrom and Bowring, 1993. Proterozoic orogenic history of Arizona; *in* J.C. Reed, Jr., M.E. Bickford, R.S. Houston, P.K. Link, D.W. Rankin, P.K. Sims, and W.R. Van Schmus (eds.), Precambrian: Conterminous U.S.; Geological Society of America, The Geology of North America, v. C-2, p. 188-211.

Kay, M., 1951. North American Geosynclines; Geological Society of America, Memoir 48, 143 p.

Kay, R.W. and Kay, S.M., 1986. Petrology and geochemistry of the lower continental crust; an overview; J.B. Dawson, D.A. Carswell, J. Hall, and K.H. Wedepohl (eds.), The Nature of the Lower Continental Crust; Geological Society (London), Special Publication 24, p. 147-159.

Kay, R.W. and Kay, S.M., 1993. Delamination and delamination magmatism; *in* A.G. Green, A. Kröner, H.J. Götze, and N. Pavlenkova (eds.), New Horizons in Strong Motion; Seismic Studies and Engineering Practice; Tectonophysics, v. 219, p. 177-189.

Kent, D.M. and Christopher, J.E., 1994. Geological history of the Williston Basin and Sweetgrass Arch; *in* G.D. Mossop and I. Shetsen (comps.), Geological Atlas of the Western Canada Sedimentary Basin; Canadian Society of Petroleum Geologists & Alberta Research Council, p. 421-430.

Kent, P.E., Satterthwaite, G.E., and Spencer, A.M. (eds.), 1969. Time and Place in Orogeny; Geological Society (London), Special Publication 3, 311 p.

Kerr, A., 1991. A decade of evolution of Archean thought: the Third International Archean Symposium; Geoscience Canada, v. 18, p. 25-27.

King, E.R. and Zietz, I., 1971. Aeromagnetic study of the midcontinent gravity high of central United States; Bulletin of the Geological Society of America, v. 82, p. 2187-2208.

King, J.E., Davis, W.J., and Relf, C., 1989. Comment on "Accretion of the Archean Slave province"; Geology, v. 17, p. 963-964.

King, P.B., 1969. Tectonic Map of North America, scale 1:5,000,000; U.S. Geological Survey.

King, P.B., 1977. The Evolution of North America; Princeton University Press, 197 p.

Kirschvink, J.L., Ripperdan, R.L., and Evans, D.A., 1997. Evidence for a large-scale reorganization of Early Cambrian continental masses by inertial interchange true polar wonder; Science, v. 277, p. 541-545.

Klasner, J.S. and King, E.R., 1986. Precambrian basement geology of North and South Dakota; Canadian Journal of Earth Sciences, v. 23, p. 1083-1102.

Klasner, J.S. and King, E.R., 1990. A model of tectonic evolution of the Trans-Hudson orogen in North and South Dakota; *in* J.F. Lewry and M.R. Stauffer (eds.), The Early Proterozoic Trans-Hudson Orogen of North America; Geological Association of Canada, Special Paper 37, p. 271-286.

Klassen, R.W., 1989. Quaternary geology of the southern Interior Plains; *in* R.J. Fulton (ed.), Quaternary Geology of Canada and Greenland; Geological Society of America, The Geology of North America, v. K-1, p. 138-173.

Klein, G.deV. and Hsui, A.T., 1987. Origin of cratonic basins; Geology, v. 15, p. 1094-1098.

Klepacki, D.W. and Wheeler, J.O., 1985. Stratigraphic and structural relations of the Milford, Kaslo and Slocan Groups, Goat Range, Lardeau and Nelson map-area, British Columbia; *in* Current Research, Part A; Geological Survey of Canada, Paper 85-1A, p. 277-286.

Klovan, J.E., 1974. Development of Western Canada Devonian reefs and comparison with Holocene analogs; Bulletin of the American Association of Petroleum Geologists, v. 58, p. 787-799.

Kober, L., 1925. Die Gestaltungsgeschichte der Erde; Bornträger Verlag, Berlin, 200 p.

Kozlovsky, Ye.A. (ed.), 1984 (translated in 1987). The Superdeep Well of the Kola Peninsula; Springer-Verlag, 558 p.

Krasnyi, L.I. (ed.), 1980. Geology of the Baikal-Amur Railway region; Nedra Publishing, Moscow, 159 p. (in Russian).

Krause, F.F. and Burrowes, O.G. (eds.), 1987. Devonian Lithofacies and Reservoir Styles in Alberta; Canadian Society of Petroleum Geologists, 13th Core Conference.

Krogstad, E.J., 1995. The general failure of whole rock isochron approaches in upper amphibolite and granulite grade terranes (abs.); American Geophysical Union, 1995 Fall Meeting, San Francisco, Program; Supplement to EOS, p. F703.

Kröner, A. and Jaeckel, P., 1995. Dating the peak of high-temperature regional metamorphism by using metamorphic zircons (abs.); American Geophysical Union, 1995 Fall Meeting, San Francisco, Program; Supplement to EOS, p. F703.

Krumbein, W.C. and Sloss, L.L., 1951. Stratigraphy and Sedimentation; W.H. Freeman & Co., 497 p.

Kurtz, R.D., DeLaurier, J.M., and Gupta, J.C., 1986. A magnetotelluric sounding across Vancouver Island sees the subducting Juan de Fuca plate; Nature, v. 321, p. 596-599.

Kusky, T.M., 1989. Accretion of the Archean Slave province; Geology, v. 17, p. 63-67.

Kuvaas, B. and Kodaira, S., 1997. The formation of the Jan Mayen microcontinent: the missing piece in the continental puzzle between the Møre-Vøring Basins and East Greenland; First Break, v. 15, p. 239-247.

Labazin, G.S., 1963. On the Geologic and Metallogenic Development of Mobile Belts in the Earth's Crust; VSEGEI, Trudy, Leningrad (St. Petersburg), v. 85 (in Russian).

Lamerson, P.R., 1982. The Fossil basin and its relationship to the Absaroka thrust system, Wyoming and Utah; *in* R.B. Powers (ed.), Geologic Studies of the Cordilleran Thrust Belt; Rocky Mountain Association of Geologists, p. 817-830.

Langenberg, C.W., 1983. Polyphase Deformation in the Canadian Shield of Northeastern Alberta; Alberta Geological Survey, Bulletin 45, 33 p.

Langenberg, W. and Kalkreuth, W., 1991. Tectonic controls on regional coalification and vitrinite-reflectance anisotropy of Lower Cretaceous coals in the Alberta Foothills, Canada; Bulletin de la Société Géologique de France, v. 162, p. 375-383.

Langenberg, C.W., Kalkreuth, W., and Dawson, R., 1989. Influences of structural setting on coal rank and thickness in the Grande Cache area, Alberta, Canada; Geologie en Mijnbouw, v. 68, p. 241-252.

Law, J., 1971. Regional Devonian geology and oil and gas possibilities, upper Mackenzie River area; Bulletin of Canadian Petroleum Geology, v. 19, p. 437-486.

Lawton, D.C., Spratt, D.A., and Hopkins, J.C., 1994. Tectonic wedging beneath the Rocky Mountain foreland basin, Alberta, Canada; Geology, v. 22, p. 519-522.

Lebel, D., Langenberg, W., and Mountjoy, E.W., 1996. Structure of the central Canadian Cordilleran fold-and-thrust belt, Athabasca-Brazeau area, Alberta: a large, complex intercutaneous wedge; Bulletin of Canadian Petroleum Geology, v. 44, p. 282-298.

Leckie, D.A., Bhattacharya, J.P., Bloch, J., Gilboy, C.F., Norris, B., Campbell, R., Plint, G., Gilders, M., Holmstrom, G., Krause, F.F., Reinson, G.E., Safton, D., Sawicki, J., and Sawicki, O., 1994. Cretaceous Colorado/Alberta Group of the Western Canada Sedimentary Basin; in G.D. Mossop and I. Shetsen (comps.), Geological Atlas of the Western Canada Sedimentary Basin, Canadian Society of Petroleum Geologists & Alberta Research Council, p. 335-354.

Leclair, A.D., Lucas, S.B., Broome, H.J., Viljoen, D.W., and Weber, W., 1997. Regional mapping of Precambrian basement beneath Phanerozoic cover in southeastern Trans-Hudson Orogen, Manitoba and Saskatchewan; Canadian Journal of Earth Sciences, v. 34, p. 618-634.

Levi, B.G., 1997. Earth's upper mantle: how low can it flow?; Physics Today, v. 50/8, p. 17-20.

Levine, J.R., 1986. Deep burial of coal-bearing strata, Anthracite region, Pennsylvania: sedimentation or tectonics; Geology, v. 14, p. 384-387.

Levorsen, A.I., 1934. Relation of oil and gas pools to unconformities in the Mid-Continent region; in W.E. Wrather and F.H. Lahee (eds.), Problems of Petroleum Geology; American Association of Petroleum Geologists, p. 761-768.

Levorsen, A.I., 1943. Discovery thinking; Bulletin of the American Association of Petroleum Geologists, v. 27, p. 887-928.

Levorsen, A.I., 1960. Paleogeological Maps; W.H. Freeman & Co., 174 p.

Levorsen, A.I., 1967. Geology of Petroleum, 2nd edition; W.H. Freeman & Co., 724 p.

Lewry, J.F. and Collerson, K.D., 1990. The Trans-Hudson Orogen: extent, subdivision and problems; in J.F. Lewry and M.R. Stauffer (eds.), The Early Proterozoic Trans-Hudson Orogen of North America; Geological Association of Canada, Special Paper 37, p. 1-14.

Lewry, J.F., Collerson, K.D., Bickford, M.E., and Van Schmus, W.R., 1986. An evolutionary model of the western Churchill province and western margin of the Superior province in Canada and the north-central United States - discussion; Tectonophysics, v. 131, p. 183-188.

Lewry, J.F., Hajnal, Z., Green, A., Lucas, S.B., White, D., Stauffer, M.R., Ashton, K.E., Weber, W., and Clowes, R.M., 1994. Structure of a Paleoproterozoic continent-continent collision zone: a Lithoprobe seismic reflection profile across the Trans-Hudson Orogen, Canada; Tectonophysics, v. 232, p. 143-160.

Lewry, J.F., Lucas, S., Stern, R., Ansdell, K., and Ashton, K.E., 1996. Tectonic assembly and orogenic closure in the Trans-Hudson Orogen (abs.); Summary of Investigations 1996, Saskatchewan Geological Survey, Miscellaneous Report 96-4, p. 174.

Lewry, J.F., Macdonald, R., Livesey, C., Meyer, M., Van Schmus, W.R., and Bickford, M.E., 1987. U-Pb geochronology of accreted terranes in the Trans-Hudson orogen in northern Saskatchewan; in T.C. Pharaoh, R.D. Beckinsale, and D. Rickard (eds.), Geochemistry and Mineralization of Proterozoic Volcanic Suites; Geological Society (London), Special Publication 33, p. 147-166.

Lewry, J.F. and Sibbald, T.I.I., 1980. Thermotectonic evolution of the Churchill province in northern Saskatchewan; Tectonophysics, v. 68, p. 45-82.

Lewry, J.F. and Stauffer, M.R. (eds.), 1990. The Early Proterozoic Trans-Hudson Orogen of North America; Geological Association of Canada, Special Paper 37, 505 p.

Lewry, J.F., Thomas, M.D., Macdonald, R., and Chiarenzelli, J., 1990. Structural relations in accreted terranes of the Trans-Hudson Orogen, Saskatchewan: telescoping in a collisional regime?; in J.F. Lewry and M.R. Stauffer (eds.), The Early Proterozoic Trans-Hudson Orogen of North America; Geological Association of Canada Special Paper 37, p. 75-94.

Lickorish, W.H. and Simony, P.S., 1995. Evidence for late rifting of the Cordilleran margin outlined by stratigraphic division of the Lower Cambrian Gog Group, Rocky Mountain Main Ranges, British Columbia and Alberta; Canadian Journal of Earth Sciences, v. 32, p. 860-874.

Lidiak, E.G., 1971. Buried Precambrian rocks of South Dakota; Bulletin of the Geological Society of America, v. 82, p. 1411-1420.

Link, P.K., Christie-Blick, N., Devlin, W.J., Elston, D.P., Horodyski, R.J., Levy, M., Miller, J.M.G., Pearson, R.C., Prave, A., Stewart, J.H., Winston, D., Wright, L.A., and Wrucke, C.T., 1993. Middle and Late Proterozoic stratified rocks of the western U.S. Cordillera, Colorado Plateau, and Basin and Range Province; in J.C. Reed, Jr., M.E. Bickford, R.S. Houston, P.K. Link, D.W. Rankin, P.K. Sims, and W.R. Van Schmus (eds.), Precambrian: Conterminous U.S.; Geological Society of America, The Geology of North America, v. C-2, p. 463-595.

Lis, M.G. and Price, R.A., 1976. Large-scale block faulting during deposition of the Windermere Supergroup (Hadrynian) in southeastern British Columbia; in Current Research, Part A; Geological Survey of Canada, Paper 76-1A, p. 135-136.

Litak, R.K. and Hauser, E.C., 1992. The Bagdad reflection sequence as tabular mafic intrusions: evidence from seismic modeling of mapped exposures; Bulletin of the Geological Society of America, v. 104, p. 1315-1325.

Locke, A., Billingsley, P.R., and Blakemore, M.E., 1940. Sierra Nevada tectonic patterns; Bulletin of the Geological Society of America, v. 51, p. 513-539.

Lucas, S.B., Green, A., Hajnal, Z., White, D., Lewry, J., Ashton, K., Weber, W., and Clowes, R., 1993. Deep seismic profile across a Proterozoic collision zone: surprises at depth; Nature, v. 363, p. 339-342.

Ludwigsen, R., 1989. The Burgess shale: not in the shadow of the Cathedral escarpment; Geoscience Canada, v. 16, p. 51-59.

Lyatsky, H.V., 1993. Basement-controlled structure and evolution of the Queen Charlotte Basin, west coast of Canada; Tectonophysics, v. 228, p. 123-140.

Lyatsky, H.V., 1994a. Formation of non-compressional sedimentary basins on continental crust: limitation on modern models; Journal of Petroleum Geology, v. 17, p. 301-316.

Lyatsky, H.V., 1994b. Book review of 'Foreland Basins and Fold Belts', American Association of Petroleum Geologists, Memoir 55; Journal of Petroleum Geology, v. 17, p. 247-248.

Lyatsky, H.V., 1996. Continental-Crust Structures on the Continental Margin of Western North America; Lecture Notes in Earth Sciences 62, Springer-Verlag, 352 p.

Lyatsky, H.V., Dietrich, J.R., and Edwards, D.J., 1998. Analysis of Gravity and Magnetic Horizontal-Gradient Vector Data Over the Buried Trans-Hudson Orogen and Churchill-Superior Boundary Zone in Southern Saskatchewan and Manitoba; Geological Survey of Canada, Open File 3614, 34 p.

Lyatsky, H.V. and Haggart, J.W., 1993. Petroleum exploration model for the Queen Charlotte Basin, offshore British Columbia; Canadian Journal of Earth Sciences, v. 30, p. 918-927.

Lyatsky, H.V. and Lawton, D.C., 1988. Application of the surface reflection seismic method to shallow coal exploration in the Plains of Alberta; Canadian Journal of Exploration Geophysics, v. 24, p. 124-140.

Lyatsky, H.V. and Lawton, D.C., 1989. Reflection seismic study of a Lower Paleocene coal deposit, Wabamun, Alberta; in L.V. Hills (gen. ed.), Geophysical Atlas of Western Canadian Hydrocarbon Pools; Canadian Society of Exploration Geophysicists & Canadian Society of Petroleum Geologists, Calgary, p. 301-310.

Lyatsky, H.V., Thurston, J.B., Brown, R.J., and Lyatsky, V.B., 1992. Hydrocarbon-exploration applications of potential-field horizontal-gradient vector maps; Recorder, Canadian Society of Exploration Geophysicists, v. XVII, no. 9, p. 10-15.

Lyatsky, V.B., 1965. Towards a Common Geochronological Scale of Absolute Age of Geologic Formations (in Russian); Proceedings of Meeting "General Regularities of Geologic Phenomena", U.S.S.R. Geographic Society, Leningrad (St. Petersburg), p. 177-181.

Lyatsky, V.B., 1967. Earthquakes and Geotectonics; U.S.S.R. Geographical Society, Leningrad (St. Petersburg), 22 p. (in Russian).

Lyatsky, V.B., 1969. Shelves - State of the Art and Prospects for Research; U.S.S.R. Geographic Society, Leningrad (St. Petersburg), 130 p. (in Russian).

Lyatsky, V.B., 1974. Theoretical principles of geological exploration and mapping of continental shelves; in Mapping of Continental Shelves; U.S.S.R. Geographic Society, Leningrad (St. Petersburg), p. 31-51 (in Russian).

Lyatsky, V.B., 1978. The System Approach as a Methodological Basis for Theory and Practice in the Studies of Shelves; U.S.S.R. Academy of Sciences, Institute of Zoology, Leningrad (St. Petersburg), 26 p. (in Russian).

Lyatsky, V.B., 1988. "Depositional sequence mapping" as a technique to outline tectonics and hydrocarbon potential of continental margins: the limitations of seismic stratigraphy in the geological basin analysis; Northeastern Geology, v. 10, p. 134-150.

Lyatsky, V.B., Brown, R.J., and Lyatsky, H.V., 1990. The Use of Potential-Field Horizontal-Gradient Vector Data in Hydrocarbon Exploration; A.P. Holder (ed.), Lyatsky Geoscience Research and Consulting Ltd., Calgary, 26 p.

Lyatsky, V.B., Fong, G., and Ha, T., 1988. Petroleum Potential in Sedimentary Basins in North-Western China: Explanatory Notes to the Atlas "Petroleum Potential of Central Asia"; Calgary, Canada, 98 p.

Lyatsky, V.B. and Lyatsky, H.V., 1990. Integrated geological basin analysis as a method of hydrocarbon exploration on continental shelves; in 22nd Offshore Technology Conference, Houston, Proceedings, p. 237-242.

Lyell, C., 1830-33. Principles of Geology, v. 1-3; John Murray (London), 511 p.

MacDonald, R., Upton, B.G.J., Collerson, K.D., Hearn, B.C.J., and James, D., 1992. Potassic mafic lavas of the Bearpaw Mountains, Montana: mineralogy, chemistry and origin; Journal of Petrology, v. 33, p. 305-346.

MacDonald, D.E., 1987. Geology and Resource Potential of Phosphates in Alberta; Alberta Research Council, Earth Sciences Report 87-2, 65 p.

Macdonald, K.C., Scheirer, D.S., and Carbotte, S.M., 1991. Mid-ocean ridges: discontinuities, segments and giant cracks; Science, v. 253, p. 986-994.

MacIntyre, D.G., 1991. SEDEX - sedimentary-exhalative deposits; in McMillan, W.J. (ed.), Ore Deposits, Tectonics and Metallogeny in the Canadian Cordillera; British Columbia Ministry of Energy, Mines and Petroleum Resources, Paper 1991-4, p. 25-70.

Mack, G.M. and Jerzykiewicz, T., 1989. Provenance of post-Wapiabi sandstones and its implications for Campanian to Paleocene tectonic history of the southern Canadian Cordillera; Canadian Journal of Earth Sciences, v. 26, p. 665-676.

Macqueen, R.W. and Leckie, D.A. (eds.), 1992. Foreland Basins and Fold Belts; American Association of Petroleum Geologists, Memoir 55, 460 p.

Mandler, H.A.F. and Clowes, R.M., 1997a. Evidence for extensive tabular intrusions in the Precambrian shield of western Canada: a 160 km long sequence of bright reflections; Geology, v. 25, p. 271-274.

Mandler, H.A.F. and Clowes, R.M., 1997b. Strong and long: crustal bright reflectors give evidence for extensive Precambrian magmatism in western Canada; Newsletter of the Lithoprobe Seismic Processing Facility, University of Calgary, v. 10, no. 1, p. 45-46.

Mariano, J. and Hinze, W.J., 1994. Structural interpretation of the Midcontinent Rift in eastern Lake Superior from seismic reflection and potential-field studies; in R.A. Gibb, W.J. Hinze, and M.D. Thomas (eds.), Potential Field Signatures of Continental Rifts; the Great Lakes Region; Canadian Journal of Earth Sciences, v. 31, p. 619-628.

Marquis, G., Jones, A.G., and Hyndman, R.D., 1995. Coincident conductive and reflective middle and lower crust in southern British Columbia; Geophysical Journal International, v. 120, p. 111-131.

Marshak, S. and Paulsen, T., 1996. Mid-continent U.S. fault and fold zones: a legacy of Proterozoic intracratonic extensional tectonism?; Geology, v. 24, p. 151-154.

Martignole, J., 1986. Some questions about crustal thickening in the central part of the Grenville Province; in J.M. Moore, A. Davidson, and A.J. Baer (eds.), The Grenville Province; Geological Association of Canada, Special Paper 31, p. 327-339.

Masters, J.A. (ed.), 1984. Elmworth - Case Study of a Deep Basin Gas Field; American Association of Petroleum Geologists, Memoir 38, 316 p.

Mathews, W.H. (comp.), 1986. Physiography of the Canadian Cordillera; Geological Survey of Canada, Map 1701A, scale 1:5,000,000.

Maughan, E.K. and Perry, Jr., W.J., 1986. Lineaments and their tectonic implications in the Rocky Mountains and adjacent Plains regions; in J.A. Peterson (ed.), Paleotectonics and Sedimentation in the Rocky Mountain Region; American Association of Petroleum Geologists, Memoir 41, p. 41-53.

McCrossan, R.G. and Glaister, R.P. (eds.), 1964. Geological History of Western Canada; Alberta Society of Petroleum Geologists, Atlas, 232 p.

McDonough, M.R., McNicoll, V.J., and Schetselaar, E.M., 1995. Age and kinematics of crustal shortening and escape in a two-sided oblique-slip collisional and magmatic orogen, Paleoproterozoic Taltson magmatic zone, northeastern Alberta; in G.M. Ross (ed.), Alberta Basement Transects Workshop, Lithoprobe Report 47, p. 264-308.

McDonough, M.R. and Simony, P.S., 1988. Structural evolution of basement gneisses and Hadrynian cover, Bulldog Creek area, Rocky Mountains, British Columbia; Canadian Journal of Earth Sciences, v. 25, p. 1687-1702.

McMechan, M.E., 1987. Stratigraphy and Structure of the Mount Selwyn Area, Rocky Mountains, Northeastern British Columbia; Geological Survey of Canada, Paper 85-28.

McMechan, M.E., 1990. Upper Proterozoic to Middle Cambrian history of the Peace River Arch: evidence from the Rocky Mountains; in S.C. O'Connell and J.S. Bell (eds.), Geology of the Peace River Arch; Bulletin of Canadian Petroleum Geology, v. 38A, p. 36-44.

McMechan, M.E., 1991. Purcell Anticlinorium; *in* H. Gabrielse and C.J. Yorath (eds.), Geology of the Cordilleran Orogen in Canada; Geological Society of America, The Geology of North America, v. G-2, p. 628-630.

McMechan, M.E., 1997. Provenance and tectonic significance of Late Jurassic Monteith conglomerates, western Canadian Foreland Basin, northeastern British Columbia and northwestern Alberta (abs.); Canadian Society of Petroleum Geologists & Society of Exploration Paleontologists and Mineralogists, Calgary; Program with Abstracts, p. 189.

McMechan, M.E. and Price, R.A., 1982. Superimposed low-grade metamorphism in the Mount Fisher area, southeastern British Columbia - implications for the East Kootenay orogeny; Canadian Journal of Earth Sciences, v. 19, p. 476-489.

McMechan, M.E. and Thompson, R.I., 1989. Structural style and history of the Rocky Mountain Fold and Thrust Belt; *in* B.D. Ricketts (ed.), Western Canada Sedimentary Basin, A Case History; Canadian Society of Petroleum Geologists, Special Publication 30, p. 47-72.

McMechan, M.E. and Thompson, R.I., 1993. The Canadian Cordilleran fold-and-thrust belt south of 66°N and its influence on the Western Interior Basin; *in* W.G.E. Caldwell and E.G. Kauffman (eds.), Evolution of the Western Interior Basin; Geological Association of Canada, Special Paper 39, p. 73-90.

McMillan, W.J., 1991. Overview of the tectonic evolution and setting of mineral deposits in the Canadian Cordillera; *in* Ore Deposits, Tectonics and Metallogeny in the Canadian Cordillera; British Columbia Ministry of Energy, Mines and Petroleum Resources, Paper 1991-4, p. 5-24.

McNutt, M.K. and Menard, H.W., 1982. Constraints on yield strength in the oceanic lithosphere derived from observations of flexure; Geophysical Journal of the Royal Astronomical Society, v. 71, p. 363-394.

Meijer Drees, N.C., 1986. Evaporitic Deposits of Western Canada; Geological Survey of Canada, Paper 85-20, 118 p.

Meijer Drees, N.C., 1994. Devonian Elk Point Group of the Western Canada Sedimentary Basin; *in* G.D. Mossop and I. Shetsen (comps.), Geological Atlas of the Western Canada Sedimentary Basin; Canadian Society of Petroleum Geologists & Alberta Research Council, p. 129-147.

Meissner, R., 1989. Rupture, creep, lamellae and crocodiles: happenings in the continental crust; Terra Review, v. 1, p. 17-28.

Menning, M., 1988. Synopsis of numerical time scales 1917-1986; Episodes, v. 12, p. 3-5.

Meyer, M.T., Bickford, M.E., and Lewry, J.F., 1992. The Wathaman batholith: an Early Proterozoic continental arc in the Trans-Hudson orogenic belt, Canada; Bulletin of the Geological Society of America, v. 104, p. 1073-1085.

Miall, A.D., 1987. Epeirogeny: is it really orogeny or theology?; Geoscience Canada, v. 14, p. 126-129.

Milici, R.C. and de Witt, W., Jr., 1988. The Appalachian Basin; *in* L.L. Sloss (ed.), Sedimentary Cover - North American Craton; U.S.; Geological Society of America, The Geology of North America, v. D-2, p. 427-470.

Milkereit, B., Eaton, D., Wu, J., Salisbury, M., Berrer, E.K., and Morrison, G., 1996. Seismic imaging of massive sulfide deposits: Part II. Reflection seismic profiling; Economic Geology, v. 91, p. 829-834.

Miller, D.M., Nilsen, T.H., and Bilodeau, W.L., 1992. Late Cretaceous to early Eocene geologic evolution of the U.S. Cordillera; *in* B.C. Burchfiel, P.W. Lipman, and M.L. Zoback (eds.), The Cordilleran Orogen: Conterminous U.S.; Geological Society of America, The Geology of North America, v. G-3, p. 205-260.

Miller, K.C., Keller, R.G., Gridley, J.M., Luetgert, J.H., Mooney, W.D., and Thybo, H., 1997. Crustal structure along the west flank of the Cascades, western Washington; Journal of Geophysical Research, v. 102, p. 17,857-17,873.

Misra, K.S., Slaney, V.R., Graham, D., and Harris, J., 1991. Mapping of basement and other tectonic features using Seasat and Thematic Mapper in hydrocarbon-producing areas of the Western Sedimentary Basin of Canada; Canadian Journal of Remote Sensing, v. 17, p. 137-151.

Mitchum, Jr., R.M., Vail, P.R., and Sangree, J.B., 1977. The depositional sequence as a basic unit for stratigraphic analysis; *in* C.E. Payton (ed.), Seismic Stratigraphy - Applications to Hydrocarbon Exploration; American Association of Petroleum Geologists, Memoir 26, p. 53-63.

Mitrofanov, G.L. and Taskin, A.P., 1994. Structural relations of the Siberian platform with its folded frame; Geotectonics, v. 28, p. 1-13.

Molenaar, C.M. and Rice, D.D., 1988. Cretaceous rocks of the Western Interior Basin; *in* L.L. Sloss (ed.), Sedimentary Cover - North American Craton; U.S.; Geological Society of America, The Geology of North America, v. D-2, p. 77-82.

Mollard, J.D., 1988. First R.M. Hardy Memorial Lecture: fracture lineament research and applications on the Western Canadian Plains; Canadian Geotechnical Journal, v. 25, p. 749-767.

Monger, J.W.H., 1989. Overview of Cordilleran geology; *in* B.D. Ricketts (ed.), Western Canada Sedimentary Basin, A Case History; Canadian Society of Petroleum Geologists, Special Publication 30, p. 9-32.

Monger, J.W.H., 1993. Canadian Cordilleran tectonics: from geosynclines to crustal collage; Canadian Journal of Earth Sciences, v. 30, p. 209-231.

Monger, J.W.H. and Irving, E., 1980. Northward displacement of north-central British Columbia; Nature, v. 285, p. 289-294.

Monger, J.W.H. and Price, R.A., 1996. Comment on "Paleomagnetism of the Upper Cretaceous strata of Mount Tatlow: evidence for 3000 km of northward displacement of the eastern Coast Belt, British Columbia" by P.J. Wynne et al., and on "Paleomagnetism of the Spences Bridge Group and northward displacement of the Intermontane Belt, British Columbia: a second look" by E. Irving et al.; Journal of Geophysical Research, v. 101, p. 13,793-13,799.

Monger, J.W.H., Price, R.A., and Tempelman-Kluit, D., 1982. Tectonic accretion and the origin of the two major metamorphic and plutonic welts in the Canadian Cordillera; Geology, v. 10, p. 70-75.

Monger, J.W.H. and Ross, C.A., 1971. Distribution of fusilinaceans in the Canadian Cordillera; Canadian Journal of Earth Sciences, v. 8, p. 259-278.

Monger, J.W.H., Souther, J.G., and Gabrielse, H., 1972. Evolution of the Canadian Cordillera: a plate-tectonic model; American Journal of Science, v. 272, p. 577-602.

Mooney, W.D. and Christensen, N.I., 1994. Composition of the crust beneath the Kenya rift; Tectonophysics, v. 236, p. 391-408.

Moore, J.M., 1986. Introduction; The "Grenville" problem then and now; *in* J.M. Moore, A. Davidson, and A.J. Baer (eds.), The Grenville Province; Geological Association of Canada, Special Paper 31, p. 1-11.

Moore, J.M., Davidson, A., and Baer, A.J. (eds.), 1986. The Grenville Province; Geological Association of Canada, Special Paper 31, 348 p.

Moore, P.F., 1989. The Lower Kaskaskia Sequence - Devonian; *in* B.D. Ricketts (ed.), Western Canada Sedimentary Basin, A Case History; Canadian Society of Petroleum Geologists, Special Publication 30, p. 139-164.

Moore, P.F., 1993. Devonian; *in* D.F. Stott and J.D. Aitken (eds.), Sedimentary Cover of the Craton in Canada; Geological Society of America, The Geology of North America, v. D-1, p. 150-201.

Morgan, W.J., 1968. Rises, trenches, great faults and crustal blocks; Journal of Geophysical Research, v. 73, p. 1959-1982.

Morris, L.K., Lund, S.K., and Bottjer, D.J., 1986. Paleolatitude drift history of displaced terranes in southern and Baja California; Nature, v. 321, p. 844-847.

Morrow, D.W., 1991. The Silurian-Devonian sequence in the northern part of the Mackenzie Shelf, Northwest Territories; Geological Survey of Canada, Bulletin 413, 121 p.

Moskaleva, V.N., 1989. Petrological and mineralogical studies of crystalline rocks; VSEGEI, Leningrad (St. Petersburg), 152 p. (in Russian).

Mossop, G.D. and Shetsen, I. (comps.), 1994. Geological Atlas of the Western Canada Sedimentary Basin; Canadian Society of Petroleum Geologists & Alberta Research Council.

Mountjoy, E.W., 1978. Upper Devonian reef trends and configuration of the western portion of the Alberta Basin; *in* I.A. McIlreath and P.C. Jackson (eds.), The Fairholme Carbonate Complex at Hummingbird and Cripple Creek; Canadian Society of Petroleum Geologists, p. 1-30.

Mountjoy, E.W., 1980. Some questions about the development of Upper Devonian carbonate buildups (reefs), Western Canada; Bulletin of Canadian Petroleum Geology, v. 28, p. 315-340.

Muehlenbachs, K., Chacko, T., and Burwash, R.A., 1993. Oxygen isotope evidence for a metamorphic core complex in the Precambrian basement of Alberta (abs.); Geological Society of America, Abstracts with Program, p. A-80.

Muehlenberger, A.W., 1996. Tectonic Map of North America; American Association of Petroleum Geologists, scale 1:5,000,000.

Mueller, P.A., Shuster, R.D., Wooden, J.L., Erslev, E.A., and Bowes, D.R., 1993. Age and composition of Archean crystalline rocks from the southern Madison Range, Montana: implications for crustal evolution of the Wyoming craton; Bulletin of the Geological Society of America, v. 105, p. 437-446.

Murphy, D.C., Walker, R.T., and Parrish, R.R., 1991. Age and geological setting of Gold Creek gneiss, crystalline basement of the Windermere Supergroup, Cariboo Mountains, British Columbia; Canadian Journal of Earth Sciences, v. 28, p. 1217-1231.

Murphy, M.A., 1988. Unconformity-bounded stratigraphic units: discussion; Bulletin of the Geological Society of America, v. 100, p. 155.

Nelson, J., 1991. The tectonic setting of alkaline porphyry suite copper-gold deposits in the Canadian Cordillera (abs.); Canadian Institute of Mining, Metallurgy and Petroleum, 93rd Annual General Meeting, Vancouver, p. 57.

Nelson, K.D., Baird, D.J., Walters, J.J., Hauck, M., Brown, L.D., Oliver, J.E., Ahern, J.L., Hajnal, Z., Jones, A.G., and Sloss, L.L., 1993. Trans-Hudson orogen and Williston basin in Montana and North Dakota: new COCORP deep-profiling results; Geology, v. 21, p. 447-450.

Nelson, S.J. and Nelson, E.R., 1985. Allochthonous Permian micro- and macrofauna, Kamloops area, British Columbia; Canadian Journal of Earth Sciences, v. 22, p. 442-451.

Newton, C.R., 1988. Significance of the "Tethyan" fossils in the American Cordillera; Science, v. 242, p. 385-391.

Nielsen, P.A., Langenberg, C.W., Baadsgaard, H., and Godfrey, J.D., 1981. Precambrian metamorphic conditions and crustal evolution, northeastern Alberta, Canada; Precambrian Research, v. 16, p. 171-193.

Nilsen, T.H. and Stewart, J.H., 1980. The Antler Orogeny - mid-Paleozoic tectonism in western North America; Geology, v. 8, p. 298-302.

Nokleberg, W.J., Brew, D.A., Grybeck, Yeend, W., D., Bundtzen, T.K., Robinson, M.S., Smith, T.E., and Berg, H.C., with contributions by Andersen, G.L., Chipp, E.R., Gaard, D.R., Burton, P.J., Dunbier, J., Schrekenbach, D.A., Foley, J.Y., Thurow, G., Warner, J.D., Freeman, C.J., Gamble, B.M., Nelson, S.W., Schmidt, J.M., Hawley, C.C., Hitzman, M.W., Jones, B.K., Lange, I.M., Maars, C.D., Puchner, C.C., Steefel, C.I., Menzie, D.W., Metz, P.A., Modene, J.S., Plahuta, J.P., Young, L.E., Nauman, C.R., Newkirk, S.R., Newberry, R.J., Rogers, R.K., Rubin, C.M., Swainbank, R.C., Smith, P.R., and Stephens, J.E., 1994. Metallogeny and major mineral deposits of Alaska; in G. Plafker and H.C. Berg (eds.), Geological Society of America, The Geology of North America, v. G-1, p. 855-903.

Norford, B.S., 1990. Ordovician and Silurian stratigraphy, paleogeography and depositional history in the Peace River Arch area, Alberta and British Columbia; *in* S.C. O'Connell and J.S. Bell (eds.), Geology of the Peace River Arch; Bulletin of Canadian Petroleum Geology, v. 38A, p. 45-54.

Norford, B.S., Haidl, F.M., Bezys, R.K., Cecile, M.P., McCabe, H.R., and Paterson, D.F., 1994. Middle Ordovician to Lower Devonian strata of the Western Canada Sedimentary Basin; *in* G.D. Mossop and I. Shetsen (comps.), Geological Atlas of the Western Canada Sedimentary Basin; Canadian Society of Petroleum Geologists & Alberta Research Council, p. 109-127.

Norman, A.R., Williams, P.F., and Ansdell, K.M., 1995. Early Proterozoic deformation along the southern margin of the Kisseynew gneiss belt, Trans-Hudson Orogen, a 30 Ma progressive deformation cycle; Canadian Journal of Earth Sciences, v. 32, p. 875-894.

North American Commission on Stratigraphic Nomenclature, 1983. North American Stratigraphic Code; Bulletin of the American Association of Petroleum Geologists, v. 67, p. 841-875.

Nur, A., 1982. The origin of tensile fracture lineaments; Journal of Structural Geology, v. 4, p. 31-40.

Nurkowski, J.R., 1984. Coal quality, coal rank variation and its relation to reconstructed overburden, Upper Cretaceous and Tertiary plains coals, Alberta, Canada; Bulletin of the American Association of Petroleum Geologists, v. 68, p. 285-295.

Obradovich, J.D., Zartman, R.E., and Peterman, Z.E., 1984. Update on the geochronology of the Belt Supergroup; *in* W.S. Hobbs (ed.), The Belt; Montana Bureau of Mines and Geology, Special Publication 90, p. 82-84.

O'Connell, S.C. and Bell, J.S. (eds.), 1990. Geology of the Peace River Arch; Bulletin of Canadian Petroleum Geology, Special Volume; v. 38A, 281 p.

O'Connell, S.C., Dix, G.R., and Barclay, J.E., 1990. The origin, history, and regional structural development of the Peace River Arch, Western Canada; *in* S.C. O'Connell and J.S. Bell (eds.), Geology of the Peace River Arch; Bulletin of Canadian Petroleum Geology, v. 38A, p. 4-24.

O'Driscoll, E.S.T., 1982. Deep tectonic foundations of the Eromanga Basin; *in* B.R. Griffith (ed.), Developing for the Future; The APEA Journal, v. 23 (Part 1), p. 35-42.

Oldale, H.S. and Munday, R.J., 1994. Devonian Beaverhill Lake Group of the Western Canada Sedimentary Basin; *in* G.D. Mossop and I. Shetsen (comps.), Geological Atlas of the Western Canada Sedimentary Basin; Canadian Society of Petroleum Geologists & Alberta Research Council, p. 149-163.

Oldow, J.S., Bally, A.W., Avé Lallemant, H.G., and Leeman, W.P., 1989. Phanerozoic evolution of the North American Cordillera; United States and Canada; *in* A.W. Bally and A.R. Palmer (eds.), The Geology of North America - An Overview; Geological Society of America, The Geology of North America, v. A, p. 139-232.

O'Neill, J.M. and Lopez, D.A., 1985. Character and regional significance of Great Falls Tectonic Zone, east-central Idaho and west-central Montana; Bulletin of the American Association of Petroleum Geologists, v. 69, p. 437-447.

Oreskes, N., Shrader-Frechette, K., and Belitz, K., 1994. Verification, validation, and confirmation of numerical models in the earth sciences; Science, v. 263, p. 641-646.

Osadetz, K.G., 1989. Basin analysis applied to petroleum geology in western Canada; in B.D. Ricketts (ed.), Western Canada Sedimentary Basin, A Case History; Canadian Society of Petroleum Geologists, Special Publication 30, p. 287-306.

Osadetz, K.G. and Haidl, F.M., 1989. Tippecanoe sequence: Middle Ordovician to lowest Devonian: vestiges of a great epeiric sea; in B.D. Ricketts (ed.), Western Canada Sedimentary Basin: A Case Study; Canadian Society of Petroleum Geologists, Special Publication 30, p. 121-137.

Pakiser, L.C. and Mooney, W.D. (eds.), 1989. Geophysical Framework of the Continental United States; Geological Society of America, Memoir 172, 826 p.

Parrish, R.R., 1991. Precambrian basement rocks of the Canadian Cordillera; in H. Gabrielse and C.J. Yorath (eds.), Geology of the Cordilleran Orogen in Canada; Geological Society of America, The Geology of North America, v. G-2, p. 89-95.

Parrish, R.R., 1995. Thermal evolution of the southeastern Canadian Cordillera; Canadian Journal of Earth Sciences, v. 32, p. 1618-1642.

Parrish, R. and Armstrong, R.L., 1983. U-Pb zircon age and tectonic significance of gneisses in structural culminations of the Omineca Crystalline Belt, British Columbia (abs.); Geological Society of America, Abstracts with Program, p. 324.

Patton, Jr., W.W., Box, S.E., Moll-Stalcup, E.J., and Miller, T.P., 1994. Geology of west-central Alaska; in G. Plafker and H.C. Berg (eds.), The Geology of Alaska; Geological Society of America, The Geology of North America, v. G-1, p. 241-270.

Payton, C.E. (ed.), 1977. Seismic Stratigraphy - Applications to Hydrocarbon Exploration; American Association of Petroleum Geologists, Memoir 26, 516 p.

Pearson, N.J. and O'Reilly, S.Y., 1991. Thermobarometry and P-t paths: the granulite-to-eclogite transition in lower crustal xenoliths from eastern Australia; Journal of Metamorphic Geology, v. 9, p. 349-359.

Pell, J., 1987. Alkaline Ultrabasic Rocks in British Columbia: Carbonatites, Nepheline Syenites, Kimberlites, Ultramafic Lamprophyres and Related Rocks; British Columbia Geological Survey Branch, Open File 1987-17.

Pell, J., 1997. Kimberlites in the Slave craton, Northwest Territories, Canada; Geoscience Canada, v. 24, p. 77-91.

Peltier, W.R., 1985. New constraint on transient lower mantle rheology and internal mantle buoyancy from glacial rebound data; Nature, v. 318, p. 614-617.

Penner, L.A. and Mollard, J.D., 1991. Correlated photolineament and geoscience data on eight petroleum and potash study projects in southern Saskatchewan; Canadian Journal of Remote Sensing, v. 17, p. 174-184.

Penner, L.A., Mollard, J.D., and Hodgson, R.A., 1987. Remote Sensing for Petroleum Exploration and Exploitation in Saskatchewan; Saskatchewan Energy and Mines, Technical Report 5, 200 p.

Percival, J.A. and Card, K.D., 1983. Archean crust as revealed in the Kapuskasing uplift, Superior province, Canada; Geology, v. 11, p. 323-326.

Percival, J.A., Mortensen, J.K., Stern, R.A., Card, K.D., and Bégin, N.J., 1992. Giant granulite terranes of northeastern Superior Province: the Ashuanipi complex and Minto block; Canadian Journal of Earth Sciences, v. 29, p. 2287-2308.

Peterman, Z.E. and Futa, K., 1987. Is the Archean Wyoming Province exotic to the Superior craton? Evidence from Sm-Nd model ages of basement cores (abs.); Geological Society of America, Abstracts with Programs, v. 19, p. 803.

Peterman, Z.E. and Hildreth, R.A., 1978. Reconnaissance Geology and Geochronology of the Precambrian of the Granite Mountain, Wyoming; U.S. Geological Survey, Professional Paper 1055, 22 p.

Peterson, J.A., 1985. Regional stratigraphy and general petroleum geology of Montana and adjacent areas; in J.J. Tonnsen (ed.), Montana Oil and Gas Fields Symposium, 1985: Billings; Montana Geological Society, p. 5-45.

Peterson, J.A., 1986. General stratigraphy and regional paleotectonics of the Western Montana Overthrust Belt; in J.A. Peterson (ed.), Paleotectonics and Sedimentation in the Rocky Mountain Region, United States; American Association of Petroleum Geologists, Memoir 41, p. 57-86.

Peterson, J.A. (ed.), 1986. Paleotectonics and Sedimentation in the Rocky Mountain Region, United States; American Association of Petroleum Geologists, Memoir 41.

Peterson, J.A., 1988. Phanerozoic stratigraphy of the northern Rocky Mountain region; in L.L. Sloss (ed.), Sedimentary Cover - North American Craton; U.S.; Geological Society of America, The Geology of North America, v. D-2, p. 83-108.

Peterson, J.A. and Smith, D.L., 1986. Rocky Mountain paleogeography through geologic time; in J.A. Peterson (ed.), Paleotectonics and Sedimentation in the Rocky Mountain Region, United States; American Association of Petroleum Geologists, Memoir 41, p. 3-19.

Pilkington, M., 1989. Variable-depth magnetization mapping: application to the Athabasca basin, northern Alberta and Saskatchewan, Canada; Geophysics, v. 54, p. 1164-1173.

Pinet, C., Jaupart, C., Mareschal, J.C., Gariepy, C., Bienfait, G., and Lapointe, R., 1991. Heat flow and structure of the lithosphere in the eastern Canadian Shield; Journal of Geophysical Research, v. 96, p. 19,941-19,963.

Piskarev, A.L. and Tchernyshev, M.Yu., 1997. Magnetic and gravity anomaly patterns related to hydrocarbon fields in northern West Siberia; Geophysics, v. 62, p. 831-841.

Pitman, W., 1997. Response to citation; EOS, Transactions of the American Geophysical Union, v. 78, p. 122.

Plafker, G. and Berg, H.C. (eds.), 1994. The Geology of Alaska; Geological Society of America, The Geology of North America, v. G-1, 1055 p.

Plint, A.G., Hart, B.S., and Donaldson, W.S., 1993. Lithospheric flexure as a control on stratal geometry and facies distribution in Upper Cretaceous rocks of the Alberta Foreland Basin; Basin Research, v. 5, p. 69-77.

Podruski, J.A., Barclay, J.E., Hamblin, A.P., Lee, P.J., Osadetz, K.G., Procter, R.M., and Taylor, G.C., 1988. Conventional Oil Resources of Western Canada (Light and Medium). Part 1: Resource Endowment; in Geological Survey of Canada, Paper 87-26, p. 1-125.

Poole, F.G., Stewart, J.H., Palmer, A.R., Sandberg, C.A., Madrid, R.J., Ross, R.J., Jr., Hintze, L.F., Miller, M.M., and Wrucke, C.T., 1992. Latest Precambrian to latest Devonian time; development of a continental margin; in B.C. Burchfiel, P.W. Lipman, and M.L. Zoback (eds.), The Cordilleran Orogen: Conterminous U.S.; Geological Society of America, The Geology of North America, v. G-3, p. 9-56.

Poulton, T.P., 1989. Upper Absaroka to Lower Zuni; the transition to the Foreland Basin; in B.D. Ricketts (ed.), Western Canada Sedimentary Basin: A Case History; Canadian Society of Petroleum Geologists, Special Publication 30, p. 233-247.

Poulton, T.P., Braun, W.K., Brooke, M.M., and Davies, E.H., 1993. Jurassic; in D.F. Stott and J.D. Aitken (eds.), Sedimentary Cover of the Craton in Canada; Geological Society of America, The Geology of North America, v. D-1, p. 321-357.

Poulton, T.P., Christopher, J.E., Hayes, B.J.R., Losert, J., Tittemore, J., and Gilchrist, R.D., 1994. Jurassic and lowermost Cretaceous strata of the Western Canada Sedimentary Basin; in G.D. Mossop and I. Shetsen (comps.), Geological Atlas of the Western Canada Sedimentary Basin; Canadian Society of Petroleum Geologists & Alberta Research Council, p. 297-316.

Powell, C.McA., Dalziel, I.W.D., Li, Z.X., and McElhinny, M.W., 1995. Did *Pannonia*, the Latest Neoproterozoic southern supercontinent, really exist? (abs.); American Geophysical Union, 1995 Fall Meeting, San Francisco, Program; Supplement to EOS, p. F577.

Powell, C.McA., Li, Z.X., McElhinny, M.W., Meert, J.G., and Park, J.K., 1993. Paleomagnetic constraints on timing of the Neoproterozoic breakup of Rodinia and the Cambrian formation of Gondwana; Geology, v. 21, p. 889-892.

Powell, D., Andersen, T.B., Drake, A.A., Jr., Hall, L., and Keppie, J.D., 1988. The age and distribution of basement rocks in the Caledonide orogen of the N Atlantic; in A.L. Harris and D.J. Fettes (eds.), The Caledonian-Appalachian Orogen; Geological Society (London), Special Publication 38, p. 63-74.

Press, F. and Siever, R., 1978. Earth, second edition; Freeman, 649 p.

Price, R.A., 1973. Large-scale gravitational flow of supracrustal rocks, southern Canadian Rockies; in K.A. De Jong and R. Scholten (eds.), Gravity and Tectonics; Wiley, p. 491-502.

Price, R.A., 1994. Cordilleran tectonics and evolution of the Western Canada sedimentary basin; in G.D. Mossop and I. Shetsen (comps.), Geological Atlas of Western Canada; Canadian Society of Petroleum Geologists & Alberta Research Council, p. 13-24.

Price, R.A. and Douglas, R.J.W. (eds.), 1972. Variations in Tectonic Styles in Canada; Geological Association of Canada; Special Paper 11, 688 p.

Pugh, D.C., 1973. Subsurface Lower Paleozoic Stratigraphy in Northern and Central Alberta; Geological Survey of Canada, Paper 72-12, 54 p.

Pugh, D.C., 1975. Cambrian Stratigraphy from Western Alberta to Northeastern British Columbia; Geological Survey of Canada, Paper 74-37, 31 p.

Ramaekers, P., 1981. Hudsonian and Helikian basins of the Athabasca region, northern Saskatchewan; in F.H.A. Campbell (ed.), Proterozoic Basins of Canada; Geological Survey of Canada, Paper 81-10, p. 219-233.

Rankin, D.W., Chiarenzelli, J.R., Drake, Jr., A.A., Goldsmith, R., Hall, L.M., Hinze, W.J., Isachsen, Y.W., Lidiak, E.G., McLelland, J., Mosher, S., Ratcliffe, N.M., Secor, Jr., D.T., and Whitney, P.R., 1993. Proterozoic rocks east and southeast of the Grenville front; in J.C. Reed, Jr., M.E. Bickford, R.S. Houston, P.K. Link, D.W. Rankin, P.K. Sims, and W.R. Van Schmus (eds.), Precambrian: Conterminous U.S.; Geological Society of America, The Geology of North America, v. C-2, p. 335-461.

Rast, N., 1969. Orogenic belts and their parts; in P.E. Kent, G.E. Satterthwaite, and A.M. Spencer (eds.), Time and Place in Orogeny; Geological Society (London), Special Publication 3, p. 197-213.

Read, P.B. and Wheeler, J.O., 1976. Geology of the Lardeau west-half map area, British Columbia; Geological Survey of Canada, Open-File Map 432, scale 1:125,000.

Redden, J.A., Peterman, Z.E., Zartman, R.E., and DeWitt, E., 1990. U-Th-Pb geochronology and preliminary interpretation of Precambrian tectonic events in the Black Hills, South Dakota; in J.F. Lewry and M.R. Stauffer(eds.), The Early Proterozoic Trans-Hudson Orogen of North America; Geological Association of Canada, Special Paper 37, p. 229-251.

Redly, P. and Hajnal, Z., 1997. Interpretation of regional seismic sections from the Williston Basin - a "Slossian" approach (abs.); Canadian Society of Exploration Geophysicists and Society of Exploration Paleontologists and Mineralogists, Joint Convention, Calgary; Program with Abstracts, p. 235.

Reed, J.C., Jr., Ball, T.T., Farmer, G.L., and Hamilton, W.B., 1993. A broader view; in J.C. Reed, Jr., M.E. Bickford, R.S. Houston, P.K. Link, D.W. Rankin, P.K. Sims, and W.R. Van Schmus (eds.), Precambrian: Conterminous U.S.; Geological Society of America, The Geology of North America, v. C-2, p. 597-636.

Reed, J.C., Jr., Bickford, M.E., Houston, R.S., Link, P.K., Rankin, D.W., Sims, P.K., and Van Schmus, W.R. (eds.), 1993. Precambrian: Conterminous U.S.; Geological Society of America, The Geology of North America, v. C-2, 657 p.

Reed, J.C., Jr. and Harrison, J.E., 1993. Introduction; in J.C. Reed, Jr., M.E. Bickford, R.S. Houston, P.K. Link, D.W. Rankin, P.K. Sims, and W.R. Van Schmus (eds.), Precambrian: Conterminous U.S.; Geological Society of America, The Geology of North America, v. C-2, p. 1-

Reesor, J.E., 1984. The Purcell Supergroup in the Purcell Mountains, British Columbia; *in* S.W. Hobbs (ed.), The Belt; Montana Bureau of Mines and Geology, Special Publication 90, p. 33-35.

Revelle, R. and Maxwell, A.E., with a discussion by E.C. Bullard, 1952. Heat flow through the floor of the North Pacific Ocean; Nature, v. 170, p. 199-200.

Reynolds, M.W., 1984. Tectonic setting and development of the Belt Basin, northwestern United States; *in* S.W. Hobbs (ed.), The Belt; Montana Bureau of Mines and Geology, Special Publication 90, p. 44-46.

Rich, J.L., 1951. Origin of compressional mountains and associated phenomena; Bulletin of the Geological Society of America, v. 62, p. 1179-1222.

Richards, B.C., 1989. Upper Kaskaskia Sequence: uppermost Devonian and Lower Carboniferous; *in* B.D. Ricketts (ed.), Western Canada Sedimentary Basin, A Case Study; Canadian Society of Petroleum Geologists, Special Publication 30, p. 165-201.

Richards, B.C., Bamber, E.W., Higgins, A.C., and Utting, J., 1993. Carboniferous; *in* D.F. Stott and J.D. Aitken (eds.), Sedimentary Cover of the Craton in Canada; Geological Society of America, The Geology of North America, v. D-1, p. 202-271.

Richards, B.C., Barclay, J.E., Bryan, D., Hartling, A., Henderson, C.M., and Hinds, R.C., 1994. Carboniferous strata of the Western Canada Sedimentary Basin; *in* G.D. Mossop and I. Shetsen (comps.), Geological Atlas of the Western Canada Sedimentary Basin; Canadian Society of Petroleum Geologists & Alberta Research Council, p. 221-250.

Ricketts, B.D., 1989. Basin framework; *in* B.D. Ricketts (ed.), Western Canada Sedimentary Basin, A Case History; Canadian Society of Petroleum Geologists, Special Publication 30, p. 1-8.

Ricketts, B.D. (ed.), 1989. Western Canada Sedimentary Basin, A Case History; Canadian Society of Petroleum Geologists, Special Publication 30.

Roberts, M.R. and Finger, F., 1997. Do U-Pb zircon ages from granulites reflect peak metamorphic conditions?; Geology, v. 25, p. 319-322.

Roberts, R.J., 1949. Structure and stratigraphy of the Antler Peak quadrangle, north-central Nevada (abs.); Bulletin of the Geological Society of America, v. 60, p. 1917.

Roberts, R.J., 1964. Stratigraphy and Structure of the Antler Peak Quadrangle, Humbolt and Lander Counties, Nevada; U.S. Geological Survey, Professional Paper 459-A, 93 p.

Robson, R.D., 1980. The Sparky sand trend and its performance in the Dulwich-Silverdale area of west-central Saskatchewan; *in* L.S. Beck, J.E. Christopher, and D.M. Kent (eds.), Lloydminster and Beyond: Geology of Mannville Hydrocarbon Reservoirs; Saskatchewan Geological Society, Special Publication 5, p. 177-196.

Rodgers, J., 1987. Chains of basement uplifts within cratons marginal to orogenic belts; American Journal of Science, v. 287, p. 661-692.

Rodgers, J., 1995. Lines of basement uplifts within the external parts of orogenic belts; American Journal of Science, v. 295, p. 455-487.

Root, K.G., 1987. Geology of the Delphine Creek Area, Southeastern British Columbia; Ph.D. thesis, Dept. of Geology & Geophysics, University of Calgary, 446 p.

Ross, G.M., 1991. Tectonic setting of the Windermere Supergroup revisited; Geology, v. 19, p. 1125-1128.

Ross, G.M., 1997. Assembly of the southwestern Laurentian craton; in G.M. Ross (comp.), Alberta Basement Transects Workshop; Lithoprobe Report 59, p. 23-34.

Ross, G.M. and Eaton, D.W., 1997. The Winagami reflector sequence: seismic evidence for post-collisional magmatism in the Proterozoic of western Canada; Geology, v. 25, p. 199-203.

Ross, G.M., Eaton, D.W., Boerner, D.E., and Clowes, R.M., 1997. Geologists probe buried craton in western Canada; EOS, Transactions of the American Geophysical Union, v. 78, p. 493-497.

Ross, G.M., Milkereit, B., Eaton, D., White, D., Kanasewich, E.R., and Burianyk, M.J.A., 1995. Paleoproterozoic collisional orogen beneath the western Canada sedimentary basin imaged by Lithoprobe crustal seismic-reflection data; Geology, v. 23, p. 195-199.

Ross, G.M., Parrish, R.R., Bowring, S.A., and Tankard, A.J., 1988. Tectonics of the Canadian Shield in the Alberta subsurface (abs.); Geological Association of Canada, Abstracts with Program, v. 13, p. A106.

Ross, G.M., Parrish, R.R., Villeneuve, M.E., and Bowring, S.A., 1991. Geophysics and geochronology of the crystalline basement of the Alberta Basin, western Canada; Canadian Journal of Earth Sciences, v. 28, p. 512-522.

Ross, G. and Stephenson, R.A., 1989. Crystalline basement: the foundations of the Western Canada Sedimentary Basin; in B.D. Ricketts (ed.), Western Canada Sedimentary Basin: A Case History; Canadian Society of Petroleum Geologists, Special Publication 30, p. 33-45.

Rudnick, R.L., 1995. Making continental crust; Nature, v. 378, p. 571-578.

Rudnick, R.L. and Fountain, D.M., 1995. Nature and composition of the continental crust: a lower crustal perspective; Reviews of Geophysics, v. 33, p. 267-309.

Ruzicka, V., 1987. Monometallic and polymetallic deposits associated with the sub-Athabascan unconformity in Saskatchewan; in Current Research, Part C; Geological Survey of Canada, Paper 89-1C, p. 67-79.

Salvador, A. (chmn.), 1987. Unconformity-bounded stratigraphic sequences; Geological Society of America, v. 98, p. 232-237.

Salvador, A. (chmn.), 1988. Unconformity-bounded stratigraphic sequences: reply; Geological Society of America, v. 100, p. 156.

Salvador, A. (ed.), 1994. International Stratigraphic Guide; a Guide to Stratigraphic Classification, Terminology, and Procedure; International Union of Geological Sciences & Geological Society of America, 214 p.

Sando, W.J., Bamber, E.W., and Richards, B.C., 1990. The rugose coral *Ankhelasma* - index to Viséan (Lower Carboniferous) shelf margin in the western Interior of North America; *in* Shorter Contributions to Paleontology and Stratigraphy; U.S. Geological Survey, Bulletin 1895, p. B1-B29.

Sarwar, G. and Friedman, G.M., 1995. Post-Devonian Sediment Cover over New York State; Lecture Notes in Earth Sciences, Springer-Verlag, Heidelberg, 113 p.

Saskatchewan Geological Survey, 1994. Geology and Mineral Resources of Saskatchewan; Saskatchewan Energy and Mines, Miscellaneous Report 94-6, 99 p.

Schermer, E.R., Howell, D.G., and Jones, D.L., 1984. The origin of allochthonous terranes; perspectives on the growth and shaping of continents; Annual Reviews of Earth and Planetary Sciences, v. 12, p. 107-131.

Schieber, J., 1989. The origin of the Neihart Quartzite, a basal deposit of the mid-Proterozoic Belt Supergroup, Montana, U.S.A.; Geological Magazine, v. 126, p. 271-281.

Schultz, S.S., 1964. Activated Zones of the Earth's Crust; Moscow (in Russian).

Schulz, K.J., Sims, P.K., and Morey, G.B., 1993. Tectonic synthesis; *in* J.C. Reed, Jr., M.E. Bickford, R.S. Houston, P.K. Link, D.W. Rankin, P.K. Sims, and W.R. Van Schmus (eds.), Precambrian: Conterminous U.S.; Geological Society of America, The Geology of North America, v. C-2, p. 60-64.

Scotese, C.R., van der Voo, R., and Barrett, S.F., 1985. Silurian and Devonian base maps; Philosophical Transactions of the Royal Society of London, v. B309, p. 57-77.

Scotese, C.R., 1997. Paleogeographic Atlas; PALEOMAP Progress Report 90-0497, University of Texas at Arlington, unpaginated.

Sears, J.W. and Price, R.A., 1978. The Siberian connection - a case for Precambrian separation of the North American and Siberian platforms; Geology, v. 6, p. 267-270.

Serpa, L., Setzer, T., Farmer, H., Brown, L., Oliver, J., Kaufman, S., Sharp, J., and Steeples, D., 1984. Structure of the souther Keweenawan rift from COCORP surveys across the Midcontinent Geophysical Anomaly in northeastern Kansas; Tectonics, v. 3, p. 367-384.

Shanmugam, G., 1997. The Bouma Sequence and the turbidite mind set; Earth-Science Reviews, v. 42, p. 201-229.

Shanmugam, G., Bloch, R.B., Mitchell, S.M., Beamish, G.W.J., Hodgkinson, R.J., Damuth, J.E., Straume, T., Syvertsen, S.E., and Shields, K.E., 1995. Basin-floor fans in the North Sea: sequence stratigraphic models vs. sedimentary facies; Bulletin of the American Association of Petroleum Geologists, v. 79, p. 477-512.

Shanmugam, G., Bloch, R.B., Damuth, J.E., and Hodgkinson, R.J., 1997. Basin-floor fans in the North Sea: sequence stratigraphic models vs. sedimentary facies; reply; Bulletin of the American Association of Petroleum Geologists, v. 81, p. 666-672.

Shanmugam, G. and Moiola, R.J., 1995. Reinterpretation of depositional processes in a classic flysch sequence (Pennsylvanian Jackfork Group), Ouachita Mountains, Arkansas and Oklahoma; Bulletin of the American Association of Petroleum Geologists, v. 79, p. 263-279.

Shanmugam, G., and Moiola, R.J., 1997. Reinterpretation of depositional processes in a classic flysch sequence (Pennsylvanian Jackfork Group), Ouachita Mountains, Arkansas and Oklahoma: reply; Bulletin of the American Association of Petroleum Geologists, v. 81, p. 476-491.

Sharpton, V.L., Grieve, R.A.F., Thomas, M.D., and Halpenny, J.F., 1987. Horizontal gravity gradient: an aid to the definition of crustal structure in North America; Geophysical Research Letters, v. 14, p. 808-811.

Shatsky, N.S. (ed.-in-chief), 1956. Tectonic Map of the U.S.S.R. and Adjacent Countries, scale 1:5,000,000; Gosgeoltechizdat Publishing, Moscow.

Sherrod, D.R. and Smith, J.G., 1990. Quaternary extrusion rates of the Cascade Range, north-western United States and southern British Columbia; Journal of Geophysical Research, v. 95, p. 19,465-19,474.

Shurr, G.W., 1994. Landsat mapping of tectonic blocks that show strike slip, Northern Great Plains, U.S.A. (abs.); Annual Meeting Abstracts, American Association of Petroleum Geologists & SEPM (Society of Economic Paleontologists and Mineralogists), p. 258.

Sikabonyi, L.A. and Rodgers, W.J., 1959. Paleozoic tectonics and sedimentation in the northern half of the West Canadian Basin; Journal of the Alberta Society of Petroleum Geologists, v. 7, p. 193-216.

Silver, P.G. and Chan, W.W., 1988. Implications for continental structure and evolution from seismic anisotropy; Nature, v. 35, p. 34-39.

Simony, P.S., 1995. Core complexes: lessons from little sheared margins of the Shuswap Complex (abs.); Geological Association of Canada/Mineralogical Association of Canada, Annual Meeting, Victoria; Final Program and Abstracts, p. A-98.

Simpson, R.W., Jachens, R.C., Blakeley, R.J., and Saltus, R.W., 1986. A new isostatic residual gravity map of the conterminous United States with a discussion of the significance of isostatic residual anomalies; Journal of Geophysical Research, v. 91, p. 8348-8372.

Sims, P.K. and Peterman, Z.E., 1986. Early Proterozoic Central Plains orogen: a major buried structure in the north-central United States; Geology, v. 14, p. 488-491.

Sims, P.K., Anderson, J.L., Bauer, R.L., Chandler, V.W., Hanson, G.N., Kalliokoski, J., Morey, G.B., Mudrey, Jr., M.G., Ojakangas, R.W., Peterman, Z.E., Schulz, K.J., Shirey, K.J., Smith, E.I., Southwick, D.L., Van Schmus, W.R., and Weiblen, P.W., 1993. The Lake Superior region and Trans-Hudson orogen; in J.C. Reed, Jr., M.E. Bickford, R.S. Houston, P.K. Link, D.W. Rankin, P.K. Sims, and W.R. Van Schmus (eds.), Precambrian: Conterminous U.S.; Geological Society of America, The Geology of North America, v. C-2, p. 11-120.

Sinclair, H.D., 1997. Tectonostratigraphic model for underfilled peripheral foreland basins: an Alpine perspective; Bulletin of the Geological Society of America, v. 109, p. 324-346.

Slatt, R.M., Weimer, P., and Stone, C.G.; Lowe, D.R.; Coleman, Jr., J.L.; Bouma, A.H., DeVries, M.B., and Stone, C.G.; D'Agostino, A.E. and Jordan, D.W., 1997. Reinterpretation of depositional processes in a classic flysch sequence (Pennsylvanian Jackfork Group), Ouachita Mountains, Arkansas and Oklahoma: discussions; Bulletin of the American Association of Petroleum Geologists, v. 81, p. 449-476.

Slind, O.L., Andrews, G.D., Murray, D.L., Norford, B.S., Paterson, D.F., Salas, C.J., and Tawadros, E.E., 1994. Middle Cambrian to Lower Ordovician strata of the Western Canada Sedimentary Basin; in G.D. Mossop and I. Shetsen (comps.), Geological Atlas of the Western Canada Sedimentary Basin; Canadian Society of Petroleum Geologists & Alberta Research Council, Calgary, p. 87-108.

Sloss, L.L., 1963. Sequences in the cratonic interior of North America; Bulletin of the Geological Society of America, v. 74, p. 93-114.

Sloss, L.L., 1988a. Tectonic evolution of the craton in Phanerozoic time; in L.L. Sloss (ed.), Sedimentary Cover - North American Craton: U.S.; Geological Society of America, The Geology of North America, v. D-2, p. 25-51.

Sloss, L.L., 1988b. Forty years of sequence stratigraphy; Bulletin of the Geological Society of America, v. 100, p. 1661-1665.

Sloss, L.L. (ed.), 1988. Sedimentary Cover - North American Craton: U.S.; Geological Society of America, The Geology of North America, v. D-2.

Sloss, L.L., 1990. Tectonics - the primary control on sequence stratigraphy: a countervailing view (abs.); Reservoir (Canadian Society of Petroleum Geologists), v. 17, no. 11, p. 1.

Sloss, L.L., Krumbein, W.C., and Dapples, E.C., 1949. Integrated facies analysis; in C.R. Longwell (chmn.), Sedimentary Facies in Geologic History; Geological Society of America, Memoir 39, p. 91-124.

Smith, G.G., 1989. Coal formation and resources in the Foreland Basin; in B.D. Ricketts (ed.), Western Canada Sedimentary Basin, A Case History; Canadian Society of Petroleum Geologists, Special Publication 30, p. 307-320.

Smith, G.G., Cameron, A.R., and Bustin, R.M., 1994. Coal resources of the Western Canada Sedimentary Basin; in G.D. Mossop and I. Shetsen (comps.), Geological Atlas of the Western Canada Sedimentary Basin; Canadian Society of Petroleum Geologists & Alberta Research Council, p. 471-482.

Smith, M.T. and Gehrels, G.E., 1991. Detrital zircon geochronology of Upper Proterozoic to lower Paleozoic continental margin strata of the Kootenay Arc: implications for the early Paleozoic tectonic development of the eastern Canadian Cordillera; Canadian Journal of Earth Sciences, v. 28, p. 1271-1284.

Snowdon, L.R., 1989. Organic matter properties and thermal evolution; in I.E. Hutcheon (ed.), Short Course in Burial Diagenesis; Mineralogical Association of Canada, Short Course Handbook, v. 15, p. 39-60.

Snyder, G.A., Taylor, L.A., Crozaz, G., Halliday, A.N., Beard, B.L., Sobolev, V.N., and Sobolev, N.V., 1997. The origins of Yakutian eclogite xenoliths; Journal of Petrology, v. 38, p. 85-113.

Sobczak, L.W. and Halpenny, J.F., 1990. Isostatic and Enhanced Isostatic Gravity Anomaly Maps of the Arctic; Geological Survey of Canada, Paper 89-16, 9 p.

Sonder, R.A., 1947. Shear patterns of the Earth's crust - a discussion; Transactions of the American Geophysical Union, v. 28, p. 939-945.

Speed, R.C. and Sleep, N.H., 1982. Antler orogeny and foreland basin: a model; Bulletin of the Geological Society of America, v. 93, p. 815-828.

Spizharsky, T.N., 1973. Overview Tectonic Maps of the U.S.S.R. (in Russian); Nedra Publishers, Leningrad (St. Petersburg), 240 p.

Sprenke, K.F., Wavra, C.S., and Godfrey, J.D., 1986. Geophysical Expression of the Canadian Shield of Northeastern Alberta; Alberta Geological Survey, Bulletin 52, 54 p.

Stauffer, M.R., 1984. Manikewan: an Early Proterozoic ocean in central Canada, its igneous history and orogenic closure; Precambrian Research, v. 25, p. 257-281.

Stauffer, M.R., 1990. The Missi Formation: an Aphebian molasse deposit in the Reindeer Lake Zone of the Trans-Hudson Orogen, Canada; in J.F. Lewry and M.R. Stauffer(eds.), The Early Proterozoic Trans-Hudson Orogen in North America; Geological Association of Canada, Special Paper 37, p. 121-141.

Stauffer, M.R. and Gendzwill, D.J., 1987. Fractures in the Northern Plains, stream patterns, and the midcontinent stress field; Canadian Journal of Earth Sciences, v. 24, p. 1086-1097.

Stephenson, R.A., Zelt, C.A., Ellis, R.M., Hajnal, Z., Morel-a-l'Huissier, P., Mereu, R.F., Northey, D.J., West, G.F., and Kanasewich, E.R., 1989. Crust and upper mantle structure and the origin of the Peace River Arch; Bulletin of Canadian Petroleum Geology, v. 37, p. 224-235.

Stern, R.A., Syme, E.C., Bailes, A.H., and Lucas, S.B., 1995. Palcoproterozoic (1.90-1.86 Ga) arc volcanism in the Flin Flon Belt, Trans-Hudson Orogen, Canada; Contributions to mineralogy and Petrology, v. 119, p. 117-141.

Stewart, J.H., 1972. Initial deposits of the Cordilleran geosyncline: evidence of Late Precambrian (<850my) continental separation; Bulletin of the Geological Society of America, v. 83, p. 1345-1360.

Stille, H., 1924. Grundfragen der Vergleichenden Tektonik; Gebruder Bornträger, Berlin.

Stille, H., 1936a. The present tectonic state of the Earth; Bulletin of the American Association of Petroleum Geologists, v. 20, p. 848-880.

Stille, H., 1936b. Wege und Ergebnisse der geologisch-tectonisch Forschung; Wissenschaftlich Verhandlung, Gesellschaft 25. Jahr Kaiser Wilhelm, v. 2, p. 84-85.

Stille, H., 1941. Einführung in den Bau Amerikas; Bornträger, Berlin.

Stoakes, F.A. and Wendte, J.C., 1987. The Woodbend Group; in F.F. Krause and O.G. Burrowes (eds.), Devonian Lithofacies and Reservoir Styles in Alberta; 13th Canadian Society of Petroleum Geologists Core Conference and Display, Second International Symposium on the Devonian System, Canadian Society of Petroleum Geologists, p. 153-170.

Stock, J.M. and Molnar, P., 1988. Uncertainties and implications of the Late Cretaceous and Tertiary position of North America relative to the Farallon, Kula, and Pacific plates; Tectonics, v. 6, p. 1339-1384.

Stockwell, C.H., 1961. Structural Provinces, Orogenies, and Time-Classification of Rocks of the Canadian Precambrian Shield; *in* J.A. Lowdon, Age Determinations of the Geological Survey of Canada; Geological Survey of Canada, Paper 61-17, p. 108-118.

Stockwell, C.H., 1964. Fourth report on structural provinces, orogenies and time-classification of rocks of the Canadian Precambrian Shield; Geological Survey of Canada, Paper 64-17.

Stockwell, C.H., 1966. Notes on the tectonic map of the Canadian Shield; *in* Scientific communications read to the Commission for the Geological Map of the World, p. 33-40.

Stockwell, C.H., McGlynn, J.C., Emslie, R.F., Sanford, B.V., Norris, A.W., Donaldson, J.A., Fahrig, W.F., and Currie, K.L., 1970. Geology of the Canadian Shield; *in* R.J.W. Douglas (ed.), Geology and Economic Minerals of Canada; Geological Survey of Canada, Economic Geology Report No. 1, p. 45-150.

Stott, D.F., 1984. Cretaceous sequences of the Foothills of the Canadian Rocky Mountains; *in* D.F. Stott and D.J. Glass (eds.), The Mesozoic of Middle North America; Canadian Society of Petroleum Geologists, Memoir 9, p. 85-107.

Stott, D.F., Caldwell, W.G.E., Cant, D.J., Christopher, J.E., Dixon, J., Koster, E.H., McNeil, D.H., and Simpson, F., 1993. Cretaceous; *in* D.F. Stott and J.D. Aitken (eds.), Sedimentary Cover of the Craton in Canada; Geological Society of America, The Geology of North America, v. D-1, p. 358-438.

Stott, D.F., Yorath, C.J., and Dixon, J., 1991. The Foreland Belt; *in* H. Gabrielse and C.J. Yorath (eds.), Geology of the Cordilleran Orogen in Canada; Geological Society of America, The Geology of North America, v. G-2, p. 335-345.

Stott, D.F. and Aitken, J.D. (eds.), 1993. Sedimentary Cover of the Craton in Canada; Geological Society of America, The Geology of North America, v. D-1, 826 p.

Struik, L.C., 1987. The Ancient Western North American Margin: An Alpine Rift Model For The East-Central Canadian Cordillera; Geological Survey of Canada, Paper 87-15, 19 p.

Struik, L.C., 1991. Cariboo Mountains and Quesnel Highlands; *in* H. Gabrielse and C.J. Yorath (eds.), Geology of the Cordilleran Orogen in Canada; Geological Society of America, The Geology of North America, v. G-2, p. 632-634.

Struik, L.C., Currie, L.D., O'Sullivan, P.B., Kung, R.B., and Jackson, L.E., Jr., 1997. Tertiary drain patterns in central British Columbia: implications for Oligocene to Miocene tectonics (abs.); *in* F. Cook and P. Erdmer (comps.), Slave-Northern Cordillera Lithospheric Evolution (SNORCLE) Transect and Cordilleran Tectonics Workshop Meeting; Lithoprobe Report 56, p. 183.

Su, W.-j. and Dziewonski, A.M., 1995. Inner core anisotropy in three dimensions; Journal of Geophysical Research, v. 100, p. 9831-9852.

Su, W.-j., Woodward, R.L., and Dziewonski, A.M., 1994. Degree 12 model of shear velocity heterogeneity in the mantle; Journal of Geophysical Research, v. 99, p. 6945-6980.

Suess, E., 1906. The Face of the Earth; Oxford, Clarendon Press, 556 p.

Sweeney, J.F., Stephenson, R.A., Currie, R.G., and DeLaurier, J.M., 1991. Crustal Geophysics; in H. Gabrielse and C.J. Yorath (eds.), Geology of the Cordilleran Orogen in Canada; Geological Society of America, The Geology of North America, v. G-2, p. 39-58.

Switzer, S.B., Holland, W.G., Christie, D.S., Graf, G.C., Hedinger, A.D., McAuley, R.J., Wierzbicki, R.A., and Packard, J.J., 1994. Devonian Woodbend-Winterburn Strata of the Western Canada Sedimentary Basin; in G.D. Mossop and I. Shetsen (comps.), Geological Atlas of the Western Canada Sedimentary Basin; Canadian Society of Petroleum Geologists & Alberta Research Council, p. 455-468.

Syme, E.C., 1995. 1.9 Ga arc and ocean floor assemblages and their bounding structures in the central Flin Flon Belt; in Lithoprobe Report 48, p. 261-272.

Symons, D.T.A., 1991. Paleomagnetism of the Proterozoic Wathaman Batholith and the suturing of the Trans-Hudson Orogen in Saskatchewan; Canadian Journal of Earth Sciences, v. 28, p. 1931-1938.

Symons, D.T.A., Radigan, S.P., and Lewchuk, M.T., 1996. Paleomagnetism of the Davin Lake granitoids, Rottenstone domain, Trans-Hudson orogen (part of NTS 64D-12 and -13); in Summary of Investigations 1996, Saskatchewan Geological Survey, Miscellaneous Report 96-4, p. 111-118.

Tagami, T. and Dumitru, T.A., 1996. Provenance and thermal history of the Franciscan accretionary complex: constraints from zircon fission track thermochronology; Journal of Geophysical Research, v. 101, p. 11,353-11,364.

Taylor, R.S., Mathews, W.H., and Kupsch, W.O., 1964. Tertiary; in R.G. McCrossan and R.P. Glaister (eds.), Geological History of Western Canada; Alberta Society of Petroleum Geologists, p. 190-200.

Teitz, M.W. and Mountjoy, E.W., 1989. The Late Proterozoic Yellowhead carbonate platform west of Jasper, Alberta; in H.H.J. Geldsetzer, N.P. James, and G.E. Tebbutt (eds.), Reefs, Canada and Adjacent Areas; Canadian Society of Petroleum Geologists, Memoir 13, p. 103-118.

Tempelman-Kluit, D.J., 1979. Transported Cataclasite, Ophiolite and Granodiorite in Yukon: Evidence of Arc-Continent Collision; Geological Survey of Canada, Paper 79-14, 27 p.

Tempelman-Kluit, D.J., Gabrielse, H., Evenchick, C.A., Mansy, J.L., Brown, R.L., Journeay, J.M., Lane, L.S., Struik, L.C., Murphy, D.C., Rees, C.J., Simony, P.S., Fyles, J.T., Höy, T., Gordey, S.P., Thompson, R.I., McMechan, M.E., and

Harms, T.A., 1991. Omineca Belt; *in* H. Gabrielse and C.J. Yorath (eds.), Geology of the Cordilleran Orogen in Canada; Geological Society of America, The Geology of North America, v. G-2, p. 603-674.

Thom, A., Arndt, N.T., Chauvel, C., and Stauffer, M., 1990. Flin Flon and western La Ronge belts, Saskatchewan: products of Proterozoic subduction-related volcanism; *in* J.F. Lewry and M.R. Stauffer (eds.), The Early Proterozoic Trans-Hudson Orogen in Canada; Geological Association of Canada, Special Paper 37, p. 163-175.

Thomas, G.E., 1974. Lineament-block tectonics: Williston-Blood Creek Basin; Bulletin of the American Association of Petroleum Geologists, v. 58, p. 1305-1322.

Thomas, M.D., Sharpton, V.L., and Grieve, R.A.F., 1987. Gravity patterns and Precambrian structure in the North American Central Plains; Geology, v. 15, p. 489-492.

Thompson, R.I., 1989. Stratigraphy, Tectonic Evolution and Structural Analysis of the Halfway River Map Area (94 B), Northern Rocky Mountains, British Columbia; Geological Survey of Canada, Memoir 425, 119 p.

Thurston, J.B. and Brown, R.J., 1994. Automated source-edge location with a new variable pass-band horizontal-gradient operator; Geophysics, v. 59, p. 546-554.

Tonnsen, J.J., 1986. Influence of tectonic terranes adjacent to the Precambrian Wyoming Province on Phanerozoic stratigraphy in the Rocky Mountain region; *in* J.A. Peterson (ed.), Paleotectonics and Sedimentation in the Rocky Mountain Region, United States; American Association of Petroleum Geologists, Memoir 41, p. 21-39.

Tóth, J., 1978. Gravity-induced cross-formational flow of formation fluids, Red Earth region, Alberta, Canada: analysis, patterns, evolution; Water Resources Research, v. 14, p. 805-843.

Tran, H.T., Lewry, J.F., and Ashton, K.E., 1996. The geology of the Medicine Rapids - Grassy narrows area; *in* Summary of Investigations 1996, Saskatchewan Geological Survey, Miscellaneous Report 96-4, p. 43-50.

Tremblay, L.P., 1972. Geology of the Beaverlodge mining area, Saskatchewan; Geological Survey of Canada, Memoir 367, 468 p.

Turner, R.J.W., Madrid, R.J., and Miller, E.L., 1989. Roberts Mountain allochthon: stratigraphic comparison with lower Paleozoic outer continental margin strata of the northern Canadian Cordillera; Geology, v. 17, p. 341-344.

Vail, P.R., 1992. The evolution of seismic stratigraphy and the global sea-level curve; *in* R.H. Dott, Jr. (ed.), Eustasy: the Historical Ups and Downs of a Major Geological Concept; Geological Society of America, Memoir 180, p. 83-92.

Vail, P.R., Mitchum, Jr., R.M., Todd, R.G., Widmier, J.M., Thompson III, S., Sangree, J.B., Bubb, J.N., and Hatlelid, W.G., 1977. Seismic stratigraphy and global changes of sea level; *in* C.E. Payton (ed.), Seismic Stratigraphy - Applications to Hydrocarbon Exploration; American Association of Petroleum Geologists, Memoir 26, p. 49-212.

Van der Heyden, P., 1992. A Middle Jurassic to Early Tertiary Andean-Sierran arc model for the Coast Belt of British Columbia; Tectonics, v. 11, p. 82-97.

Van der Velden, A.J. and Cook, F.A., 1994. Displacement of the Lewis thrust sheet in southwestern Canada: new evidence from seismic reflection data; Geology, v. 22, p. 819-822.

Van Hees, H., 1964. Cambrian. Part I - Plains; *in* R.G. McCrossan and R.P. Glaister (eds.), Geological History of Western Canada; Alberta Society of Petroleum Geologists, p. 20-28.

Van Schmus, W.R. and Bickford, M.E., 1993. Introduction; *in* J.C. Reed, Jr., M.E. Bickford, R.S. Houston, P.K. Link, D.W. Rankin, P.K. Sims, and W.R. Van Schmus (eds.), Precambrian: Conterminous U.S.; Geological Society of America, The Geology of North America, v. C-2, p. 1-9.

Van Schmus, W.R., Bickford, M.E., Anderson, J.L., Bender, E.E., Anderson, R.R., Bauer, P.W., Robertson, J.M., Bowring, S.A., Condie, K.C., Denison, R.E., Gilbert, M.C., Grambling, J.A., Mawer, C.K., Shearer, C.K., Hinze, W.J., Karlstrom, K.E., Kisvarsanyi, E.B., Lidiak, E.G., Reed, J.C., Jr., Sims, P.K., Tweto, O., Silver, L.T., Treves, S.B., Williams, M.L., and Wooden, J.L., 1993. Transcontinental Proterozoic provinces; *in* J.C. Reed, Jr., M.E. Bickford, R.S. Houston, P.K. Link, D.W. Rankin, P.K. Sims, and W.R. Van Schmus (eds.), Precambrian: Conterminous U.S.; Geological Society of America, The Geology of North America, v. C-2, p. 171-334.

Van Schmus, W.R., Bickford, M.E., Lewry, J.F., and Macdonald, R., 1987. U-Pb geochronology in the Trans-Hudson orogen, northern Saskatchewan; Canadian Journal of Earth Sciences, v. 24, p. 407-424.

Van Wagoner, J.C., Posamentier, H.W., Mitchum Jr., R.M., Vail, P.R., Sarg, J.F., Loutit, T.S., and Hardenbol, J., 1988. An overview of the fundamentals of sequence stratigraphy and key definitions; *in* C.K. Wilgus, B.S. Hastings, C.G. St. C. Kendall, H.W. Posamentier, C.A. Ross, and J.C. van Wagoner (eds.), Sea Level Changes: An Integrated Approach; Society of Economic Paleontologists and Mineralogists, Special Publication 42, p. 39-46.

Vassoyevich, N.B., 1959. Bedding in light of sedimentary differentiation; International Geology Review, v. 1, p. 59-71.

Veevers, J.J. and Powell, C.M., 1987. Late Paleozoic glacial episodes in Gondwanaland reflected in transgressive-regressive depositional sequences in Euroamerica; Bulletin of the Geological Society of America, v. 98, p. 475-487.

Vening Meinesz, F.A., 1947. Discussion of "Shear patterns of the earth's crust"; Transactions of the American Geophysical Union, v. 28, p. 939-946.

Viele, G.W., 1989. The Ouachita orogenic belt; *in* R.D. Hatcher, Jr., W.A. Thomas, and G.V. Viele (eds.), The Appalachian-Ouachita Orogen in the United States; Geological Society of America, The Geology of North America, v. F-2, p. 555-561.

Villeneuve, M.E., Ross, G.M., Theriault, R.J., Miles, W., Parrish, R.R., and Broome, J., 1993. Tectonic Subdivision, U-Pb Geochronology and Sm-Nd Isotope Geochemistry of the Crystalline Basement of the Alberta Basin, Western Canada; Geological Survey of Canada, Bulletin 447, 86 p.

Vine, F.J. and Matthews, D.H., 1963. Magnetic anomalies southwest of Vancouver Island; Nature, v. 199, p. 947-949.

Vinogradov, A.P. (ed.), 1974. Paleogeography of the U.S.S.R., 4 volumes; Nedra Publishing, Moscow (in Russian).

Walcott, R.I., 1968. The Gravity Field of Northern Saskatchewan and Northeastern Alberta; Gravity Map Series, Publications of the Dominion Observatory, Ottawa, No. 16 to 20.

Walker, R.G. and Eyles, C.H., 1991. Topography and significance of a basinwide sequence-bounding erosion surface in the Cretaceous Cardium Formation, Alberta; Journal of Sedimentary Petrology, v. 61, p. 473-496.

Wanless, R.K. and Eade, K.E., 1975. Geochronology of Archean and Proterozoic rocks in the southern District of Mackenzie; Canadian Journal of Earth Sciences, v. 12, p. 95-114.

Wannamaker, P.E., Johnson, J.M., Stodt, J.A., and Booker, J.R., 1997. Anatomy of the southern Cordilleran hingeline, Utah and Nevada, from deep electrical resistivity profiling; Geophysics, v. 62, p. 1069-1086.

Warner, L.A., 1978. The Colorado lineament: a middle Precambrian wrench fault system; Bulletin of the Geological Society of America, v. 89, p. 161-171.

Watts, A.B., 1982. Tectonic subsidence, flexure and global changes in sea level; Nature, v. 297, p. 469-474.

Watters, B.R. and Pearce, J., 1987. Metavolcanic rocks of the La Ronge Domain in the Churchill province, Saskatchewan; in T.C. Pharaoh, R.D. Beckinsale, and D. Rickard (eds.), Geochemistry and Mineralization of Proterozoic Volcanic Suites; Geological Society (London), Special Publication 33, p. 167-182.

Waync, W.J., Aber, J.S., Agard, S.S., Bergantino, R.N., Bluemle, J.P., Coates, D.A., Cooley, M.E., Madole, R.F., Martin, J.E., Mears, B., Jr., Morrison, R.B., and Sutherland, W.M., 1991. Quaternary geology of the Northern Great Plains; in R.B. Morrison (ed.), Quaternary Nonglacial Geology: Conterminous U.S.; Geological Society of America, The Geology of North America, v. K-2, p. 441-476.

Weber, W., 1990. The Churchill-Superior Boundary zone, southeast margin of the Trans-Hudson orogen: a review; in J.F. Lewry and M.R. Stauffer (eds.), The Early Proterozoic Trans-Hudson Orogen of North America; Geological Association of Canada, Special Paper 37, p. 41-55.

Wegener, A., 1915. Die Entstehung der Kontinente und Ozeane; Vieweg, Braunschweig.

Wegmann, C.E., 1935. Zur Deutung der Migmatite; Geologische Rundschau, v. 26, p. 305-350.

Wegmann, C.E., 1956. Stockwerktektonik und Modelle von Gesteinsdifferentiation; Geotektonisches Symposium zu Ehren von Hans Stille, Stuttgart, p. 3-19.

Weimer, R.J., 1980. Recurrent movements on basement faults, a tectonic style for Colorado and adjacent areas; in H.C. Kent and K.W. Porter (eds.), Colorado Geology; Symposium of the Rocky Mountain Association of Geologists, Denver, p. 23-35.

Weimer, R.J., 1984. Relations to unconformities, tectonics, and sea-level changes, Cretaceous of Western Interior, USA; *in* J.S. Schlee (ed.), Inter-Regional Unconformities and Hydrocarbon Migration; American Association of Petroleum Geologists, Memoir 36, p. 7-36.

Weller, J.M., 1960. Stratigraphic Principles and Practice; Harper & Row, 725 p.

Wellman, P., 1985. Block structure of continental crust derived from gravity and magnetic maps, with Australian examples; *in* W.J. Hinze (ed.), The Utility of the Regional Gravity and Magnetic Anomaly Maps; Society of Exploration Geophysicists, p. 102-108.

Westermann, G.E.G. (ed.), 1984. Jurassic-Cretaceous Paleogeography of North America; Geological Association of Canada, Special Paper 27.

Wheeler, J.O. and McFeely, P. (comps.), 1991. Tectonic Assemblage Map of the Canadian Cordillera and Adjacent Parts of the United States of America; Geological Survey of Canada, Map 1712A, scale 1:2,000,000.

White, R.S., 1988. The Earth's crust and lithosphere; Journal of Petrology, special issue, p. 1-10.

White, W.H., 1959. Cordilleran tectonics in British Columbia; Bulletin of the American Association of Petroleum Geologists, v. 43, p. 60-100.

White, D.J., Lucas, S.B., Hajnal, Z., Green, A.G., Lewry, J.F., Weber, W., Bailes, A.H., Syme, E.C., and Ashton, K., 1994. Paleo-Proterozoic thick-skinned tectonics: Lithoprobe seismic reflection results from the eastern Trans-Hudson Orogen; Canadian Journal of Earth Sciences, v. 31, p. 458-469.

Whitney, P.R., 1983. A three-stage model for the tectonic history of the Adirondack region, New York; Northeastern Geology, v. 5, p. 61-72.

Whittaker, A., Cope, J.C.W., Cowie, J.W., Gibbons, W., Hailwood, E.A., House, M.R., Jenkins, D.G., Rawson, P.F., Rushton, A.W.A., Smith, D.G., Thomas, A.T., and Wimbledon, W.A., 1991. A Guide to Stratigraphic Procedure; Geological Society (London), Special Report 20, 824 p.

Wilson, J.A., 1986. Geology of the basement beneath the Athabasca Basin in Alberta; Alberta Research Council, Bulletin 65, 61 p.

Wilson, J.T., 1966. Did the Atlantic close and then reopen?; Nature, v. 207, p. 343-347.

Winston, D., 1986. Sedimentation and tectonics of the Middle Proterozoic Belt Basin and their influence on Phanerozoic compression and extension in western Montana and northern Idaho; *in* J.A. Peterson (ed.), Paleotectonics and Sedimentation in the Rocky Mountain Region, United States; American Association of Petroleum Geologists, Memoir 41, p. 87-118.

Wise, D.U., 1982. Linesmanship and the practice of linear geo-art; Bulletin of the Geological Society of America, v. 93, p. 886-888.

Wold, R.J. and Hinze, W.J. (eds.), 1982. Geology and Tectonics of the Lake Superior Basin; Geological Society of America, Memoir 156, 280 p.

Woodsworth, G.J., Anderson, R.G., and Armstrong, R.L., 1991. Plutonic regimes; *in* H. Gabrielse and C.J. Yorath (eds.), Geology of the Cordilleran Orogen in Canada; Geological Society of America, The Geology of North America, v. G-2, p. 491-531.

Workum, R.H. and Hedinger, A.S., 1987. Geology of the Devonian Fairholme Group Cline Channel, Alberta; Second International Symposium on the Devonian System, Field Excursion A6, 41 p.

Woussen, G., Roy, D.W., Dimroth, E., and Chown, E.H., 1986. Mid-Proterozoic extensional tectonics of the core zone of the Grenville Province; *in* J.M. Moore, A. Davidson, and A.J. Baer (eds.), The Grenville Province; Geological Association of Canada, Special Paper 31, p. 297-311.

Wu, P., 1991. Flexure of the lithosphere beneath the Alberta foreland basin: evidence of an eastward stiffening continental lithosphere; Geophysical Research Letters, v. 18, p. 451-454.

Wu, P., 1993. Postglacial rebound in a power-law medium with axial symmetry and the existence of the transition zone in relative sea-level data; Geophysical Journal International, v. 114, p. 417-432.

Wyllie, P.J., 1971. The Dynamic Earth; John Wiley & Sons, 416 p.

Yañez, G.A. and LaBrecque, J.L., 1997. Age-dependent three-dimensional magnetic modeling of the North Pacific and North Atlantic oceanic crust at intermediate wavelengths; Journal of Geophysical Research, v. 102, p. 7947-7961.

Yarger, H.L., Robertson, R., Martin, J., Ng, K., Sooby, R., and Wentland, R., 1981. Aeromagnetic Map of Kansas; Kansas Geological Survey, Map M-16, scale 1:500,000.

Yoos, T.R., Potter, C.J., Thigpen, J.L., and Brown, L.D., 1991. The Cordilleran foreland thrust belt in northwestern Montana and northern Idaho from COCORP and industry seismic reflection data; Bulletin of the American Association of Petroleum Geologists, v. 75, p. 1089-1106.

Young, F.G., 1981. The Amundsen Embayment, Northwest Territories; relevance to the Upper Proterozoic evolution of North America; *in* F.H.A. Campbell (ed.), Proterozoic Basins of Canada; Geological Survey of Canada, Paper 81-10, p. 203-218.

Zaleski, E., Eaton, D.W., Milkereit, B., Roberts, B., Salisbury, M., and Petrie, L., 1997. Seismic reflections from subvertical diabase dikes in an Archean terrane; Geology, v. 25, p. 707-710.

Zartman, R.E. and Stacey, J.S., 1971. Lead isotopes and mineralization ages in Belt supergroup rocks, northwestern Montana and northern Idaho; Economic Geology and Bulletin of the Society of Economic Geologists, v. 66, p. 849-860.

Zhamoida, A.I. (ed.), 1979. Stratigraphic Code of the U.S.S.R. (English edition); VSEGEI, Leningrad (St. Petersburg).

Zhu, C. and Hajnal, Z., 1993. Tectonic development of the northern Williston basin: a seismic interpretation of an east-west regional profile; Canadian Journal of Earth Sciences, v. 30, p. 621-630.

Ziegler, P.A., 1969. The Development of Sedimentary Basins in Western and Arctic Canada; Alberta Society of Petroleum Geologists, 89 p.

Ziegler, P.A., 1982. Geological Atlas of Western and Central Europe; Elsevier, 130 p.

Ziegler, P.A. (ed.), 1987. Compressional Intraplate Deformations in the Alpine Foreland; Tectonophysics, special issue, v. 137, 420 p.

Zietz, I., Bond, K.R., Gilbert, F.P., Kirby, J.R., Riggle, F.E., and Snyder, S.L., 1982. Composite magnetic anomaly map of the Unites States, Part A: Conterminous United States; U.S. Geological Survey, Map GP-954A, and accompanying report.

Zoback, M.L. and Zoback, M.D., 1989. Tectonic stress field of the continental United States; in L.C. Pakiser and W.D. Mooney (eds.), Geophysical Framework of the Continental United States; Geological Society of America, Memoir 172, p. 523-540.

Zwanzig, H.V., 1990. Kisseynew gneiss belt in Manitoba: stratigraphy, structure, and tectonic evolution; in J.F. Lewry and M.R. Stauffer(eds.), The Early Proterozoic Trans-Hudson Orogen of North America; Geological Association of Canada, Special Paper 37, p. 95-120.

Zwanzig, H.V., 1997. Kisseynew metasedimentary gneiss belt, Trans-Hudson orogen (Canada): back-arc origin and collisional inversion: comment; Geology, v. 25, p. 90-91.

Zwanzig, H.V., Schledewitz, D.C.P., Ashton, K.E., Froese, E., and Dohar, V., 1996. Geological compilation map of the Flin Flon Belt-Kisseynew Belt transition zone, Manitoba-Saskatchewan: a preliminary NATMAP product (abs.); Summary of Investigations 1996, Saskatchewan Geological Survey, Miscellaneous Report 96-4, p. 178.

Lecture Notes in Earth Sciences

For information about Vols. 1–19
please contact your bookseller or Springer-Verlag